TREATISE ON BASIC PHILOSOPHY

Volume 4

ONTOLOGY II: A WORLD OF SYSTEMS

TREATISE ON BASIC PHILOSOPHY

MARIO BUNGE

Treatise on Basic Philosophy

VOLUME 4

Ontology II:

A WORLD OF SYSTEMS

D. REIDEL PUBLISHING COMPANY
DORDRECHT : HOLLAND / BOSTON : U.S.A.
LONDON : ENGLAND

Library of Congress Cataloging in Publication Data (Revised)

CIP

Bunge, Mario Augusto
Ontology.

(His Treatise on basic philosophy ; v. 3–4)
Includes bibliographies and indexes.
CONTENTS: 1. The furniture of the world. 2. A world of systems.
1. Ontology–Collected works. I. Title.
BD311.B84 111 77-1750
ISBN 90-277-0534-8 (v. 1)
ISBN 90-277-0572-0 pbk. (v. 1)
ISBN 90-277-0535-6 (v. 2)
ISBN 90-277-0573-9 pbk. (v. 2)
ISBN 90-277-0780-4 (v. 3)
ISBN 90-277-0785-5 pbk. (v. 3)
ISBN 90-277-0944-0 (v. 4)
ISBN 90-277-0945-9 pbk. (v. 4)

Published by D. Reidel Publishing Company,
P.O. Box 17, Dordrecht, Holland

Sold and distributed in the U.S.A., Canada, and Mexico
by D. Reidel Publishing Company, Inc.
Lincoln Building, 160 Old Derby Street, Hingham,
Mass. 02043, U.S.A.

Printed in The Netherlands

GENERAL PREFACE TO THE *TREATISE*

This volume is part of a comprehensive *Treatise on Basic Philosophy*. The treatise encompasses what the author takes to be the nucleus of contemporary philosophy, namely semantics (theories of meaning and truth), epistemology (theories of knowledge), metaphysics (general theories of the world), and ethics (theories of value and of right action).

Social philosophy, political philosophy, legal philosophy, the philosophy of education, aesthetics, the philosophy of religion and other branches of philosophy have been excluded from the above *quadrivium* either because they have been absorbed by the sciences of man or because they may be regarded as applications of both fundamental philosophy and logic. Nor has logic been included in the *Treatise* although it is as much a part of philosophy as it is of mathematics. The reason for this exclusion is that logic has become a subject so technical that only mathematicians can hope to make original contributions to it. We have just borrowed whatever logic we use.

The philosophy expounded in the *Treatise* is systematic and, to some extent, also exact and scientific. That is, the philosophical theories formulated in these volumes are (*a*) formulated in certain exact (mathematical) languages and (*b*) hoped to be consistent with contemporary science.

Now a word of apology for attempting to build a system of basic philosophy. As we are supposed to live in the age of analysis, it may well be wondered whether there is any room left, except in the cemeteries of ideas, for philosophical syntheses. The author's opinion is that analysis, though necessary, is insufficient – except of course for destruction. The ultimate goal of theoretical research, be it in philosophy, science, or mathematics, is the construction of systems, i.e. theories. Moreover these theories should be articulated into systems rather than being disjoint, let alone mutually at odds.

Once we have got a system we may proceed to taking it apart. First the tree, then the sawdust. And having attained the sawdust stage we should move on to the next, namely the building of further systems. And this for three reasons: because the world itself is systemic, because no idea can become fully clear unless it is embedded in some system or other, and because sawdust philosophy is rather boring.

The author dedicates this work to his philosophy teacher

Kanenas T. Pota

in gratitude for his advice: "Do your own thing. Your reward will be doing it, your punishment having done it".

CONTENTS OF *ONTOLOGY II*

MARIO BUNGE

PREFACE TO *ONTOLOGY II*

This volume continues and concludes the task begun in Part I, titled *The Furniture of the World* – namely the building of an exact and systematic ontology consistent with contemporary science. However, it can be read independently by anyone willing to take for granted the basic notions analyzed and systematized in the companion volume, namely those of substance, property, thing, possibility, change, space, and time.

The three main themes of this book are wholeness (or systemicity), variety, and change. These three notions are analyzed and systematized, and they occur in some of the main assumptions of our ontology. One of these hypotheses is that the universe is not a heap of things but a thing composed of interconnected things – i.e. a system. This supersystem is composed of subsystems of various kinds: physical, biological, social, etc. Only physical systems may be composed of things that are not themselves systems, such as elementary particles and field quanta. However, even nonsystems are components of some system or other, and every system but the universe is a subsystem of some system: there are no strays. Ours is, in sum, a world of interconnected systems. Moreover it is the only one.

Another postulate of this system of ontology is that concrete systems are not all alike except insofar as they are systems and therefore tractable with the help of a unifying systems-theoretic framework. There are different kinds of system and each is characterized by its own peculiar properties and laws. Surely we sometimes succeed in accounting for the emergence and the history of a system in terms of its composition, environment, and structure. Nevertheless, explaining need not be explaining away: explained systems are not heaps, explained emergence is no mere resultant, and explained novelty is not old. Systemicity, emergence, and qualitative novelty and variety are as genuine as they are capable of being explained. Far from being incompatible with reason, wholeness and emergence can be understood.

A third major thesis of this work is that no system, except for the world as a whole, lasts forever. Systems get assembled, change, and break down. If natural, systems emerge as a result of self-assembly processes – often from debris of former systems. Even modest accretion processes can ensue in systems possessing emergent properties. Order can thus emerge from randomness, systems from physical precursors, living systems from nonliving ones, and so on. (Entropy need not increase in open systems.)

All three theses are by now common knowledge or nearly so. Now they – jointly with many others – have become part and parcel of a science-oriented ontological system couched in a fairly exact language. Thus the novelty of this system resides sometimes in its components, and at other times in their organization.

This volume covers the following ground. Chapter 1 defines the notions of system and subsystem, and emergence and level, and it lays down a handful of laws supposed to be shared by systems of all kinds. Chapter 2 defines the notions of chemical system and biochemical system, regarding them as emergent relative to the physical level. Chapter 3 attempts to catch the elusive notion of life, and tackles the problems of biovalue, health, adaptation, and bioprogress. Chapter 4 is devoted to the mind-body problem from a point of view that combines the systemic approach with the psychobiological one. (Individual neurons cannot think; but certain systems of neurons, if connected in special plastic ways, can.) Chapter 5 defines the concept of a sociosystem and analyzes every society into four main subsystems – the kinship system, the economy, the culture, and the polity. Chapter 6 presents the world view that emerges from the preceding chapters – one which is naturalistic (but not physicalist), systemic (but not holistic), pluralistic (with respect to properties, not substances), and dynamicist. Finally, the two appendices review some mathematical models of systems and of change which, because of their generality, belong to ontology as well as to science. These appendices are helpful but not essential for the understanding of the text.

ACKNOWLEDGMENTS

I am grateful for the advice and criticism offered by Professors Lina Bettucci (Departamento de Biología, Universidad Autónoma Metropolitana, México), Ernesto M. Bravo (Instituto Superior de Medicina, La Habana), A. Brito da Cunha (Instituto de Biociências, Universidade de São Paulo), Sir Francis Crick (The Salk Institute), Bernardo Dubrovsky (Department of Psychiatry, McGill University), Máximo García-Sucre (Sección Física, Instituto Venezolano de Investigaciones Científicas), Bernd Küppers (Max-Planck-Institut für Biophysikalische Chemie, Göttingen), Rodolfo Llinás (Department of Physiology and Biophysics, New York University Medical Center), Roger Palfree (Department of Biology, McGill University), Rafael Pérez Pascual (Instituto de Física, Universidad Nacional Autónoma de México), Osvaldo A. Reig (Departamento de Estudios Ambientales, Universidad Simón Bolívar, Caracas), Daniel Seni (Institut d'Urbanisme, Université de Montréal), and René Zayan (Centre de psychologie expérimentale et comparée, Université de Louvain). I am also indebted to my former research assistant Robert Blohm for suggesting a number of corrections. My former students David Conter, Adolfo García de la Sienra, Gordon Graham, and John Hawthorne made several remarks for which I am grateful. I am thankful to Dr Ricardo Peralta Fabi for drawing all the diagrams in the book. Last, but not least, I thank Professors D. R. Axelrad and Walter Hitschfeld, as well as the Canada Council and the McGill Faculty of Graduate Studies and Research, for supporting my research.

MARIO BUNGE

SPECIAL SYMBOLS

$\varnothing$	the *empty set*
$a \in A$	individual *a belongs* to set *A*
$A \subset B$	set *A* is *included* in set *B*
$A \cup B$	the set of objects in *A or* in *B*
$A \cap B$	the set of objects in *A and* in *B*
$A - B$	the set of objects *in A but not in B*
$A \triangle B$	the set of objects in *A or* in *B* but *not in both*
$\langle a, b \rangle$	the *ordered pair* of *a* and *b*
$A \times B$	the *cartesian product* of *A* and *B*
$\|A\|$	the *cardinality* (numerosity) of *A*
$a \dotplus b$	the *association* of things *a* and $b = sup\ \{a, b\}$
$\mathscr{C}(x)$	the *composition* of system *x*
$\mathscr{E}(x)$	the *environment* of system *x*
$\mathscr{S}(x)$	the *structure* of system *x*
$f: A \to B$	*function f* maps set *A* into set *B*
$f(x)$	the *value* of function *f* at *x*
$\mathbb{F} = \langle F_1, F_2, \ldots, F_n, \ldots \rangle$	*state function* for some system
$h(x, \tau)$	the *history* of thing *x* during time interval τ
iff	*if and only if* (necessary and sufficient condition)
$\mathbb{L}(x)$	the *set of laws* of thing *x*
$\mathbb{N}$	the set of *natural numbers*
$x \sqsubset y$	thing *x* is a *part* of thing *y*
$x \prec y$	system *x* is a *subsystem* of system *y*
2^S	the *power set* (family of subsets) of set *S*
Pr	*probability* function
$\mathbb{R}$	the *real line*
$\mathbb{R}^+$	the set of *non-negative real numbers*
$S_L(x)$	*lawful state space* of thing *x*
T	set of *time* instants

CHAPTER 1

SYSTEM

Every science studies systems of some kind, whether natural (physical, chemical, biological, or social) or artificial (technical). Moreover most sciences study nothing but systems. Thus biology studies biosystems, sociology sociosystems, and technology technosystems. Physics seems to be the only science that investigates not only systems, such as atoms and large scale fields, but also putatively simple or elementary things such as electrons and photons. Even so, physicists acknowledge that every such basic thing is a component of some system or other

Until recently every species of system was studied separately. About four decades ago a number of specialists joined efforts to launch various cross-disciplinary ventures, such as operations research and cybernetics. Their success suggested to some workers that a unified approach to problems in various fields was possible. They pointed out that (*a*) there are some concepts and structural principles that seem to hold for systems of many kinds, and (*b*) there are some modeling strategies – in particular the state space approach – that seem to work everywhere.

The discipline that purports to develop such a unified approach is often called 'general systems theory' (Bertalanffy, 1950, 1958; Boulding, 1956). Paradoxically enough, this is not a single theory but a whole set of theories – automata theory, linear systems theory, control theory, network theory, general Lagrangian dynamics, etc. – unified by a philosophical framework (Bunge, 1974c, 1977c). We shall call *systemics* this set of theories that focus on the structural characteristics of systems and can therefore cross the largely artificial barriers between disciplines.

Systemics has two related motivations, one cognitive and one practical. The cognitive or theoretical rationale of systemics is, of course, the wish to discover similarities among systems of all kinds despite their specific differences – e.g. between body temperature control systems and furnace thermostats. The practical motivation for systemics is the need to cope with the huge and many-sided systems characteristic of industrial societies – such as communications networks, factories, hospitals, and armies. This complexity, in particular the variety of components of such systems, violates the traditional borders among disciplines and calls for a cross-disciplinary approach.

Note the differences between the standard scientist, engineer or social scientist on the one hand, and the systems "specialist" (actually a generalist) on the other. Whereas the former do or apply some particular science, the systemics expert de-emphasizes the physics (chemistry, biology, or sociology) of his systems, focusing instead on their structure and behavior. Moreover he is interested particularly in duplicating or imitating (modeling or simulating) the behavior of any given system (e.g. a person) by one of a different kind (e.g. a pattern recognition automaton). This holds not only for the mathematician who takes systemics as an honorable pretext for playing with abstract structures but without any serious concern for practical problems in engineering or management: it also holds for the systemist intent on solving practical problems, such as modeling and simulating a grazing land system or a university.

The method employed by the system theorist is mathematical modeling and the experimental (or at least the computer) testing of system models. Both are of course part of the scientific method. What is peculiar to the way the systemics expert proceeds is that, far from incorporating any specific (e.g. chemical) laws into his model, he aims at building a black box, a grey box, or a kinematical model free from details concerning the materials composing the system, and noncommittal enough to cover some of the global aspects of the organization and behavior of the system on some of its levels. The scientific method is thus taken for granted: what is emphasized is the general or cross disciplinary approach in contrast to the specific or disciplinary one. In other words, the systemics expert is a jack of all trades – a quasi-philosopher if not a full blown one.

Systemics is not quite the same as *systems analysis*, a much advertised, ill defined, and sometimes controverted thing. Systems analysis too, when serious, uses the scientific method but, unlike systemics, it is not particularly interested in de-emphasizing the peculiarities of the components of the system concerned. What it does emphasize is that, because it studies many-sided and multi-level systems – such as ecosystems and transportation systems – it must adopt various points of view on different levels. For example, hospitals are not just buildings with medical equipment but social systems as well – whose components include medical personnel and patients – and moreover subsystems of a larger social system, namely a health-care system, which is in turn a subsystem of a society. The novelty of systems analysis resides less in its methods than in the objects it studies, namely complex man-artifact systems never before approached in a scientific manner. Unlike systemics, systems analysis is hardly interested in building extremely general models: it aims instead at drawing flow charts, network diagrams, and occasionally specific mathematical models accounting if possible not just for the structure and kinematics of the system but also for its dynamics, and thus enabling one to understand how it operates and malfunctions, hence how it can be repaired. (For a hilarious account of systems antics, see Gall (1977).)

Systemics, or general system theory, is a field of scientific and technological research and one of considerable interest to philosophy. Because of its generality it has a sizable overlap with ontology or metaphysics construed in the traditional, pre-Hegelian sense as well as in our own sense of scientific ontology (Bunge, 1973a, 1977a). Both sytemics experts and ontologists are interested in the properties common to all systems irrespective of their particular constitution, and both are intrigued by the peculiarities of extremely general theories, which are methodologically quite different from specific theories (Bunge, 1973a, 1977c).

The main differences between systemics and ontology seem to be these: (*a*) while systems theorists take certain concepts for granted – e.g., those of property, possibility, change, and time – ontologists take nothing for granted except logic and mathematics; (*b*) while systems theorists are often interested in the details of the couplings of the components of a system, ontologists seldom are; (*c*) while systems theorists focus their attention on input–output models of systems that are largely at the mercy of their environment, ontologists are interested in free systems as well (in which respect they do not differ from physicists); (*d*) while systems theorists are mainly interested in deterministic (or rather nonstochastic) models – partly because theirs are large scale things – ontologists are also interested in stochastic ones; and (*e*) while some systems theorists focus their attention on the search for analogies among systems of different kinds, and particularly on different levels, ontologists are primarily interested in analyzing and systematizing concepts referring to all kinds of system.

In this chapter we shall propose a number of definitions and principles concerning concrete systems in general. These ideas will be used in succeeding chapters, where certain system genera will be studied. Details on mathematical models of systems are found in the two appendices.

1. Basic Concepts

1.1. *Aggregate and System*

An *aggregate* or assemblage is a collection of items not held together by bonds, and therefore lacks integrity or unity. Aggregates can be either conceptual or concrete (material). A conceptual aggregate is a set. (But not every set is a conceptual aggregate: a set equipped with a structure is a conceptual system.) A concrete or material aggregate, on the other hand, is a compound thing, the components of which are not coupled, linked, connected, or bonded, such as a field constituted by two superposed fields, a celestial constellation, and a random sample of a biological population.

Because the components of an aggregate do not interact – or do not interact appreciably – the behavior of each is independent of the behavior of the others. Consequently the history of the aggregate is the union of the histories of its members. On the other hand the components of a concrete system are linked, whence the history of the whole differs from the union of the histories of its parts. We shall take the last statement to be an accurate version of the fuzzy slogan of holistic metaphysics, namely *The whole is greater than the sum of its parts.* But we shall go far beyond this characterization of wholeness or systemicity. To this end we shall make use of a few elementary mathematical concepts as well as of a number of common notions – such as those of thing, property, and time – that have been clarified in our companion volume (Bunge, 1977a).

A system, then, is a complex object, the components of which are interrelated rather than loose. If the components are conceptual, so is the system; if they are concrete or material, then they constitute a concrete or material system. A theory is a conceptual system, a school a concrete system of the social kind. These are the only system kingdoms we recognize: conceptual and concrete. We have no use for mixed systems, such as Popper's "world 3", allegedly composed of conceptual objects, such as theories, as well as concrete objects, such as books (Popper, 1968; Popper and Eccles, 1977). We do not because, in order to be able to speak of the association or combination of two items, we must specify the association bond or operation. And, while mathematical theories specify the way conceptual items combine, and ontological and scientific theories take care of the combination of concrete items, no known theory specifies the manner whereby conceptual items could combine with concrete ones – and no experience suggests that such hybrids exist.

Whatever its kingdom – conceptual or concrete – a system may be said to have a definite composition, a definite environment, and a definite structure. The composition of a system is the set of its components; the environment, the set of items with which it is connected; and the structure, the relations among its components as well as among these and the environment. For example, a theory is composed of propositions or statements; its environment is the body of knowledge to which it belongs (e.g. algebra or ecology); and its structure is the entailment or logical consequence relation. The merger of these three items is a propositional system, i.e. a system $\mathcal{T}$ composed of a set P of propositions, embedded in a certain conceptual body B, and glued together by the relation $\vdash$ of entailment: in short, $\mathcal{T} = \langle P, B, \vdash \rangle$. And the composition of a school is the union of its staff and pupils; the environment is the natural and social milieu, and the structure consists of the relations of teaching and learning, managed and being managed, and others. The environment must be included in the description of a system because the behavior of the latter depends critically on the nature of its milieu. But of course in the case of the universe the environment is empty, and so it is in the case of the important fiction known as the free particle (or field).

One way of characterizing the general concept of a system is this. Let T be a nonempty set. Then the ordered triple $\sigma = \langle C, E, S \rangle$ is (or represents) a *system over* T iff C and E are mutually disjoint subsets of T (i.e. $C \cap E = \varnothing$), and S is a nonempty set of relations on the union of C and E. The system is conceptual if T is a set of conceptual items, and concrete (or material) if $T \subseteq \Theta$ is a set of concrete entities, i.e. things. However, the preceding is not a definition proper, because it does not tell us what exactly is the membership of the coordinates C, E and S of the ordered triple. We must therefore define the notions of composition, environment, and structure of a thing.

1.2. *Concrete System: Definition*

Let us start by defining the composition of a system. A social system is a set of socially linked animals. The brains of such individuals are parts of the latter but do not qualify as members or components of a social system because they do not enter independently into social relations: only entire animals can hold social relations. In other words, the composition of a social system is not the collection of its parts but just the set of its atoms, i.e. those parts that are socially connectible. This particular notion of composition is that of atomic composition or A-composition for short. It will be defined thus: The *A-composition* (or *composition at the A level*) of a thing x is the set of parts of x that belong to A. In symbols: let $A \subseteq \Theta$ be a class of things and let x be a thing (i.e. $x \in \Theta$). Then the (absolute) *composition* of x is the set of its parts, i.e.

$$\mathscr{C}(x) = \{y \in \Theta \mid y \sqsubset x\},$$

where '$y \sqsubset x$' designates "y is a part of x". And the *A-composition* of x is the set of its A-parts:

$$\mathscr{C}_A(x) = \mathscr{C}(x) \cap A = \{y \in A \mid y \sqsubset x\}.$$

Let us introduce next the concept of link, connection, or coupling among the components of a thing. We must distinguish between a mere relation, such as that of being older, and a connection, such as that of exerting pressure. Unlike a mere relation, a connection makes some difference to its relata. That is, two things are connected just in case at least one of them acts upon the other – where the action need not consist in eventuating something but may consist in either cutting out or opening up certain possibilities.

In turn, we say that one thing *acts* upon another if it modifies the latter's behavior line, or trajectory, or history. The acting of thing a on thing b is symbolized

$$a \rhd b.$$

If a thing acts upon another, and the latter does not react back, the former is called the *agent* and the latter the *patient*. If neither action nor reaction are nil, the things are said to *interact*. Finally, two things are *connected* (or *coupled*, or *linked*, or *bonded*) if at least one of them acts on the other. The *bondage* of a set $A \subseteq \Theta$ of things is the set $\mathbb{B}_A$ of bonds (or couplings or links or connections) among them. So, the total set of relations among the components of a complex entity may be decomposed into its bondage $\mathbb{B}_A$ and the set $\bar{\mathbb{B}}_A$ of nonbonding relations.

We can now introduce the notion of the *A-environment* of a thing x with A-composition $\mathscr{C}_A(x)$. It will be defined as the set of all things, other than those in $\mathscr{C}_A(x)$, that act on or are acted upon the latter:

$$\mathscr{E}_A(x) = \{y \in \Theta \mid \neg(y \in \mathscr{C}_A(x)) \,\&\, (\exists z)(z \sqsubset x \,\&\, (y \rhd z \vee z \rhd y))\}.$$

Finally the *structure* of a thing will be defined as the set of all the relations among the thing's components as well as among these and the things in the thing's environment.

We now have all we need to define the notion of a concrete system:

DEFINITION 1.1 An object is a *concrete system* iff it is composed of at least two different connected things.

(a) (b)

Fig. 1.1. Two systems with the same composition but different structures and environments.

Example A molecule, a coral reef, a family and a factory are systems. On the other hand a set of states of a thing and a collection of events, even if ordered, are not concrete systems. Symbol: Σ.

And now the three characteristics of any system:

DEFINITION 1.2 Let $\sigma \in \Sigma$ be a concrete system and $A \subset \Theta$ a class of things. Then

(i) the *A-composition* of σ at a given time t is the set of its A-parts at t:

$$\mathscr{C}_A(\sigma, t) = \{x \in A \mid x \sqsubset \sigma\};$$

(ii) the *A-environment* of σ at time t is the set of all things of kind A, not components of σ, that act or are acted on by components of σ at t:

$$\mathscr{E}_A(\sigma, t) = \{x \in A \mid x \notin \mathscr{C}_A(\sigma, t) \;\&\; (\exists y)(y \in \mathscr{C}_A(\sigma, t) \;\&\; (x \rhd y \vee y \rhd x)\};$$

(iii) the *A-structure* (or *organization*) of σ at time t is the set of relations, in particular bonds, among the components of σ, and among them and the things in the environment of σ, at t:

$$\mathscr{S}_A(\sigma,t) = \{R_i \in \mathbb{B}_A(\sigma,t) \cup \bar{\mathbb{B}}_A(\sigma,t) \mid \mathbb{B}_A(\sigma,t) \neq \varnothing \;\&\; 1 \leq i \leq n,\}$$

where $\mathbb{B}_A(\sigma, t)$ is the set of bonding relations, and $\bar{\mathbb{B}}_A(\sigma, t)$ that of non-bonding relations, defined on $\mathscr{C}_A(\sigma, t) \cup \mathscr{E}_A(\sigma, t)$.

Example The simplest possible system is one composed of two connected things, a and b, in an environment lumped into a single thing c. That is, $\mathscr{C}(\sigma) = \{a, b\}$, $\mathscr{E}(\sigma) = \{c\}$. This system can have either of the following internal structureses: $a \rhd b$, $b \rhd a$, or $a \bowtie b$: see Figure 1.2. (These are the conceivable internal structures. But some of them may not be nomologically possible, let alone technically feasible or even desirable.) As for

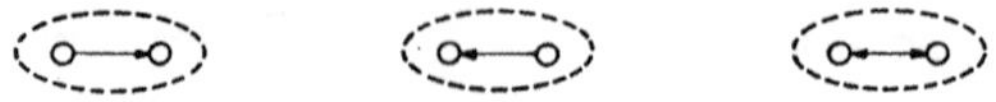

Fig. 1.2. A two component system with three possible internal structures.

the external structures, they can be any of these or their unions: $\{a \rhd c\}$, $\{b \rhd c\}$, $\{c \rhd a\}$, $\{c \rhd b\}$.

An exhaustive knowledge of a system would comprise the following items: (*a*) the composition, the environment and the structure of the system; (*b*) the history of the system (particularly if this is a biosystem or a sociosystem), and (*c*) the laws of the system. Such complete knowledge is seldom attainable, particularly with reference to complex systems. But in order to be able to speak of systems at all we should know at least their composition, their environment and their structure. Thus we can say that the model constituted by the ordered triple

$$s_A(\sigma, t) = \langle \mathscr{C}_A(\sigma, t), \mathscr{E}_A(\sigma, t), \mathscr{S}_A(\sigma, t)\rangle$$

is the *minimal model* of system σ at time t. Obviously, this qualitative model will not suffice for quantitative purposes, such as predicting the rate of formation or breakdown of a system. We shall therefore supplement the above minimal model with a quantitative model to be introduced in Sec. 2.2. However, before doing so we shall use the minimal model to clarify a number of issues that are often obscure in the literature on systems.

1.3. *More of the Same*

Before continuing our study of systems and their models we must make sure that the concept of a concrete system is not idle – i.e. that some things are systems while others are not. That some things are not systems follows from the assumption that there are basic or elementary things, i.e. things without parts (Vol. 3, Postulate 1.4). And that other things are systems follows from the ontological hypothesis that every thing – except the universe as a whole – acts on, and is acted upon by, other things (Vol. 3, Postulate 5.10). In sum, we have proved the nontrivial

THEOREM 1.1 (i) There are concrete systems. (ii) Every thing is a component of at least one system.

Surely the identification and modeling of a concrete system may be an extremely difficult task. Thus it is not always clear what the composition, hence also the environment, of a system is, particularly if it is strongly coupled to other systems – as is the case with the economic and the political systems of a society. However, this is a scientific problem not an ontological one.

Note that actions and the corresponding connections have been defined for things not for properties. The latter can be *interdependent* but not *interacting*. That is, the common phrase 'Properties P and Q interact' should be understood either as "Properties P and Q (of a given thing) are interdependent", or "Things with property P interact with things with property Q".

Connections can be permanent or temporary, static or dynamic. In the latter case they are often called *flows* – of energy, as in heat transfer, of matter, as in migrations, or of fields, as in a television network. If a physical flow happens to carry information, the connection is called *informational* and the entire system an 'information system'. However, the physical/informational distinction is one of emphasis, not a dichotomy, for every information flow rides on some energy flow. (See Appendix A, Sec. 1.4.)

Our definition of the environment of a system as the set of all things coupled with components of the system makes it clear that it is the *immediate* environment, not the *total* one – i.e. the set of all the things that are not parts of the system. Except in extragalactic astronomy and in cosmology, we are interested not in the transactions of a system with the rest of the universe but only in that portion of the world that exerts a significant influence on the thing of interest. This immediate environment or *milieu* is the cell in the case of chromosomes, the rest of the organism in the case on an organ, the ecosystem in the case of an organism, the solar system in the case of a biosphere, and so on. In other words, the immediate environment of a thing is the composition of its next supersystem. (More in Sec. 1.4.)

A system that neither acts on nor is acted upon by any other thing is said to be closed. In other words, we make

DEFINITION 1.3 Let σ be a system with environment $\mathscr{E}(\sigma, t)$. Then σ is *closed* at t iff $\mathscr{E}(\sigma, t) = \varnothing$ – otherwise σ is *open*.

Since every thing but the universe interacts with some other things, we infer

COROLLARY 1.1 The universe is the only system closed at all times.

This holds whether or not the universe turns out to be spatially infinite, for the universe may be defined as that thing which has a void environment (i.e. which is self-contained).

So much for the concept of total closure. We need also the notion of partial closure, or closure relative to a given property, since a system may be open in some respects and closed in others. (Thus all systems are gravitationally open, but some are electrically closed, others are closed to the exchange of matter, still others to cultural influences, and so on.) We make then

DEFINITION 1.4 Let P be a property of a system σ in an environment $\mathscr{E}(\sigma, t)$. Then σ is *open with respect to P at t* iff P is related, at t, to at least one property of things in $\mathscr{E}(\sigma, t)$ – otherwise σ is *closed* in the respect P.

Comparing this definition with the previous one we realize that a system is closed iff it is closed in every respect.

Finally some comments on the concept of structure. Our use of it is common in mathematics and in the social sciences. Thus a famous anthropologist: for the biochemist an organism "is a complexly integrated system of complex molecules. The system of relations by which these units are related is the organic structure. As the terms are here used the organism is *not* itself a structure; it is a collection of units (cells or molecules) arranged in a structure, i.e. in a set of relations; the organism *has* a structure" (Ratcliffe–Brown, 1935). Biologists use 'structure' sometimes in the above sense and at other times as a synonym for 'anatomic component'. In the latter case they run the risk of talking about the structure of a structure.

It is sometimes useful to distinguish a system's internal structure from its external one. The former is the subset of the total structure formed by the relations (in particular connections) among the system components. And the external structure is of course the complement of the internal structure to the total structure. Though distinct, the internal and the external structure are interdependent. Thus the internal structure of a molecule, far from being a permanent and intrinsic property of the molecule, depends critically upon its external structure – i.e. the interactions between the molecule and its milieu (e.g. the solvent).

Another distinction worth making is that between total structure and spatial structure, or set of spatial relations among the parts of a thing. (Spatial structure or configuration should not be mistaken for shape. The great majority of systems in the universe, i.e. the hydrogen and helium atoms, are shapeless. Nor do social systems have a shape although they have a spatial configuration since they are made up of living beings that stand in definite spatial relations to one another.) Every system has both a system structure (or bondage) and a spatial structure (or configuration). On the other hand aggregates or assemblages have spatial structures but no system structures.

For ease of reference we collect some of the above elucidations in

DEFINITION 1.5 Let σ be a concrete system with A-structure $\mathcal{S}_A(\sigma, t)$ at time t. Then

(i) the *internal A-structure* of σ at t is the subset of $\mathcal{S}_A(\sigma, t)$ composed of the relations among the A-parts of σ at t;

(ii) the *configuration* (or *spatial structure*) of σ at t is the subset of $\mathcal{S}_A(\sigma, t)$ composed of the spatial relations among the A-parts of σ at t.

1.4. *Subsystem*

A system component may or may not be a system itself. If it is we call it a 'subsystem'. More explicitly, we make

DEFINITION 1.6 Let σ be a system with composition $\mathcal{C}(\sigma, t)$, environment $\mathcal{E}(\sigma, t)$ and structure $\mathcal{S}(\sigma, t)$ at time t. Then a thing x is a *subsystem* of σ at t, or $x \prec \sigma$, iff

(i) x is a system at time t, and

(ii) $\mathcal{C}(x, t) \subseteq \mathcal{C}(\sigma, t)$ & $\mathcal{E}(x, t) \supseteq \mathcal{E}(\sigma, t)$ & $\mathcal{S}(x, t) \subseteq \mathcal{S}(\sigma, t)$.

By definition, the subsystem relation $\prec$ is an order relation, i.e. it is reflexive, asymmetric, and transitive. So, in particular, if $\sigma_1 \prec \sigma_2$ and $\sigma_2 \prec \sigma_3$, then $\sigma_1 \prec \sigma_3$. We shall make use of this property when defining the notion of a system of nested systems (Definition 1.7).

Example 1 Factories, hospitals and schools constitute subsystems of any modern society. On the other hand the persons composing them are not themselves social systems: they are biosystems. *Example 2* A foetus is a subsystem of its mother; it becomes a system in its own right after birth: before that it does not fall under any of the laws, natural or social, that hold for independent systems.

Systems of different kinds have different compositions or different structures. (A difference in composition induces a structural difference but not conversely, as shown by the existence of isomers, i.e. systems with the same composition but different structures.) However, all systems of the same genus seem to have the same overall structure or "general plan" –

pardon the anthropomorphism. For example, all atoms consist of nuclei surrounded by electrons, all solids are atomic or ionic lattices inhabited by wandering electrons, and even the overall structure of the skeleton and organs is the same for all vertebrates. (However, the precise characterization of the notion of overall structure is an open problem.)

Structures are often said to come superposed or nested like systems of Chinese boxes. Thus a polypeptide is said to have two structures, one primary or basic (the linear sequence of amino acids), the other secondary and consisting in the configuration of the entire coil. The helical configuration of the DNA molecule is an example of a secondary structure. In turn the secondary structure may determine a tertiary structure, e.g. the folding of the whole double strand into a regular configuration. See Figure 1.3.

In our view there is no such thing as a hierarchy of structures. (Etymologically 'hierarchy' means a set of sacred components ordered by a power or domination relation.) What we do have here is a system of nested systems, i.e. a collection of systems each of which is a subsystem of a larger system (or supersystem). And what molecular biologists call 'primary structure' is the structure of the innermost or core system, the secondary structure is the structure of the next supersystem, and so on. This notion is elucidated by

DEFINITION 1.7. Let σ be a system and call Σ the totality of systems, and

$$N_\sigma = \{\sigma_i \in \Sigma \mid \sigma \prec \sigma_i \;\&\; 1 \leq i \leq n\}$$

a collection of supersystems of σ partially ordered by the subsystem relation $\prec$. Then

(i) N_σ is a system of *nested* systems with core σ;

Fig. 1.3. An imaginary system of Chinese boxes or hierarchy of systems. The primary structure, i.e. the amino acid sequence, is not shown. The secondary structure is the helix, the tertiary the *Z* shape. And the quaternary structure is the way the individual *Z*'s are assembled together, i.e. the double staircase.

(ii) the *primary structure* of σ is the structure of σ itself; the *secondary structure* of σ is the structure of the smallest supersystem of σ in N_σ, i.e. σ_1; in general, the *n-ary structure* of σ is the structure of σ_{n-1}.

1.5. *Level*

Talk of levels of organization (or complexity, integration, or evolution) and of a hierarchy of such has been rampant in science, particularly in biology, for the last half century or so. Unfortunately there is no consensus on the significance of the terms 'level' and 'hierarchy', which are used in a variety of ways and seldom if ever defined (Bunge, 1959b, 1959c). This fuzziness must be blamed not only on scientists but also on philosophers – on the inexact philosophers who despise clarity and on the exact ones who are not aware of the philosophical problems raised by scientific research. Let us attempt to remedy this situation by clarifying one concept of level, and the corresponding one of hierarchy, that are widely used in contemporary science.

The intuitive idea is simple: the things at any given level are composed of things belonging to preceding levels. Thus biospheres are composed of ecosystems, which are composed of populations, which are composed of organisms, which are composed of organs, which are composed of cells, which are composed of organelles, which are composed of molecules, which are composed of atoms, which are composed of so-called elementary particles. One way of exactifying this notion is by means of

DEFINITION 1.8 Let $L = \{L_i \mid 1 \leq i \leq n\}$ be a family of n nonempty sets of concrete things. Then

(i) one level *precedes* another iff all the things in the latter are composed of things in (some or all of) the latter. I.e. for any L_i and L_j in L,

$$L_i < L_j =_{df} (\forall x)[x \in L_j \Rightarrow (\exists y)\,(y \in L_i \;\&\; y \in \mathscr{C}(x))];$$

(ii) a thing *belongs to a given level* iff it is composed of things in (some or all of) the precedings levels. I.e. for any $L_i \in L$:

For any x in L_i: $x \in L_i =_{df} \mathscr{C}(x) \subset \bigcup_{k=1}^{i-1} L_k$;

(iii) $\mathscr{L} = \langle L, < \rangle$ is a *level structure*.

Note the following points. First, a level is not a thing but a set and therefore a concept, though not an idle one. Hence levels cannot act upon one another. In particular the higher levels cannot command or even obey the lower ones. All talk of interlevel action is elliptical or metaphorical, not literal. Second, the relation between levels is neither the part–whole relation nor the set inclusion relation but a *sui generis* relation definable in terms of the former. Third, there is nothing obscure about the notion of level precedence as long as one sticks to the above definition instead of construing '$L_i < L_j$' as "the L_i's are inferior to the L_j's" or in similar guise. Fourth, it is mistaken to call a level structure $\mathscr{L} = \langle L, < \rangle$ a *hierarchy*, because the level order $<$ is not a dominance relation (Bunge, 1973a). Fifth, our concept is so far static: we are not assuming anything about the origin or mode of composition of systems in terms of evolution.

1.6. *Systems Association*

Whether or not two things form a system, they can be assumed to associate (or add physically) to form a third thing. Thus thing a and thing b, no matter how distant and indifferent, may be assumed to form thing $c = a + b$. In other words, the set of things is closed under the operation $+$ of *association*, physical addition, or juxtaposition (Vol. 3, Ch. 1, Sec. 1).

Not so with systems: two systems may or may not associate to form a third. Thus two molecules may not combine to form a system, and two social systems may not merge to form a third. In general the physical addition or association of two things will be a thing but not a system: *systemicity is not conserved.* The environment, the structure and perhaps even the composition of the resulting thing are different from the mere union of the partial compositions, environments, and structures. See Figure 1.4. In short, the set of all systems has no algebraic structure – not even the rather modest one of a semigroup. But, of course, since systems are things, they do comply with the algebra of things. In particular they associate to form further things.

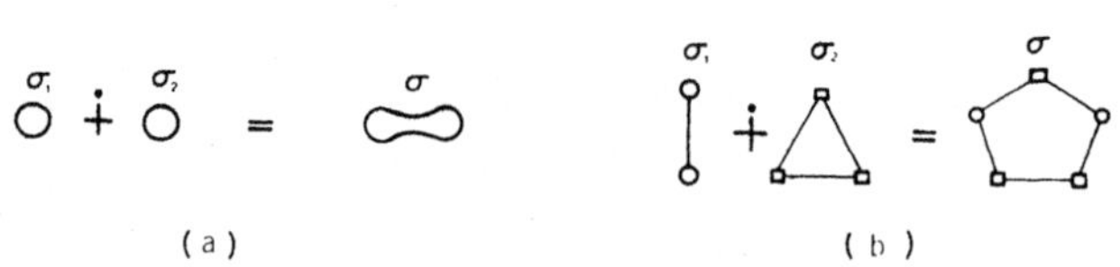

Fig. 1.4. (*a*) Before fusion, $\mathscr{E}(\sigma_1) = \{\sigma_2\}$, $\mathscr{E}(\sigma_2) = \{\sigma_1\}$. After fusion, $\mathscr{E}(\sigma_1 + \sigma_2) = \varnothing$. (*b*) Before merger, $\mathscr{S}(\sigma_1) = \{\text{Linear link}\}$ and $\mathscr{S}(\sigma_2) = \{\text{Triangular link}\}$. After merger, $\mathscr{S}(\sigma_1 + \sigma_2) = \{\text{Pentagonal link}\}$.

1.7. *Other Kinds of System: Property and Functional*

We are interested not only in concrete systems but also in property systems, or sets of interrelated properties, as well as in functional systems, or sets of coupled processes. For example, most of the properties of a thing that is either simple (basic) or a system hang together, i.e. a change in one of them is accompanied by changes in others. As a consequence most of the changes occurring in either a simple thing or a system are coupled, so that if one of them starts or stops, others change. (The cautious prefix 'most' is intended to exclude superficial properties, such as position and color, which can often change considerably, within bounds, without dragging changes in other properties.) Although every thing other than an aggregate or conglomerate has properties and undergoes processes that constitute systems, property systems and functional systems are particularly conspicuous among organisms. In particular the mental abilities of an animal form a system.

We shall use the following conventions:

DEFINITION 1.9 Let $p(x)$ be the set of properties of a thing x, and $\pi(x)$ the set of processes occurring in x. Then

(i) the subset $p_0(x) \subset p(x)$ is a *property system* of x iff every property in $p_0(x)$ is lawfully related to at least one other property in $p_0(x)$;

(ii) the subset $\pi_0(x) \subset \pi(x)$ is a *functional system* of x iff every process in $\pi_0(x)$ is lawfully related to at least one other process in $\pi_0(x)$.

Because in a system all properties and processes are lawfully interrelated we conclude that, for every x, x is a concrete system iff $p(x)$ is a property system or $\pi(x)$ a functional system.

1.8. *Concluding Remarks*

The literature on systems is vast, rapidly growing, and somewhat bewildering. (Cf. Klir and Rogers, 1977.) However, the field is still immature and its reputation is jeopardized by a fringe of charlatans. Suffice it to mention three indicators of immaturity.

Firstly, the very definition of the concept of a system is still in doubt, so that many a paper starts by spending time defining or redefining the concept. Yet, so much effort spent on definitions has yielded only three which are as popular as they are incorrect. According to the first definition, a system is a set of interrelated elements – which is fine for conceptual systems but not for concrete ones since sets, no matter how structured, are sets, hence concepts not things. The second definition equates a system with a black box equipped with inputs and outputs, which is fine in a few cases but useless when the internal structure of the system is relevant. And the third widely used definition is a generalization of the preceding, namely this: a system is a binary relation – again a conceptual object.

Secondly, some writers claim that everything imaginable is a system, and that a general theory of systems should deal with every possible thing (without thereby becoming part of philosophy) and every possible problem, theoretical or practical, concerning the behavior of systems of all kinds. Some have even asserted that such a theory should cover not only concrete systems but also conceptual ones, so that it would be a thoroughly unified science of everything.

Thirdly, some enthusiasts of general theories of systems have seen in these a vindication of holistic philosophies, hence a condemnation of the analytic method characteristic of science. However, most of those who approve of general systems theories for their alleged holistic virtues either misuse the term 'holistic' to designate "systemic", or are interested in instant wisdom rather than painstaking scientific or philosophical research.

Such confusions and wild claims, which linger because of insufficient foundational research in the field of systemics, have elicited some entirely negative reactions to it (e.g. Berlinski, 1976). While there is some legitimacy in such reactions, there is no denying that systemics abounds in good theories – such as automata theory and general lagrangian dynamics – serviceable in a number of fields, and that it provides an inspiring framework for posing problems and building models. Rather than throw out the baby together with the bath water we ought to change the latter once in a while.

2. System representations

2.1. *Coupling Graphs and Matrices*

We shall presently review two standard and equivalent ways of representing a system with a denumerable composition, be it a molecule or an industrial plant. They are the graph and the matrix representations. (See Klir and Valach (1967).) The following examples show how to proceed.

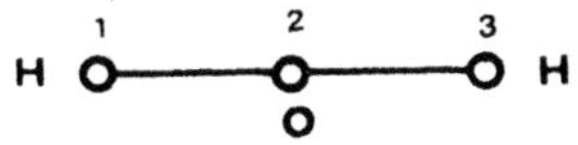

$$\sigma_1 = \begin{Vmatrix} 0 & 1 & 0 \\ 1 & 0 & 1 \\ 0 & 1 & 0 \end{Vmatrix}$$

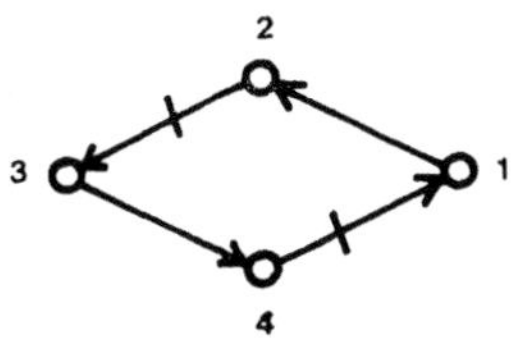

$$\sigma_2 = \begin{Vmatrix} 0 & 1 & 0 & 0 \\ 0 & 0 & -1 & 0 \\ 0 & 0 & 0 & 1 \\ -1 & 0 & 0 & 0 \end{Vmatrix}$$

The arrows indicate excitation, the crossed arrows inhibition.

$$\sigma_3 = \begin{Vmatrix} 0 & 1 & 0 \\ 0 & 1 & 1 \\ 0 & 1 & 1 \end{Vmatrix}$$

The loops indicate self action or feedback.

We generalize the foregoing into the following semantic assumptions. Let σ be a system with m components and n different kinds of connection among them (e.g. mechanical, chemical, informational, social, etc.). Then σ is representable by

(i) a set of n directed graphs over the composition of σ, one for each kind of connection, with a total of m nodes (vertices), in such a way that (*a*) the nodes represent the components and (*b*) the edges represent the connections; or

(ii) a set of n $m \times m$ matrices ${}^pM$, where $1 \leq p \leq n$, such that (*a*) the matrix element ${}^pM_{rs}$ of the pth matrix represents the strength of the action of component r, in the pth respect, upon component s, and (*b*) the matrix element ${}^pM_{rr}$ stands for the action of kind p of the rth component on itself (feedback).

The off-diagonal elements ${}^pM_{rs}$, with $r \neq s$, represent connections other than self connections. There are $m^2 - m = m(m - 1)$ such elements per matrix, and a total of $nm(m - 1)$ per system with n different kinds of connection. This number is called the *coupling capacity* of the system.

So far we have represented the composition and the internal structure of a system while neglecting its environment, hence its external structure. An open system, i.e. one connected with its environment, can be represented as follows. Instead of building an $m \times m$ matrix for an m component system, as directed by the previous semantic postulate, we form an $(m + 1) \times (m + 1)$ matrix for each kind of connection, letting 0 stand for the environment *en bloc*. Any system component r for which $M_{0r} \neq 0$ is an input or receiver component, whereas s is an output or donor component of the system if $M_{s0} \neq 0$. For example, a two component open system with a single kind of connection can be represented by the matrix

$$M = \begin{Vmatrix} 0 & M_{01} & M_{02} \\ M_{10} & M_{11} & M_{12} \\ M_{20} & M_{21} & M_{22} \end{Vmatrix}.$$

The elements M_{01} and M_{02} are the inputs (to the first and second components respectively) and the entries M_{10} and M_{20} the outputs (of the first and second components respectively). The other entries represent the internal (or internuncial) connections among the system's components.

We generalize the preceding into the following semantic assumption. Let σ be a system with m components and n different kinds of connections among them. Moreover, let the environment of σ be construed as a single entity labelled 0. Then σ is representable by n $(m + 1) \times (m + 1)$ matrices ${}^pM$, where $1 \leq p \leq n$, such that

(i) the *internal connectivity* of σ in the pth respect is representable by the matrix obtained from ${}^pM$ by striking off the M_{r0} and M_{0s} elements;

(ii) the *input* to σ in the respect p is represented by the row of input entries of ${}^pM$, i.e.

$${}^p\mathcal{I}(\sigma) = \| {}^pM_{01} \quad {}^pM_{02} \ldots {}^pM_{0m} \|;$$

(iii) the *output* of σ in the respect p is represented by the column of output entries of ${}^pM$, i.e.

$${}^p\mathcal{O}(\sigma) = \| {}^pM_{10} \quad {}^pM_{20} \ldots {}^pM_{m0} \|^t,$$

where t designates the transposition operation (conversion of row matrix into column matrix);

(iv) the *behavior* (or *observable performance*) of σ in the respect p is the ordered pair

$${}^p\beta(\sigma) = \langle {}^p\mathcal{I}(\sigma), {}^p\mathcal{O}(\sigma) \rangle;$$

(v) the (total) *behavior* of σ is the set of its partial behaviors:

$$\beta(\sigma) = \{{}^p\beta(\sigma) \mid 1 \leq p \leq n\}.$$

Example In the simplest case, of a two component system interacting with its environment in a single way, we have

$$\mathscr{I}(\sigma) = \|M_{01} \quad M_{02}\|, \qquad \mathscr{O}(\sigma) = \left\| \begin{matrix} M_{10} \\ M_{20} \end{matrix} \right\|.$$

In the absence of any data or hypotheses concerning the internal structure (i.e. the full coupling matrix) of such a system, we must restrict our attention to its behavior. The best we can do is to guess that the latter is linear, i.e. that there exists a matrix T transforming inputs into outputs: $\mathscr{O} = T\mathscr{I}^t$, where $\mathscr{I}^t$ is the transpose of $\mathscr{I}$. We set then

$$T = \left\| \begin{matrix} T_{11} & T_{12} \\ T_{21} & T_{22} \end{matrix} \right\|$$

with unknown T_{ij}'s and perform the indicated operations:

$$T\mathscr{I}^t = \left\| \begin{matrix} T_{11} & T_{12} \\ T_{21} & T_{22} \end{matrix} \right\| \cdot \left\| \begin{matrix} M_{01} \\ M_{02} \end{matrix} \right\| = \left\| \begin{matrix} T_{11}M_{01} + T_{12}M_{02} \\ T_{21}M_{01} + T_{22}M_{02} \end{matrix} \right\| = \left\| \begin{matrix} M_{10} \\ M_{20} \end{matrix} \right\|,$$

thus obtaining the algebraic system

$$\begin{aligned} T_{11}M_{01} + T_{12}M_{02} &= M_{10} \\ T_{21}M_{01} + T_{22}M_{22} &= M_{20}. \end{aligned}$$

This system of equations has no unique solution when only the behavior of the concrete system (i.e. the M_{0i}'s and the M_{j0}'s) are given, for in this case there are only two conditions (equations) for four unknowns (the T_{ij}'s). Even hitting upon a solution by trial and error won't advance us a single step in the process of finding out the structure of the system, i.e. the complete coupling matrix M. The only procedure that could bring success is to guess and try out alternative assumptions about the system's structure and check whether they do yield the observed (or conjectured) behavior. That is, the way to theoretical knowedge is not from behavior to inferred structure but from hypothesized structure to behavior. This shows that behaviorism, phenomenalism, and inductivism are incapable, not just unwilling, to explain behavior.

Obviously neither the graph nor the matrix representation of a system suffices for all purposes. It represents only the composition, structure, and environment of a system with neglect of its dynamics. A more complete representation can only be obtained by setting up a full fledged dynamical theory incorporating and expanding the information contained in the graph or the matrix representation. We turn next to the common core of such dynamical representations, namely the state space representation. (See the Appendices for a number of particular, yet also cross-disciplinary, mathematical models of systems.)

2.2. *The State Space Representation*

Every system of a given kind K has a finite number n of general properties, such as age, number of components, connectivity among them, inputs, and outputs. And each such general property is representable by a function $F_i: A \rightarrow V_i$, where $1 \leq i \leq n$. Collecting all such property-representing functions into a single ordered n-tuple or list

$$\mathbb{F} = \langle F_1, F_2, \ldots, F_n \rangle : A \longrightarrow V_1 \times V_2 \times \ldots \times V_n,$$

we form the *state function* of systems of the given kind. Just as $\mathbb{F}$ represents the totality of general properties of the K's, so each value $\mathbb{F}(a) = \langle F_1(a), F_2(a), \ldots, F_n(a) \rangle$ represents the totality of individual properties of a particular system, such as its age and composition at a given time.

The domain A of the state function $\mathbb{F}$ of systems of kind K is the cartesian product of certain sets, such as K, the family 2^E of sets of environmental items with which the members of K are coupled, the set F of reference frames, the set T of time instants, and so on. (2^E is the power set of the set E of environmental things, so the environment e of a particular system is a member of that family, i.e. $e \in 2^E$.) And the codomain V_i of the ith component F_i of the state function is usually taken to be some subset of the real line $\mathbb{R}$. (If a property is represented by a complex valued function each component of the latter counts as a component of $\mathbb{F}$.) In short,

$$\mathbb{F} : K \times 2^E \times F \times T \times \cdots \longrightarrow \mathbb{R}^n.$$

The value $\mathbb{F}(k, e, f, t, \ldots) = \langle a, b, \ldots, n \rangle \in \mathbb{R}^n$ of the state function of system $k \in K$ interacting with environmental items $e \in 2^E$, relative to reference frame $f \in F$ at time $t \in T$, is the *state* of k at t. The collection of all such possible states, which is a subset of $\mathbb{R}^n$, is the (conceivable) *state space* of systems of kind K, or $S(K)$ for short. However, since the components of $\mathbb{F}$ are lawfully interrelated, and thus mutually restricted, not every n-tuple of real numbers represents a really (or nomologically) possible state of a system. I.e. the *lawful state space* of systems of kind K, or $S_L(K)$ for short, is a proper subset of the conceivable state space $S(K)$.

In short, every really possible state of a K is a point in some region $S_{\mathbb{L}}(K)$ of the cartesian space $\mathbb{R}^n$. See Figure 1.5.

Example 1 In the elementary kinetic theory of gases, the state function is the triple consisting of the pressure, volume, and temperature functions. The corresponding state space is a cube contained in $(\mathbb{R}^+)^3$. *Example 2* In Hamiltonian dynamics the state (or phase) vector is $\langle q(k, f, t), p(k, f, t)\rangle$, where q is the canonical coordinate and p the corresponding momentum – neither of which need be mechanical properties. *Example 3* In chemical kinetics the instantaneous state of a chemical system is described by the values of the partial concentrations of reactants and reaction products. Therefore the state space of the system is a hypercube contained in $(\mathbb{R}^+)^n$, where n is the number of system components (reactants, catalyzers, and products). If there is diffusion, further axes must be added to the state space, in particular temperature and position coordinates. *Example 4* In the genetics of populations three commonly used state variables are the size of a population, the probability (incorrectly called "frequency') of some particular gene or gene constellation, and the latter's adaptive value. Hence for a system composed of two interacting populations, A and B, the state space is the region of $\mathbb{R}^6$ spanned by the sextuples $\langle N_A(t), N_B(t), P_A(t), P_B(t), v_A(t), v_B(t)\rangle$ in the course of time.

The concept of state space can be used to clarify that of system. The state space of an aggregate or conglomerate of non-interacting things is uniquely determined by the partial states spaces. Moreover, since the

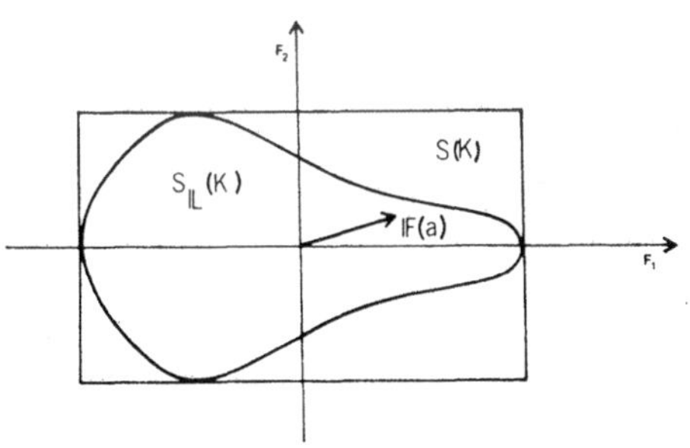

Fig. 1.5. The lawful state space $S_{\mathbb{L}}(K)$ of systems of kind K is a subset of the cartesian product of the ranges of the components of the state function. Only two such components, F_1 and F_2, are depicted here. $\mathbb{F}(a) = \langle F_1(a), F_2(a)\rangle$ is a (really) possible state of a particular system of the given kind. As time 'goes by', the tip of $\mathbb{F}(a)$ moves about within $S_{\mathbb{L}}(K)$.

contributions of the latter are all on the same footing, we may take the total state space to equal the union of the partial state spaces. In particular, let $S_L(K)$ and $S_L(M)$ be the lawful states spaces of things of kinds K and M respectively. Then the state space of the association $k + m$ of two non-interacting things of kinds K and M respectively, relative to the same reference frame, is $S_L(K) \cup S_L(M)$. Not so in the case of a system: here the state of every component is determined, at least partly, by the states other system components are in, so that the total state space is no longer the union of the partial state spaces. Thus in Example 4 above the state space of the two component system must be constructed *ab initio* rather than on the sole basis of the state spaces for the individual biopopulations. In sum, a thing is an *aggregate* if and only if its state space equals the union of the state spaces of its components – otherwise it is a (concrete) *system*. (Cf. Bunge, 1977a, 1977b.)

Every event occurs in or to some concrete thing and it consists in a change of state of the thing – the change being merely quantitative, as in the case of motion, or qualitative as well, as in the case of the coming into being or the metamorphosis of a thing. A light flash, the dissociation of a molecule, a storm, the growth of a bud, the learning of a trick, and the fall of a government are events – or rather processes, for they are complex and therefore analyzable into further events. Being changes in the states of things, events and processes are representable as trajectories in the state spaces of changing things. (An unchanging thing, if there were one, would have a state space consisting of a single point.) Different trajectories in a state space may have the same end points. That is, there are cases in which one and the same net change can be effected along alternative routes. See Figure 1.6.

The functions g and g' occurring in Figure 1.6 are not supposed to be arbitrary: they must be lawful if we are to allow only lawful events and discard lawless ones, i.e. miracles. In other words, the g occurring in the event representation $e = \langle s, s', g \rangle$ must be compatible with the laws of the system(s) concerned. Equivalently: a *lawful event or process* occurring in a system of kind K, with end points s and s', is representable by a triple $\langle s, s', g \rangle$, where $g : S_L(K) \to S_L(K)$ is compatible with the laws of the K's. If we disregard the intermediate states between the end points of the processes, we are left with arrows or ordered pairs $\langle s, s' \rangle \in S_L(K) \times S_L(K)$. The collection of all such pairs of states, i.e. the set of all net events (for a given g) constitutes the *event space* of the K's (for g). Symbol:

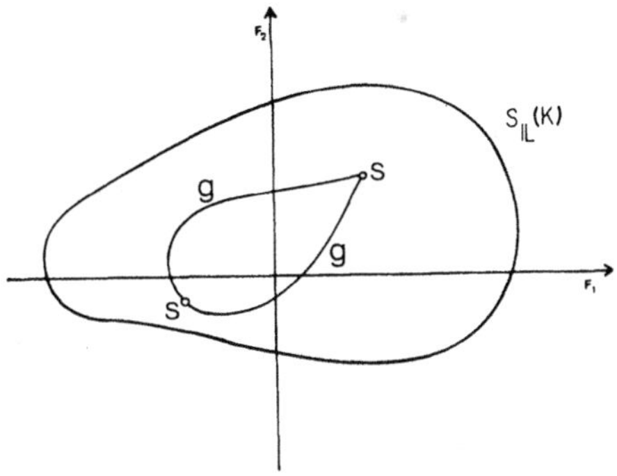

Fig. 1.6. Two different processes resulting in the same net change. The net change from state s to state s' can be represented as the ordered pair (or arrow) $\langle s, s' \rangle$. Since the change along curve g may be distinct from the change along curve $g' \neq g$, we must represent the full events (or processes) by $\langle s, s', g \rangle$ and $\langle s, s', g' \rangle$ respectively.

$E_L(K) \subseteq S_L(K) \times S_L(K)$. In general the inclusion is proper: not all state transitions are lawful.

Since we shall make ample use of the state space representation in this work, we may as well compress the preceding into the following semantic assumption. For each kind K of system possessing n properties there is a property-representing function $\mathbb{F}: A \to V_1 \times V_2 \times \ldots \times V_n$ with n components, called a *state function* for systems of the kind. Furthermore

(i) the *totality of general properties* of systems of kind K is representable by the set of all the components (or coordinates) of $\mathbb{F}$, i.e. $p(K) = \{F_i \mid 1 \leq i \leq n\}$;

(ii) every *particular property* of a system of kind K is representable by a value of a component of $\mathbb{F}$, i.e. by $F_i(a)$ for $a \in A$ and some $1 \leq i \leq n$;

(iii) the *state* of systems of kind K at $a \in A$ is representable by the value of $\mathbb{F}$ at a, i.e. $s = \mathbb{F}(a) = \langle F_1(a), F_2(a), \ldots, F_n(a) \rangle$;

(iv) the collection of all such states of things of kind K, i.e. the range of $\mathbb{F}$, is the *lawful state space* of the K's, or $S_L(K)$ for short;

(v) every *event* occurring in a system of kind K is representable by an ordered triple $\langle s, s', g \rangle$, where $s, s' \in S_L(K)$ and g is a lawful map of $S_L(K)$ into itself;

(vi) the collection of all really possible (lawful) events occurring in systems of kind K is the *event space* of K, or $E_L(K)$ for short;

(vii) for a system in a given environment, and relative to a given reference frame, the state function takes often the form of a time-dependent function $\mathbb{F}: T \to \mathbb{R}^n$, where $T \subseteq \mathbb{R}$ is the set of instants relative to the given frame;

(viii) If $\mathbb{F}: T \to \mathbb{R}^n$, then the totality of *processes* occurring in a system x of kind K during the time interval $\tau \subseteq T$ is representable by the set of states x is in during τ:

$$\pi(x, \tau) = \{\mathbb{F}(t) \mid t \in \tau\};$$

(ix) the *history* of a system x of kind K representable by a state function $\mathbb{F} : T \to \mathbb{R}^n$, during the interval $\tau \subseteq T$, is representable by the trajectory

$$h(x) = \{\langle t, \mathbb{F}(t)\rangle \mid t \in \tau\};$$

(x) the *total action* (or *effect*) of a thing x on a thing y equals the difference between the forced trajectory and the free trajectory of the patient y:

$$A(x, y) = h(y \mid x) \cap \overline{h(y)}.$$

A detailed treatment of these concepts is given elsewhere (Vol. 3, Ch. 5). We shall presently use them to advance a handful of general principles concerning systems.

3. Basic assumptions

3.1. *Structural Matters*

So far we have made only a few definitions and semantic assumptions but no substantive hypotheses on the nature of systems. (Theorem 1.1 on the existence of systems and the nonexistence of stray things, as well as Corollary 1.1 on the openness of systems, followed from our definition of a concrete system in conjunction with certain general postulates, about the nature of things, laid down in Volume 3.) We shall presently wager a handful of basic assumptions concerning systems of all kinds, the first being certain postulates of a structural kind. Since these assumptions will concern the transactions of a system with its environment, we may as well define the concepts of input and of output in more general terms than we did in Sec. 2.1. We start then with

DEFINITION 1.10 Let σ be a system with an (immediate) environment $\mathscr{E}(\sigma)$. Then

(i) the totality of *inputs* of σ is the set of all the environmental actions on σ:

$$U(\sigma) = \bigcup_{x \in \mathscr{E}(\sigma)} A(x, \sigma);$$

(ii) the totality of *outputs* of σ is the set of all the actions of the system on its environment:

$$V(\sigma) = \bigcup_{y \in \mathscr{E}(\sigma)} A(\sigma, y);$$

(iii) the *activity of the environment* of σ is

$$E(\sigma) = \bigcup_{x, y \in \mathscr{E}(\sigma)} A(x, y) \cup U(\sigma) \cup V(\sigma).$$

Our first hypothesis is that all systems receive inputs and are selective, i.e. accept only a (small) subset of the totality of environmental actions impinging on them. More precisely, we lay down

POSTULATE 1.1. Let σ be a system with total input $U(\sigma)$. Then
(i) $U(\sigma) \neq \varnothing$;
(ii) $U(\sigma) \subset E(\sigma)$ or, equivalently, the (in-selection) function

$$\mu : U(\sigma) \longrightarrow E(\sigma)$$

is the inclusion (or embedding) map of $U(\sigma)$ into $E(\sigma)$.

Example Talking to plants is ineffectual – except insofar as it nourishes them with water and carbon dioxyde.

A second, equally pervasive, feature of concrete systems is that they react on their environment, i.e. that their output is never nil. (The so-called machines without output, studied in automata theory, are of course fictions.) Moreover, in every system there is spontaneous activity, i.e. not elicited by any inputs. Thus we make

POSTULATE 1.2. Let $V(\sigma)$ be the total output of a system σ. Then
(i) $V(\sigma) \neq \varnothing$;
(ii) the (out-selection) function

$$\nu : V(\sigma) \longrightarrow E(\sigma)$$

assigns each system output an environmental action but not conversely.

Example In every neuron there is spontaneous activity that is superposed on the activity elicited by afferent stimulation.

The previous hypotheses are typical metaphysical principles in as much as they can be confirmed but not refuted, for they hinge on an item that is only partially knowable, namely the set E of environmental actions. Any evidence unfavorable to our postulates may be blamed on our ignorance of most of E.

The in-selection function μ and the out-selection function ν are joined in

DEFINITION 1.11 The function f that composes with the out-selection function ν of a system to yield its in-selection function μ, i.e. such that $\mu = f \circ \nu$, is called the *transfer* (or *transducer*) function f: $U(\sigma) \to V(\sigma)$ of σ:

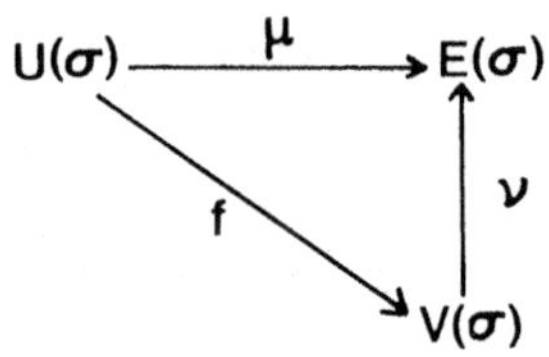

Example The retina transforms (or maps or codes) light stimuli into nervous signals.

We close this subsection with a batch of general principles that will be stated in an informal way:

POSTULATE 1.3 Let σ be an arbitrary system other than the universe. Then

(i) every input to σ is an output of some other system (i.e. there are no inputs out of the blue);

(ii) σ receives inputs of several kinds (i.e. at some time or other every one of the components of the state function of σ is bound to be affected by environmental changes);

(iii) for every action upon σ there is a threshold below which σ does not respond;

(iv) the total input of σ has a nonvanishing random component;

(v) there is a delay, however small, between every input and the corresponding output if any.

So much for our general structural assumptions. Let us now look at systems from an evolutionary perspective.

3.2. *Assembly and Emergence*

Any process whereby a system is formed out of is components is called 'assembly'; if the process is spontaenous, it is called 'self-assembly'. An assembly process can occur either in a single step or, more likely, in a number of steps: see Figure 1.7. We can express the idea formally with the help of the concept of bondage, or set of bonds among the components of a system, introduced in Sec. 1.2. Indeed we make

DEFINITION 1.12 Let x be a concrete thing composed initially of uncoupled parts (possibly systems themselves), i.e. such that $\mathbb{B}(x, t) = \varnothing$. Then

(i) x *assembles into* y at time $t' > t$ iff y is a system with the same composition as x but a nonempty bondage, i.e.

$$\mathscr{C}(y, t') = \mathscr{C}(x, t) \;\&\; \mathbb{B}(y, t') \neq \varnothing;$$

(ii) the assembly process is one of *self-assembly* iff the aggregate x turns by itself [i.e. naturally rather than artificially] into the system y;

(iii) the self-assembly process is one of *self-organization* iff the resulting system is composed of subsystems that were not in existence prior to the onset of the process.

Assembly processes can be natural or artificial, and those of the latter kind in turn experimental (or laboratory) or industrial. Artificial assembly processes are of course man-guided. However, there are degrees of control. It is one thing to assemble a machine from its parts and another to assemble a molecule from its precursors. In most cases the latter process

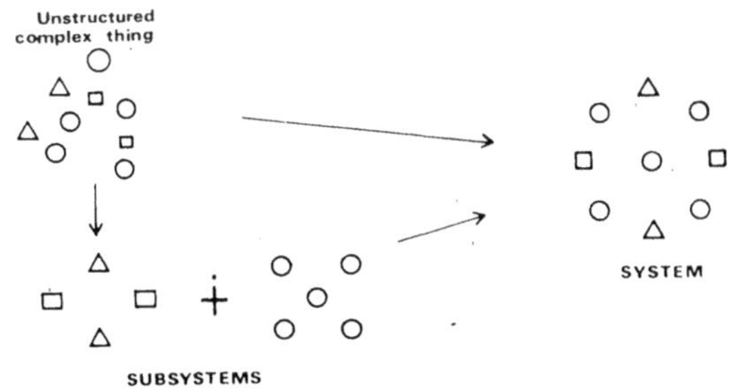

Fig. 1.7. Assembly of a system out of previously unconnected units, either directly or step-wise.

proceeds on its own – i.e. by virtue of its inner dynamics – once suitable reactants and physical conditions have been supplied. For example, proteins and even ribosome units will self-assemble *in vitro* in a matter of minutes if suitable precursors and physical conditions are provided. In this case man only lends a hand to nature, by reproducing conditions that have or could have occurred spontaneously.

All chemical and biochemical syntheses are of course assembly processes and moreover processes accompanied by the emergence of new properties. However, self-assembly occurs on all levels in a variety of manners. Perhaps the most conspicuous of all self-assembly process is the aggregation or clumping of atoms and cosmic dust brought about by gravitation. This is how planets and other celestial bodies are believed to have formed. Moreover, far from always leading to disorganized lumps, gravitational attraction can give rise to complex clustering on all astronomical scales, in particular star clusters, galaxies, and galaxy clusters. (See, e.g. de Vaucouleurs, (1970).) Similar examples of self-assembly by condensation of units of the same kind are the formation of molecular complexes (such as the haemocyanine polymolecule), of polymers such as the polypeptides and of crystals out of solutions. (See Calvin (1969); Eigen (1971); Lehninger (1975).) Needless to say, self-assembly, and in particular self-organization, occurs also on the social level: witness the formation of families, bands, communities, and social organizations of various kinds. In sum self-assembly and self-organization are not exclusive to life. What is peculiar to biotic self-organization is that it happens to ensue in living systems rather than in biochemical systems or in systems of some other kind. (In other words, it is an emergence mechanism leading to a new level of organization.) We shall assume that all systems short of the entire world have formed by assembly:

POSTULATE 1.4 All systems, except the universe, originate by assembly.

Remark 1 Natural systems originate by self-assembly, and artificial systems by artificial or man-made assembly. *Remark 2* An exception is made for the universe because (*a*) in naturalistic cosmologies the universe has neither an origin nor an end, and (*b*) in religious cosmologies it makes no sense to say that the universe originates by assembly, as this presupposes the prior existence of its components. *Remark 3* Surely a system may be formed by the breakdown of some supersystem. However, this is no counterexample to our postulate, which requires only that the original supersystem was the result of some assembly process. *Remark 4* Our axiom is far from obvious and it could not have been formulated before systems-theoretic thinking became pervasive, if only because the general concept of a system had not been elucidated until recently. *Remark 5* Our axiom is of particular relevance to the problem of the origin of biomolecules and biosystems. Until recently both were assumed either to have been created by divine *fiat* or to have existed from all eternity. Darwin's *Origin of Species* appeared only in 1859, and scientific research into the evolution of molecules started only one century later.

The components of a self-assembled system are called its *precursors*, an apt name suggesting that the system had not always been around but emerged from pre-existing things. Interestingly enough, (*a*) the precursors of a system are not blended but keep their individuality to some extent, and yet (*b*) they give rise to a thing possessing emergent properties. For example, the two hydrogen atoms that combine into a hydrogen molecule are distinct components of the latter, yet the spectrum of the molecule is radically different from that of its components. The first characteristic is formalized by saying that the composition of the self-assembled system equals the set of parts of its ancestors – i.e. the set of its precursors. And the notion of emergence can be elucidated as follows.

Call x a thing and $t \in T$ an instant of time, and introduce a function p that assigns the ordered couple $\langle x, t \rangle$ the set $p(x, t)$ of all the properties of x at t. That is, p is a function $p: \Theta \times T \to 2^{\mathbb{P}}$, where Θ is the set of all things, T the set of all instants, and $2^{\mathbb{P}}$ the family of subsets of the set $\mathbb{P}$ of all general properties of things. A change in thing x can be viewed as a certain change of state of x. Since x is held fixed throughout that change of state, we can use the simpler function

$$p_x: T \to 2^{\mathbb{P}} \quad \text{such that } p_x(t) = p(x, t).$$

In short, $p_x(t)$ is the collection of properties of thing x at time t. (For details on p_x, see Vol. 3, Ch. 2.)

Now let t and t' be two distinct instants, such that t precedes t'. The corresponding values of p_x are of course $p_x(t)$ and $p_x(t')$. If these two sets of properties of x are the same, then the thing has not changed qualitatively. If they are different, then the thing has gained or lost some properties. If the latter is the case, the newly gained properties will be said to be *emergent* relative to the given thing, even though they may be possessed by other things as well. In sum, we make

DEFINITION 1.13 Let x be a thing with properties $p_x(t)$ at time t, and properties $p_x(t')$ at a later time $t' > t$. Then

(i) the *total qualitative novelty* occurring in x during the interval $[t, t']$ is the symmetric difference

$$n_x(t, t') = p_x(t) \triangle \bigcup_{t<\tau\leq t'} p_x(\tau);$$

(ii) the *emergent properties* that appear in x during the interval $[t, t']$ are those in

$$e_x(t, t') = \bigcup_{t<\tau\leq t'} p_x(\tau) - p_x(t).$$

Example 1 Every nuclear and every chemical reaction results in the emergence of things endowed with emergent properties. *Example 2* The breakdown (dismantling) of a system, and the substitution of some of its components, are emergence processes.

We make bold and generalize:

POSTULATE 1.5 Every assembly process is accompanied by the emergence of some properties and the loss of others. I.e. let the parts of a thing x self-assemble into a system during the interval $[t, t']$. Then the system lacks some of the properties of its precursors – i.e. $p_x(t) - p_x(t') \neq \varnothing$–but on the other hand it possesses some new properties – i.e. $p_x(t') - p_x(t) \neq \varnothing$.

Thus far we have been concerned with qualitative novelty in a particular thing regardless of whether the given emergent properties are possessed by any other things. We now elucidate the concept of emergence for the first time, or absolute emergence:

DEFINITION 1.14 The *absolutely emergent properties* (or firsts) appearing in a thing x during the lapse $[t, t']$ are those in

$$e_x^a(t, t') = e_x(t, t') - \bigcup_{y\in\Theta} \bigcup_{\tau\in T} p_y(\tau) \quad \text{with } y \neq x \quad \text{and} \quad \tau \leq t'.$$

3.3. *Selection*

New systems are being formed all the time, but not all of them are viable in the environment in which they emerge. In fact many are unfit, hence ephemeral, and this either because they are internally unstable or because they cannot cope with environmental aggression. In the latter case we have to do with selection. In other words, the environment selects the fittest systems, be they molecules or men. We can formulate this idea with more precision in the following way.

We assume that every environment performs a selective or filtering action on any population of systems of some kind. This selective action consists in the shrinking of the original population S to some subset A of S, namely the collection of viable or adapted members of S. A permissive environment is such that A is almost as big as the original S, whereas a harsh environment will reduce A to a very small, possibly empty, subset A of S. We condense these ideas into a definition and a postulate.

DEFINITION 1.15 Let S be a set of systems of a given kind K, assembled during some time interval in a given environment $E = \mathscr{E}(\sigma)$ common to all $\sigma \subset S$. Further, call $i_E: S \rightarrow A_E$ the inclusion function from S into A_E, where $A_E \subseteq S$. (I.e. $i_E(x) = x$ for any x in A_E.) Then

(i) the environment E exerts the *selectivw action*

$$i_E: S \longrightarrow A_E$$

on the population S iff, during the next time interval, only the members of A_E remain in S;

(ii) A_E is the set of systems of kind K *selected by* (or *adapted to*) the environment E, and $\overline{A_E} = S - A_E$ the collection of systems of the same kind *eliminated by* (or *ill adapted to*) E, during the given time interval;

(iii) the *selection pressure* exerted by E on S is the number

$$p(S, E) = \frac{|\overline{A_E}|}{|S|} = 1 - \frac{|A_E|}{|S|},$$

where '$|X|$' designates the numerosity of the set X.

Clearly, the p values range between 0 (maximal fitness) and 1 (maximal unfitness).

Our promised hypothesis reads thus:

POSTULATE 1.6 All systems are subject to environmental selection (or to a nonvanishing selection presssure). I.e. for every set S of any kind K and every environment E common to the members of S, there is a selective action function $i_E: S \rightarrow A_E$ with $A_E \subset S$.

An immediate consequence of this axiom together with Postulate 1.4 is

COROLLARY 1.2 All self-assembled systems are subject to environmental selection (or to a nonvanishing selection pressure).

Different environments may exert different selective pressures on one and the same system population. (This is why for every S we have as many inclusion maps i_E as possible environments of the S's.) For example, two different habitats, or two different seasons in the same habitat, may exert different selective actions upon one and the same population of organisms. And such actions compose in accordance with

THEOREM 1.2 Let E and E' be two different consecutive environments of members of a population S of systems of some kind, and let i_E and $i_{E'}$ be their respective selective actions (during the corresponding consecutive periods). Then the resulting selective action is the composition of the two partial selective actions, i.e.

$$i_{EE'} = i_{E'} \circ i_E,$$

and the corresponding selection pressure equals

$$p(S, EE') = p(S, E) \cdot p(A_E, E') = \left(1 - \frac{|A_E|}{|S|}\right) \cdot \left(1 - \frac{|A_{E'}|}{|A_E|}\right).$$

It should be obvious that the selective actions of two environments do not commute in general. So much so that, if the first environment is totally hostile, there is nothing left for the second to select from. The last environment always has the last word. See Figure 1.8.

It is often maintained that the environment is not creative, for all it is capable of is weeding out the maladapted. This is false. Firstly, the environment of every system includes further systems, some of which are capable of acquiring new properties. Secondly, every new system is as-

Fig. 1.8. The selective action of E followed by E' (diagram (*a*)) differs from the resulting selective action in the reverse order (diagram (*b*)).

sembled from units supplied by the environment: the latter provides the opportunity for self-assembly, hence for emergence. In sum, the environment of any system is creative – only, it is selective and exclusive rather than permissive.

3.4. *Evolution*

Self-assembly may result in evolution – and this at the molecular, the organismic, the populational, and other levels. Let us then examine the general concept of evolution. To this end let us begin by elucidating the general concept of descent:

DEFINITION 1.16 Let S be a collection of systems of some kind. Then for any x and y in S,

(i) x is an *immediate ancestor* of y (or y *descends immediately* from x) iff x and y belong to the same species and x or a part of x is a precursor in the assembly of y;

(ii) x is a *mediate ancestor* of y (or y *descends mediately* from x) iff there is a z in S such that x is an immediate ancestor of z, and z an immediate ancestor of y;

(iii) x is an *ancestor of* y (or y *descends* from x) iff x is either an immediate or a mediate ancestor of y. Symbol: $x < y$;

(iv) the *ancestry* of x is the collection of ancestors of x:

$$A(x) = \{y \in S \mid y < x\};$$

(v) the *progeny* of x is the collection of things of which x is an ancestor:

$$P(x) = \{y \in S \mid x < y\};$$

(vi) the *lineage* of x is the union of the ancestry and the progeny of x:

$$L(x) = \{y \in S \mid y < x \quad \text{or} \quad x < y\}.$$

Clearly, the ancestry relation $<$ is a strict partial order. The graph of $<$ is then directed with single edges and no loops.

Example 1 Hydrocarbons descend from C and H, polymers from the respective monomers, ribosomes from RNA molecules and proteins, animals from their parents, etc. *Example 2* The progeny of a bacterium reproducing by binary division looks like this:

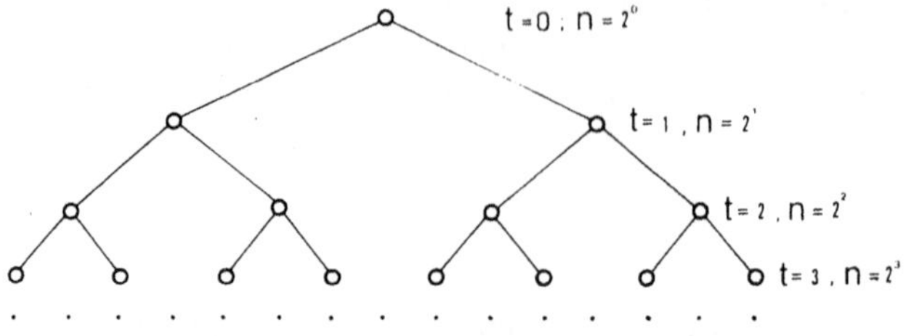

$t = 0, \quad n = 2^0$

$t = 1, \quad n = 2^1$

$t = 2, \quad n = 2^2$

$t = 3, \quad n = 2^3$

While in many cases the members of the lineage of a system belong to the same species, sometimes they do not. When they don't, the lineage constitutes an evolutionary line. More precisely, we make

DEFINITION 1.17 Let $L(x)$ be the lineage of a system x of kind K. Then $L(x)$ is an *evolutionary lineage* iff at least one ancestor or one descendant of x belongs to a kind K' different from K.

The notions of descent, lineage and evolutionary lineage carry over to collections of systems:

DEFINITION 1.18 Let S be a family of collections of systems: $S = \{S_i$ is a collection of systems $| \; 1 \leq i \leq n\}$. Then for all S_j and S_k in S,

(i) S_k *descends* from S_j iff every member of S_k descends from some members of S_j: $S_j \prec S_k$;

(ii) the *lineage* of S_j is the family of collections descending from S_j or from which S_j descends:

$$\Lambda(S_j) = \{X \in S \mid X \prec S_j \quad \text{or} \quad S_j \prec X\};$$

(iii) $\Lambda(S_j)$ is an *evolutionary lineage* iff at least one of the ancestors or the descendants of S is included in a species different from that which includes S_j.

Recall now that, according to Postulate 1.4, every system is formed by assembly. Since a system and its precursors belong to different species, it follows that assembly ensues in speciation, i.e. the formation of new species. In other words, we have

THEOREM 1.3 Every concrete system belongs to some evolutionary lineage.

In closing this subsection let us emphasize that the preceding hypotheses are assumed to hold for systems of all kinds – physical, chemical, biological, or social. The peculiarities of systems of these system genera will be studied in succeeding chapters.

4. Systemicity

4.1. *Integration, Cohesion, Coordination*

Whatever is a system is also a whole but not conversely: an aggregate of independent components is a whole but not an integrated or unitary one. (Compare a living being with its ashes.) Now, systemicity or integration comes in degrees: some systems are more tightly knit than others. The degree of integration depends on the connections or links among a system's components relative to the disintegrating actions of the environment. If the inner couplings are "positive" (or "attractive") and strong, the degree of integration is high; if the links are still positive but weak, the degree of integration is low; and if the links are "negative" (or "repulsive"), there is no systemicity or integration at all. Finally, if some of the links are "positive" while others are "negative", the degree of integration depends on which of them are overriding. For example, a stable atomic nucleus is held together by nuclear forces that overcome the electrical repulsions; and a stable human community is held together by participation in enterprises of common interest, the value of which is greater than that of rivalry or competition – until of course the latter gets the upper hand.

In the case of physical, chemical, and perhaps also biological systems, what measures their degree of integration is their binding energy or, what comes to the same, their dissociation energy. This is the minimal energy required to dissociate the system into its components. It is zero for an aggregate. But such a measure is not universal: it fails to apply to systems where information links play at least as important an integrating role as forces proper – as is the case with social systems. In short, there is no universal measure of the degree of integration or cohesion of a system. However, we can assume the methodological postulate that a measure can be set up for each system genus, or even for each class of models, regardless of the nature of the components of the systems represented by such models.

Example Two things, labeled 1 and 2, form a *linear system* if the components of the state function $\mathbb{F} = \langle F_1, F_2 \rangle$ of the latter satisfy the rate equations

$$\dot{F}_1 = a_{11}F_1 + a_{12}F_2, \qquad \dot{F}_2 = a_{21}F_1 + a_{22}F_2,$$

where the a_{ij} are in general complex numbers. The *degree of integration* or *cohesion* of the system can then be defined as

$$w = \frac{|a_{12}|}{|a_{11} + a_{12}|} + \frac{|a_{21}|}{|a_{21} + a_{22}|}.$$

If the a_{ij} are not constant but time dependent, w itself will depend on time. In general, $w: T \to [0, 1]$.

If $a_{12} = a_{21} = 0$, the components do not form a system; in every other case they do. In particular, if component 1 controls component 2, then $a_{12} = 0$ and $a_{21} \neq 0$, whence $0 < w \leq 1$. And if there is symmetric interaction, $a_{12} = a_{21} \neq 0$. Finally, if all the a_{ij} equal unity, $w = 1$, i.e. the system is maximally cohesive.

Assume then that it is possible to define in each case a measure $w: T \to [0, 1]$ of the degree of integration of systems of any given kind. Then by charting the course of w values we can follow the history of the system from buildup to breakdown through its stable stage if any. In other words, we can introduce

DEFINITION 1.19 Let σ be a system with degree of cohesion or integration $w(t)$ at time t. Then σ is *stable* during the time interval τ iff $w(t)$ is constant for all $t \in \tau$ or at most fluctuates within bounds around a constant value. Otherwise σ is *unstable* and, in particular,

(i) *builds up* (*integrates* or *assembles*) iff its degree of integration increases in time;

(ii) *breaks down* (*disintegrates* or *dismantles*) iff its degree of integration decreases in time.

The degree of integration or cohesion of a system is related to its size or number of components as well as to the nature of the latter. A system with an extremely large number of components may be unstable and ultimately break up into a number of subsystems: there is always some upper limit to the size of a system – a limit to growth. See Figure 1.9. We summarize this empirical generalization into

POSTULATE 1.7 For every kind of system there is an optimal size, i.e. a number of components that maximizes the degree of integration (cohesion) of the system in the given environment. That number is called the *critical size.*

An immediate consequence of this assumption is that, for every kind of system, there is (*a*) a *threshold size*, i.e. a number of components below

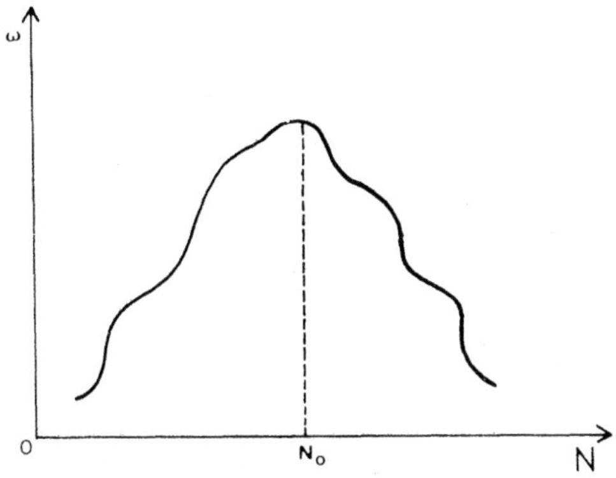

Fig. 1.9. Degree of integration (cohesion) per number of components vs. total number of components. N_0 = optimal or critical size.

which the aggregate does not form a system, and (*b*) a *maximum size*, i.e. a number above which the system breaks down.

Remark 1 Unlike most other axioms of our theory, Postulate 1.7 is empirically testable. For one thing it might well be that for systems of certain kinds there is more than a single critical size. If this were the case we would have to make a small adjustment in Postulate 1.7. *Remark 2* Accretion by gravitational pull would seem to refute our postulate. It does not, for the same process increases the density of matter and radiation energy, and this eventually sets up nuclear reactions that may lead to either explosion or collapse.

The preceding axiom concerns the overall integrity or cohesion or a system without regard for the integration of its subsystems. If a system does have subsystems, not just components, then the cohesion of the subsystems competes with that of the overall system. *Example 1* A molecule with all of its valences saturated is stable, hence it coheres poorly with similar molecules, e.g. its fellow components in a polymer. In general, the greater the complexity of a molecule the lower its overall binding energy. *Example 2* A large human family has weaker interactions with the rest of society, per family member, than a small family: the members of a large family spend a larger amount of energy interacting among themselves than with their social environment. Hence while they fare well in cases of disaster they may not make good citizens.

We compress these observations into

POSTULATE 1.8 The more cohesive each subsystem the less cohesive the total system.

A problem for the systems designer, be he an engineer or an applied social scientist, is to hit on a structure that will maximize overall integrity. He cannot maximize the cohesion of each subsystem because then it becomes self-sufficient instead of serving the whole. And he cannot minimize the partial integrities because then the subsystems would become unstable (unreliable). A compromise solution is to choose medium cohesion subsystems and to have more than one perform a given function or role. Such design enhances the reliability of the system regardless of its nature. See Figure 1.10.

Finally, another concept relevant to that of systemicity is the notion of coordination, which must be distinguished from that of integration. If integration fails the system undergoes *structural breakdown.* On the other hand coordination concerns the relation among either components or functions resulting in functional maintenance. If coordination fails the system undergoes *functional breakdown.* There can be integration without coordination but not conversely. A complex machine out of kilter is integrated but not coordinated. On the other hand organisms are coordinated and *a fortiori* integrated as long as they live. A possible characterization of the concept of coordination is given by

DEFINITION 1.19 If x and y are either in the composition or in the structure of a system, then x and y are said to be *coordinated* iff they contribute jointly to the integrity of the system.

Coordination does not exclude inhibition. Quite on the contrary: when

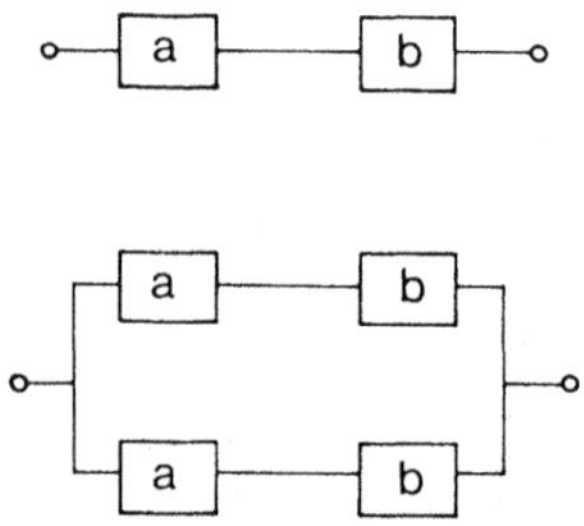

Reliability $= R = P(a) \cdot P(b)$
Probability of failure $= 1 - R$.

Probability of failure of one branch $= 1 - R$.
Probability of failure of both branches $=$
$(1 - R)^2 \leq 1 - R$
Reliability of system $= 1 - (1 - R)^2 = 2R - R^2$
$\cong 2R$ for small R.

Fig. 1.10. Increasing overall reliability (or degree of integration), by increasing redundancy, i.e. the number of subsystems performing the same function. $P(a)$ is the probability that component a will perform its regular function(s).

coordination is a result of control it includes feedback, which, when negative, is a kind of inhibition. Indeed without such a control excitation might destroy the system. But of course there can be coordination without the intervention of a control system. For example the corpus callosum bridges the two brain hemispheres in vertebrates and thus renders their coordination possible but is not a control system itself. On the other hand the entire central nervous system, which is a controller, coordinates all the subsystems that make up the vertebrate organism.

So much for the concept of wholeness. We now turn to the three main philosophical doctrines concerning wholes.

4.2. *Holism, Atomism, Systemism*

There are three possible doctrines concerning wholes: holism, atomism, and systemism. Holism is the ontological view that stresses the integrity of systems at the expense of their components and the mutual actions among them. It is characterized by the following theses.

H1 The whole precedes its parts. At first blush this thesis looks true. Thus before going for a hair cut we must have grown some hair. But of course the hair grew gradually, not all of a sudden: it became a whole in the course of a process of multiplication of parts (cells). Before making a general pronouncement concerning which precedes which we should examine the actual process in question. A system precedes its components only during a breakdown process; it succeeds them during the synthesis or formation process. In any case the existence of a system may not be obvious: it may call for an explanation in terms of both the mutual actions of parts and environment. No such explanation will be sought if the whole is taken for granted and regarded as the ultimate ground for the existence of its parts.

H2 The whole acts on its parts. For example, it will be said that the needs of the organism (or the society) as a whole dictate the functioning of its parts. But of course there would be no whole were it not for the coordination of its parts. There is no action of the whole on ist parts; rather, there are actions of some components upon others. Thus the vibration modes of any single particle in an elastic body are influenced by the motion of the other particles; likewise the behavior of any person is partially determined by that of his fellow society members. In all these cases we have not the whole acting on its parts but some or even all of the remaining components of the system acting on the given component, or the behavior of the latter being partly determined by the place it occupies in the system, in particular by its function or role.

H3 The whole is more than the sum of its parts. As it stands the thesis is hardly intelligible. It becomes intelligible if by 'sum' one means the juxtaposition (physical sum or association $\dot{+}$) that we met in Sec. 1.6, and if by 'more' one means that the whole, provided it is a system, has emergent properties that its components lack (cf. Sec. 3.2). Reformulated in this way, i.e. in a nonholistic fashion, H3 acquires a definite sense – but it proves to be only partially true. Indeed, although every system is a whole, not every totality is a system; thus the mere aggregation of things need not result in an integrated whole or system (cf. Sec. 4.1.). What make a whole into a system are precisely the actions exerted by some of its parts upon others. But the holist does not care for the disclosure of such couplings, i.e. the system structure: he despises analysis.

H4 Wholes emerge under the action of agents that transcend both the actions among the components and the environmental influences. For example, morphogenesis is guided by an entelechy, or *élan vital,* or morphogenetic field external to the components. In sum, the formation of totalities transcends their components and is traceable to inscrutable entities. Thus far the holistic account of the formation of wholes. Needless to say, science has no use for such secret hence incontrollable principles of organization. Instead, science acts on a principle of immanence, not transcendence, namely this: Only the components, the way they come together, and the environment determine what kind of thing a totality shall be. (Hence our representing a system by the ordered triple: composition-environment-structure.)

H5 Totalities cannot be explained by analysis: they are irrational. This thesis is trivially true if by 'analysis' is meant only decomposition into parts, since then only the composition of a system but not its structure is revealed. If the latter is left out then of course it becomes impossible to account for the systemic or gestalt properties of a totality. But the physicist does not claim that water is just an aggregate of H_2O molecules, and the sociologist does not assert that society is just a collection of persons. In either case the links among the components (hydrogen bonds, work relations, or what not) must be disclosed or hypothesized to understand the formation, cohesion, and eventual breakdown of a totality. Such an analysis is the conceptual basis for any effective synthesis or upward building, as well as for the effective analysis or disintegration of a system.

H6 The whole is better than any of its parts. This value judgment has been brandished as an ideological hatchet aimed at suppressing the rights of individuals and groups in the name of the good of the whole or higher system, in particular the economico-political establishment, regardless of whether the latter does in fact work for the common good. We shall waste no time on this fig leaf.

Holism, in a nutshell, is antianalytic and therefore antiscientific. In effect, it has been responsible for the backwardness of the nonphysical sciences. And it has contributed precious little to serious systemics precisely because (*a*) it has not engaged in a study of the links that hold any system together, and (*b*) rather than constructing conceptual systems (theories) to account for concrete systems, it has spent itself in attacking the analytical or atomistic approach and praising totality as such. Whatever truth there is in holism – namely that there are totalities, that they have properties of their own, and that they should be treated as wholes – is contained in systemism, or the philosophy underpinning systemics or the general theory of systems (cf. Bunge, 1977d).

In opposing holism we do not espouse its opposite, namely atomism – the thesis that the whole is somehow contained in its parts, so that the study of the latter should suffice for understanding the former. Certain wholes, to wit systems, do have collective or systemic properties not possessed by their components, and this is why they must be studied as systems. Consider the celebrated though ill understood example of a so-called contingent identity, namely Water = H_2O. This is no identity at all for, whereas the LHS is short for 'water body' (e.g. a lake), the RHS describes a property of its molecular components. (No identity is possible between a thing and a property.) What is true, of course, is that the molecular composition of water is a set of H_2O molecules, but this is no identity statement. (In other words, the correct statement is this: For any water body w, $\mathscr{C}(w) \subset$ The set of H_2O molecules.) Moreover, specifying the composition of a system does not suffice to characterize it as a system: we must add a description of the system structure. And it so happens that water, as a system composed of myriads of H_2O molecules, has properties that none of its components has – e.g. transparency, a high dielectric power (hence a high dissolving power), freezing at 0°C, and so on and so forth. Some such properties must be included in any realistic model of water.

The ontological differences between a body of water and a H_2O molecule are such that, to account for the behavior of the former, we need not only all the knowledge we have about the individual H_2O molecule but also a host of hypotheses and data concerning the structure of water (i.e. the relative configuration of H_2O molecules in the lattice) as well as hypotheses and data about the dynamics of water bodies – hypotheses and data which vary of course according to whether water is in the gaseous, liquid, or solid phase. In sum, to describe, explain or predict the properties of water we use both microlaws and macrolaws.

Atomism, an ontological doctrine, is usually, though not necessarily, allied with *reductionism*, the epistemological doctrine according to which the study of a system is reducible to the study of its components. (The converse is false: one can be an epistemological reductionist, yet acknowledge wholes, emergence, and levels.) The reductionist will claim of course that we may use macrolaws and, in general, laws of systems, as a convenience, though in principle we should be able to get along with microlaws (or laws of components) alone, since the former are reducible to (deducible from) the latter. This thesis contains a grain of truth but is not the whole truth. No theory T_2 of water as a body follows solely from a microphysical theory T_1 of the H_2O molecule – not even by our adjoining what some philosophers call the bridge laws relating macrophysical concepts (e.g. pressure) to microphysical ones (e.g. molecular impact). Much more than this must added to the primary or reducing theory T_1 in order to obtain the secondary or reduced theory, namely hypotheses concerning the interactions among the system components.

The extreme case of reduction is that of straightforward deduction from a given set of premises, or *strong reduction*. Examples: reduction of particle mechanics to continuum mechanics, and of ray optics to wave optics. (The converse reductions are impossible.) These are rather exceptional cases. In general we must resort to a more complex strategy, namely *weak reduction*, or deduction from a primary theory in conjuction with a set of conjectures and data congenial with but alien to the former. The structure of this inference is:

$$T_1 \cup \text{Subsidiary hypotheses and data concerning interactions among components} \vdash T_2.$$

The subsidiary hypotheses constitute a model of the composition and structure of the system. Since this model, though couched in the language of T_1, is not included in T_1, we are in the presence not of straight (or strong) but of partial (or weak) reduction. (Cf. Bunge, 1977f.)

Note that we are not stating that the properties of water, or of any other macrosystem, are mysterious. On the contrary, they can be explained at least in outline. For example, the exceptionally high boiling point and evaporation heat of water are explainable in terms of the hydrogen bonds linking together all the H_2O molecules in a water body, bonds which are in turn explained by the composition and the structure of the H_2O molecule. But the point is that the intermolecular hydrogen bonds do not occur in the study of the individual H_2O molecule. In other words, although water is composed of H_2O molecules it does not reduce to H_2O – *pace* the efforts of enlightened philosophers to crush the holistic monster (e.g. Kemeny and Oppenheim, 1956; Oppenheim and Putnam, 1958, Putnam, 1969).

In short, atomism is almost as false as holism, the difference lying in that, whereas the former stimulates research, the latter blocks it. Each of these views has a grain of truth that the systemic view preserves and expands.

5. Concluding remarks

The idea of a system, as distinct from a loose aggregate, is very old. Yet it has only recently been elucidated and exploited systematically. The mere suggestion that the thing we happen to be looking at, manipulating, or investigating might be a system rather than an unstructured object or a mere heap, will guide our study and handling of it. Indeed, if we suspect that a certain thing is a system then we shall endeavor to identify its composition, its environment, and its structure.

The order in which the three coordinates of the system concept appear is natural rather than accidental. Indeed, listing a system's components must precede any question regarding its environment and its structure; and identification of the environment comes before display of the structure, because the latter is the collection of relations among the components and among these and environmental items. True, when coming across certain systems, such as a plant, a watch, or a galaxy, we often begin our search with the totality and its environment, ending up by disclosing its composition and its structure. But when investigating a forest, a social system, and *a fortiori* a social supersystem such as a nation, we first meet its components (or atomic parts) in their milieu and try to figure out the structure of the whole by studying the behavior of the individual components. In either case, i.e. whatever our mode of perception may be, the conceptual analysis of a system must proceed in the way indicated – identification of composition, environment, and structure – if only for mathematical reasons.

Indeed it makes no sense to hypothesize any relations without knowing what the relata (system components and environmental units) may be. Therefore the holistic claim that atomistic analysis or the part method is incapable of grasping totalities, is groundless. On the contrary, holism is incapable of accounting for any totalities precisely because it refuses to disclose the components that are held together in the system: no components, no bonds among them. This is not to condone the antithesis of holism, namely atomism, and its usual epistemological partner, reductionism, according to which wholes – in particular, systems – are artifacts, emergents are identical with resultants, and levels are just convenient methodological categories.

The world view that emerges from this chapter is systemic: it holds that the universe is a system composed of subsystems. More precisely, the universe is the supersystem of all other systems. The world is thus neither a solid block nor a heap of disconnected items. It is held together by a number of links, from intermolecular bonds to gravitation to information. The world is material but not just a lump of physical entities: it is composed of systems of a number of qualitatively different kinds. Although all systems are physical not all of them are just physical. The universe is enormously variegated: its components can be grouped into a number of levels, such as the physical, the chemical, the biological, and the social levels. Moreover the world is restless and all of its changes are patterned (lawful). The world is, in sum, a coherent or integrated system of systems, and one that is varied, changeable, and regular.

CHAPTER 2

CHEMISM

All concrete systems are physical entities, in that they possess energy – a measure of changeability (Vol. 3, Postulate 5.3). However, some concrete systems are more than just physical entities, in the sense that the categories of physics are insufficient to account for them. For example, some are social systems. And although some of the concepts of physics – in particular the notion of energy – are necessary to account for any society, most are unnecessary. On the other hand social science requires new concepts, such as those of artifact, social group, justice, and culture. The peculiar concepts of social science are supposed to represent emergent properties of a society, i.e. properties that are not possessed by its components. Likewise the study of the components of any society, i.e. social animals, calls for a number of distinctly biological categories, such as reproduction, sex, and death. And if the animals happen to have a highly evolved nervous system, we shall need still other categories, such as the concepts of altruism and planning, which are usually regarded as psychological. Finally, organisms are composed of chemical systems, which are physical entities of a very special kind, namely composed of atoms and molecules engaged in chemical reactions.

Thus the totality of concrete entities may be grouped into five genera: the set of physical things (e.g. atoms and fields), the set of chemical systems (e.g. protein-synthesizing units), the set of biosystems (organisms and their organs), the set of sociosystems (societies and their subsystems and supersystems), and the set of artificial things. (The set of psychosystems, or animals endowed with mental abilities, is assumed to be a subset of that of biosystems.) The first three genera precede the other two in the sense that they supply the components of the higher order things or take part in some other way in their assembly. Since the lower a genus the more populous it is, we may depict the structure of reality as a pyramid: see Figure 2.1. This suggests the plan of this volume. (More on this structure in Ch. 6.)

In this chapter we shall hardly touch on physical systems: first, because they are the best understood of all; second, because we have actually dealt with them in Vol. 3. We shall focus instead on certain key chemical categories of crucial importance in the study of life, such as those of che-

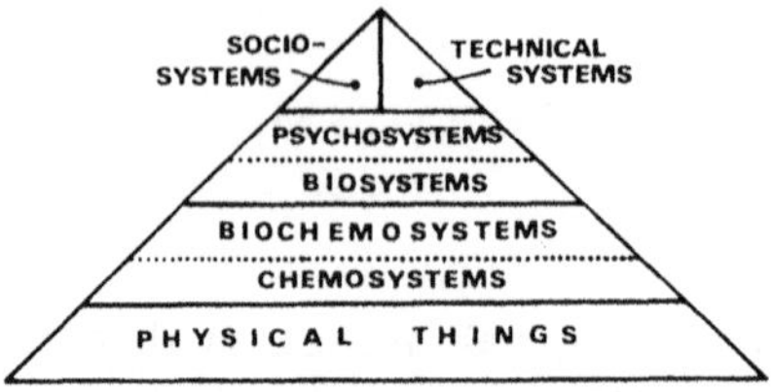

Fig. 2.1. The five main genera of things constituting the furniture of the world.

mical reaction and chemosystem, protein and nucleic acid. These categories are certainly more specific (less universal) than those of thing, property, system, and others we have encountered so far. On the other hand our chemical categories are far more universal than some of those that have dominated ontology for centuries, such as those of mind and purpose.

1. Chemical System

1.1. *Atom*

Since we shall make ample use of the concept of a chemical system, we must characterize it, the more so since no definition of it in systems theoretic terms seems to be available. And, since a chemical system is one whose components are reacting chemicals, we must start by characterizing the notions of a chemical and of a reaction.

The components of a chemical system are chemical atoms or molecules. A detailed characterization of things of these kinds calls for a number of specific scientific concepts. Luckily atoms have an essential or root property from which many another depends, namely atomic number. And molecules have an essential property of their own, namely atomic composition. These two properties will suffice for our limited purposes.

Our starting point is then the notion of chemical atom of a kind, or elementary chemical species, or element for short. There are 100 odd elements, among them hydrogen (H), carbon (C), nitrogen (N), and oxygen (O), the atomic numbers of which are 1, 6, 7, and 8 respectively. Every member of such a natural kind or species is a material system formed by at least two things of different genera, namely a system of nucleons (protons or neutrons), and an aggregate of electrons. In turn the nucleons are probably composed of subunits (quarks?), and the atoms can join forming molecules. See Figure 2.2. (This diagram suggests that molecules and their

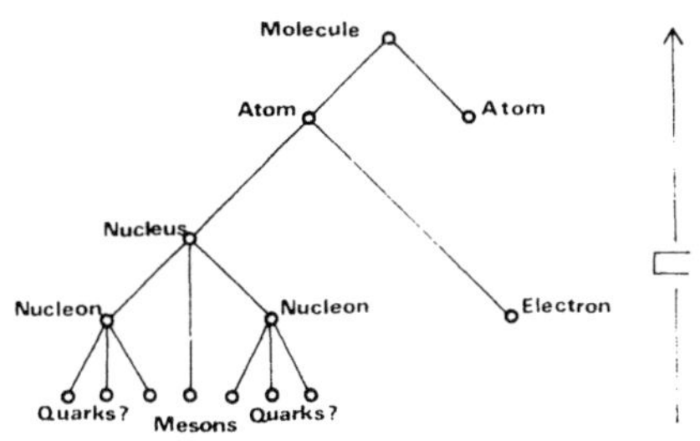

Fig. 2.2. A molecule is composed of two or more atoms. Each of these is in turn composed of a nucleus and one or more electrons. The number of protons in the nucleus equals the atomic number of the atom. Nucleons are possibly held together in the nucleus by meson fields. In turn every nucleon (proton or neutron) is presumably composed of smaller units (quarks?). Altogether six different kinds of thing are involved in a molecule.

components constitute a *sup* semilattice the order of which is that of the part-whole relation $\sqsubset$. Thus if a_1 and a_2 are two atoms forming a molecule m, then $sup\ \{a_1, a_2\} = m$. However, we shall not pursue this point here.)

In short, we steal from physics and chemistry

POSTULATE 2.1 Let A be a nonempty subset of the collection of all concrete things (or entities), and $\mathbb{N}_0$ the set formed by the first N positive integers. There exists a many to one relation (function) $Z : A \to \mathbb{N}_0$ of A onto $\mathbb{N}_0$(i.e. a surjection) that assigns each thing in A a natural number in $\mathbb{N}_0$, representing a property of the thing.

A is of course the set of atoms and Z the atomic number. The atoms with the same Z value are Z-equivalent:

If x and y are in A, then $x \sim_Z y =_{df} Z(x) = Z(y)$. The quotient of A by $\sim_Z$, or $A/\sim_Z$, is the family of classes of atoms with the same Z value. (We do not need to make the finer distinction among the isotopes of an element.) In other words, we can make

DEFINITION 2.1 Let A be the domain of the function $Z\colon A \to \mathbb{N}_0$. Then

(i) if x is in A, x is an *atom*;

(ii) the property represented by the function Z is called the *atomic number*, and the value $Z(a)$ of Z at $a \in A$, the *atomic number* of a;

(iii) the nth set

$$^{n}A = \{x \in A \mid Z(x) = n\}, \quad \text{for } 1 \leq n \leq N,$$

is the nth *atomic kind* or *chemical element*;

(iv) the partition

$$\mathscr{A} = A/\sim_Z = \{{}^nA \subset A \mid 1 \leq n \leq N\}$$

is the *family of atomic kinds* or *chemical elements*.

Since by definition $\mathscr{A}$ is the partition of A induced by $\sim_Z$,

$$A = \bigcup_{n=1}^{N} {}^nA, \quad \text{and} \quad {}^mA \cap {}^nA = \varnothing \quad \text{for } m \neq n.$$

In other words, the totality of atoms is the disjoint union of all the atomic kinds. The latter are those displayed in the periodic table. It is not known whether the number of atoms of each atomic kind is finite.

So much for the (atomic) components of a molecule.

1.2. *Molecule*

A molecule is a system composed of atoms – or, if preferred, of nuclei and electrons. Molecules range from hydrogen or H_2 (molecular weight = 2) to deoxyribonucleic acids or DNA (molecular weight up to 1 billion). There are about 4 million known species of molecules, 96 percent of which contain carbon. The combinations of atoms to form molecules are brought about by either ionic bonds or covalent bonds. In turn molecules stick together by virtue of hydrogen bonds or van der Waals forces (e.g., hydrophobic forces). All of these bonds are ultimately electric forces and they are studied by quantum chemistry, an application of quantum mechanics.

In ontology we take such bonds for granted and let chemists inquire into their nature. (Remember the fate of Hegel's alleged refutation of Berzelius' brilliant hypothesis that chemical bonds are electrical: cf. Hegel (1830), Sec. 330). The metaphysician accepts gratefully what the chemist is willing to teach him about the spontaneous formation (self-assembly) of molecules out of atoms, or of other molecules, as well as about the breakdown or dissociation of such systems as a result of their interaction with their environment (in particular thermal motion). The philosopher's task is not to compete with the chemist but to try and discover the structure and peculiarities of chemical processes *vis à vis* other kinds of process. If in the course of his work he comes up with results helping to clarify the foundations of chemistry, so much the better.

The statement that all molecules are composed of atoms can be exactified in a number of ways. We shall proceed as follows. Consider for example the first atomic species or

$$^{1}A = \mathrm{H} = \text{The set of all hydrogen atoms} = \{x \in \Theta \mid Z(x) = 1\}.$$

Since we disregard the individual differences among the hydrogen atoms, to concentrate on their shared atomic number, we may regard the species H as formed by an unspecified number of copies of a single individual. Assume now that each hydrogen atom can combine with any number of other H atoms forming hydrogen molecules – possible ones like the diatomic ones, and impossible ones like the icosiatomic ones. Since we are disregarding idiosyncrasies and concentrate on atomic composition, we can speak of the m-fold concatenation of an unspecified hydrogen atom x with itself, or $x \cdot x \cdot \ldots \cdot x = x^m$, as representing the combination of m hydrogen atoms. In this form we fashion

$$^{1}A^* = \mathrm{H}^* = \text{The set of finite concatenations of hydrogen atoms.}$$

Imagine now that the same construction is carried out for every atomic species, so that we have

$$^{n}A^* = \text{The set of finite concatenations of atoms of kind } {}^{n}A.$$

This set is the union of the following sets:

$$^{n}A_1 = \{z \in \Theta \mid z \in {}^{n}A\}$$
$$^{n}A_2 = \{z \in \Theta \mid z = x^2 \;\&\; x \in {}^{n}A\},$$
$$\cdots\cdots\cdots\cdots\cdots$$
$$^{n}A_m = \{z \in \Theta \mid z = x^m \;\&\; x \in {}^{n}A\}.$$

In other words, the set of all conceivable ^{n}A compounds is

$$^{n}A^* = \bigcup_m {}^{n}A_m = \{z \in \Theta \mid z = x^m \;\&\; x \in {}^{n}A \;\&\; 1 \leq m < \infty\}.$$

But of course not all conceivable compounds are really possible, i.e. lawful. (For the definition of real possibility in terms of lawfulness, see Vol. 3, Ch.4, Sec. 2,3.) The really possible compounds from ^{n}A are a proper subset of $^{n}A^*$, i.e.

$$^{n}A^{**} = \{z \in \Theta \mid z = x^m \;\&\; x \in {}^{n}A \;\&\; 1 \leq m < \infty \;\&\; z \text{ is lawful}\}.$$

(Do not think just of the stable compounds satisfying the valence requirements but also of all the transient molecules and free radicals.) These are all the homogeneous compounds, i.e. the molecules composed of atoms of a single atomic species. A heterogenous compound, such as the member of the chemical species CH_4, or methane, is a member of the set

$$^{1}A_4{}^{6}A_1 = \{z \in \Theta \mid z = x^4 y \;\&\; x \in {}^{1}A \;\&\; y \in {}^{6}A\}.$$

The totality of heterogeneous compounds is

$$A^* = \{z \in \Theta \mid z = x_1^{m_1} x_2^{m_2} \ldots x_N^{m_N} \;\&\; x_i \in {}^{i}A \;\&\; 1 \leq i \leq N \;\&\; 0 \leq m_i < \infty\}.$$

This is the set of all conceivable molecules, in turn a superset of the set of all really possible molecules.

The structure of A^* is obviously that of a semigroup under the binary operation $\cdot$ of combination or concatenation. More exactly, the structure $\langle A^*, \cdot \rangle$ is the *free semigroup generated by the set A* of atoms, and this because every element of A^* is either a member of A or a string of members of A. Now, as noted before, A^* contains not only the really possible molecules but also the unlawful ones. If we shrink A^* to the subset A^{**} of really possible molecules, we face the difficulty that A^{**} is not closed under the operation $\cdot$ of combination or concatenation (i.e. it does not contain all the concatenates). Equivalently: $\cdot$ is a *partial* operation on A^{**}, i.e. one not defined for every pair of members of A^{**}. We shall therefore say that the system $\mathscr{A}^{**} = \langle A^{**}, \cdot \rangle$ is a *partial* semigroup *embedded* in the free semigroup $\mathscr{A}^* = \langle A^*, \cdot \rangle$ generated by A. In other words, we make

POSTULATE 2.2 Let $\mathscr{A}^* = \langle A^*, \cdot \rangle$ be the free semigroup generated by the set A of all atoms of all kinds (Postulate 2.1), and call

$$A^{**} = \{z \in \Theta \mid z = x_1^{m_1} x_2^{m_2} \ldots x_N^{m_N} \;\&\; x_i \in {}^{i}A \;\&\; 1 \leq i \leq N \;\&\; 0 \leq m_i < \infty \;\&\; z \text{ is lawful}\}.$$

Then

(i) the partial semigroup $\mathscr{A}^{**} = \langle A^{**}, \cdot \rangle$ is embedded in $\mathscr{A}^*$;

(ii) for any two things x and y in A^{**}, if xy is in A^{**}, then xy represents the combination of x with y (or the xy compound).

Note that we are not postulating that $\mathscr{A}^{**}$ is a commutative semigroup. Hence the isomers xyz and yxz will be distinct molecules.

The previous axiom supports the following conventions that will come in handy in the sequel.

DEFINITION 2.2 Let $\mathscr{A}^{**}$ be the set defined in Postulate 2.2 Then

(i) the set of *compounds* is the complement of A^{**} to A, i.e.

$$M = A^{**} - A = A^{**} \cap \bar{A};$$

(ii) any subset of M, of the form

$$M_j = \{z \in \Theta \mid z = x_1^{m_1} x_2^{m_2} \dots x_N^{m_N} \ \& \ x_i \in {}^iA \ \& \ 1 \leq i \leq N\},$$

with fixed values of the m_i for $0 \leq m_i < \infty$, is a *molecular species*;

(iii) every member of a molecular species is a *molecule*.

(Note the strategy of first defining a set and then naming its members.)

DEFINITION 2.3 The *composition* of a molecule is the set of its constituents. I.e. let

$$\mathscr{C}^{**} : A^{**} \longrightarrow 2^{A^{**}}$$

be the function from the set of atoms and molecules to its power set, such that, if $z = x_1^{m_1} x_2^{m_2} \dots x_N^{m_N} \in A^{**}$, where $x_i \in {}^iA$ and $0 \leq i < \infty$,

$$\mathscr{C}^{**}(z) = \{x_1^{m_1}, x_2^{m_2}, \dots, x_N^{m_N}\}.$$

Then $\mathscr{C}^{**}(z)$ is the *composition* of the molecule z.

Example The composition of glucose, or $C_6H_{12}O_6$, is given by:

If $z \in C_6H_{12}O_6$, then $\mathscr{C}^{**}(z) = \{6$ carbons, 12 hydrogens, 6 oxygens$\}$.

The next definition requires a generalization, to an arbitrary set T of things, of the notion of physical addition (or association) occurring in the formula '$a + b = c$.' For an arbitrary set T of things, the association or addition of its members is the thing $[T] = \sup T =$ least upper bound of T. (Clearly, for any x in T, $x \sqsubset [T]$.)

DEFINITION 2.4 Let x be a thing. Then

(i) x is a *chemical* iff x is either a member of A^{**} (i.e. an atom or a molecule) or the aggregation of a finite set of atoms or molecules – i.e. if there is a $T \subset A^{**}$ such that $x = \sup T = [T]$;

(ii) x is a *compound* iff x is either a molecule or the aggregation of a finite set of molecules, i.e. $x \in M = A^{**} - A$ or $x = [T]$, with $T \subset M$;

(iii) x is a *polymer* iff x is either a string of molecules of a single molecular species, or the aggregation of a finite set of strings of different molecular species.

Polymers are of particular interest to us because all biomolecules are polymers. In particular, the DNA molecules are extremely long chain polymers.

Finally we make

DEFINITION 2.5 Let A^{**} be the totality of compounds and $l: A^{**} \rightarrow \mathbb{N}$ a function assigning each atom or molecule a natural number. Then the *length* of a member z of A^{**} equals the number of atoms in z. More precisely,

$$\text{If } z = x_1^{m_1} \cdot x_2^{m_2} \ldots x_N^{m_N} \in A^{**}, \quad \text{then } l(z) = m_1 + m_2 + \cdots + m_N.$$

This concept of length applies to molecules of all shapes, not just the linear ones. Thus the length of the planar molecule CH_4 is, according to our definition, equal to 5.

So much for structural matters. (For somewhat different treatments see Rosen (1959) and Chiaraviglio (1965).) Let us now turn to chemical processes.

1.3. *Chemical System*

We shall characterize a chemical system as a system whose components are chemicals varying in number (or in concentration) for being engaged in reactions. To render this idea more precise we shall avail ourselves of the state representation of systems explained in Ch. 1, Sec. 2.2. Consider a concrete system represented by a time-dependent state function $\mathbb{F} = \langle F_1, F_2, \ldots, F_n \rangle$, each component (coordinate) of which represents a property of the system. The state function spans the total lawful state space of the system. Now single out those components of $\mathbb{F}$ that represent the numbers (or else the concentrations) of the chemicals in the system. The resulting state function, and the accompanying state subspace, deserve names of their own:

DEFINITION 2.6 Let T be the time span (relative to a given reference frame) and let

$$\mathbb{F} = \langle F_1, F_2, \ldots, F_n \rangle : T \longrightarrow \mathbb{R}^n$$

be a state function for an arbitrary system σ of a certain kind, satisfying the laws in a set $\mathbb{L}$. Further, let the first $m < n$ coordinates of $\mathbb{F}$ (at a given time) equal the number of components of each kind of chemical in the system (at that time). Then

(i) the function

$\mathbb{F}_c = \langle F_1, F_2, \ldots, F_m \rangle$, with $F_i(t)$ = Number of atoms or molecules of kind $K_i \subset A^{**}$ in σ at time $t \in T$, is called the *composition state function* of σ;

(ii) the value of the composition state function at time $t \in T$, i.e. $\mathbb{F}_c(t)$, is called the *instantaneous composition* of σ;

(iii) the set of all possible instantaneous compositions of σ, i.e.

$$S_c(\sigma) = \{\mathbb{F}_c(t) \in \mathbb{R}^n \mid t \in T \ \& \ \mathbb{F}_c \text{ satisfies } \mathbb{L}_c\},$$

is called the *composition state space* of σ.

Example The composition state function of a system where water is formed out of hydrogen and oxygen, according to the reaction $2H_2 + O_2 \rightarrow 2H_2O$, has

$$\text{the } \textit{initial value } \mathbb{F}_c(t_i) = \langle 2, 1, 0 \rangle$$

and (on completion of the reaction)

$$\text{the } \textit{final value } \mathbb{F}_c(t_f) = \langle 0, 0, 2 \rangle,$$

where the integers are the numbers of molecules of the chemicals in the system. If we wish to represent the entire process from start to end, rather than only the initial and the final states, we can use the partial concentrations, to obtain

$$\mathbb{F}_c = \langle C_{H_2}, C_{O_2}, C_{H_2O} \rangle,$$

where each coordinate of the triple is a function from T to $\mathbb{R}$. A condition for being able to use these functions is that the system be rather bulky, as is tacitly assumed in chemical kinetics.

Finally we can make

DEFINITION 2.7 Let σ be a concrete system with an n component state function $\mathbb{F}$ (in a given representation). Then σ is a *chemical system*, or *chemical reactor*, during the period $\tau \subseteq T$, iff

(i) the first m coordinates of $\mathbb{F}$, where $0 < m < n$, form the composition state function $\mathbb{F}_c$ of σ (i.e., σ is composed, among other things, of chemicals: $\mathscr{C}(\sigma) \cap A^{**} \neq \varnothing$);

(ii) the instantaneous value of the ith component F_i of $\mathbb{F}_c$, for $1 \leq i \leq m$, equals the number of atoms or molecules (or else the concentration) of ith chemical natural kind or species included in A^{**} at the time concerned;

(iii) the chemical composition of σ (i.e. the value of $\mathbb{F}_c$) changes in the course of τ, and this change is not just a transport from or to environment of σ.

According to this definition there can be talk of a chemical system only where and while chemical reactions are in progress. Thus a dead battery is no longer a chemical system. Before the reactions start, and after they have reached completion, the system is a physical one and only potentially a chemical system. (A chemical system is then an essentially dynamical system that can be at most in a state of dynamical equilibrium, or steady state, which is the one at which every reaction proceeds at the same pace as its converse, so that there is no net variation of the composition of the system.) This suggests

DEFINITION 2.8 A concrete system σ is a *physical system* during the period τ iff the composition state function of σ is constant (e.g. zero) throughout τ or, if it changes, it does so by transport from or to the environment of σ.

In sum a chemical system is a system whose chemical composition changes in the course of time, however slowly, and does so not just by importation or exportation of chemicals, but by virtue of processes involving its components. These processes are of course the chemical reactions. Let us look into these.

1.4. *Chemical Reaction*

Every really possible (lawful) process in a chemical system is representable as a curve in the total state space of the system. Each m-tuple of initial concentrations of the chemicals in the system corresponds to a distinct trajectory of the tip of the state vector $\mathbb{F}$. ($\mathbb{F}$ is not really a vector, for its product with an arbitrary real number may lie outside the bounds set by the laws.) And the bundle of all such possible trajectories, for all variations of initial conditions and all systems of the same kind, represents the totality $\mathbb{L}$ of laws "at work" in chemical systems. Since a system is a chemical system only as long as there are chemical transformations going on in it, the set $\mathbb{L}$ includes every one of the laws concerning such transformations, in particular the reaction formulas and the corresponding kinetic equations. But, of course, $\mathbb{L}$ includes also every other law satisfied (possessed) by the chemical system of the given kind. See Figure 2.3.

We spell out the foregoing in

DEFINITION 2.9 Let $\mathbb{F}$ be a state function for a concrete system σ of kind K, and let $S_{\mathbb{L}}(K)$ be the corresponding lawful state space. Then

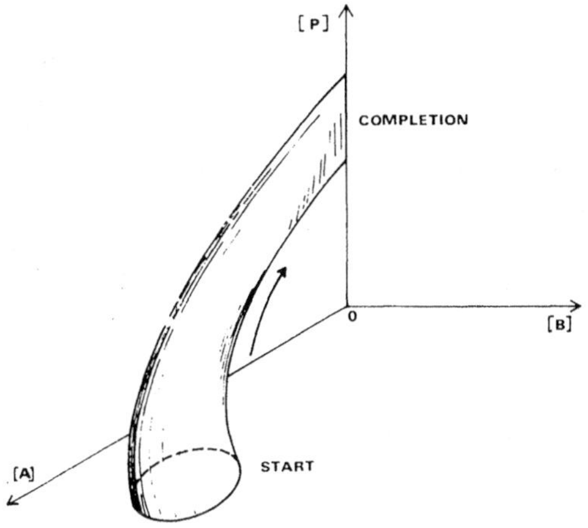

Fig. 2.3. Composition state space of a chemical system in which only reactions of the form $A + B \rightharpoonup P$ occur. $[X]$ designates the concentration of X. Each line within the conoid, from beginning to end, represents one possible complete reaction compatible with the reaction schema as well as with the formula for the rate of formation of P from A and B.

(i) the full trajectory of the state point $\mathbb{F}(t)$ in $S_{\mathrm{L}}(K)$ is called the total *history* of σ during $\tau \subseteq T$:

$$h(\sigma, \tau) = \{\langle t, \mathbb{F}(t)\rangle \mid t \in \tau\};$$

(ii) the set of *laws* of concrete systems of kind K is the family of all the histories of all the systems in K over all periods:

$$\mathbb{L} = \{h(\sigma, \tau) \mid \sigma \in K \;\&\; \tau \in 2^T\},$$

where 2^T is the family of all time periods included in T.

We are interested in certain segments of the history of an arbitrary chemical system called 'chemical reactions'. Equivalently: we shall pay particular attention to the chemical law statements (formulas) that focus on chemical transformations, especially those that disregard all the concomitant physical processes. Consider for example the substitution schema

$$r = \ulcorner A + B \rightharpoonup C + D \urcorner,$$

usually interpreted as "One unit of species A reacts with another of species B to produce one unit of species C and another of species D". (The interpretation of the more general reaction schema $m\mathrm{A} + n\mathrm{B} \rightarrow p\mathrm{C} + q\mathrm{D}$ is obvious.) This statement says nothing about the reaction mechanism or about the environmental conditions for it. (The latter are taken care of in the antecedent of the conditional statement of which r is the consequent. The entire conditional is then of the form: "If the pressure, temperature, etc., are such and such, then r".) Nor does the statement inform us about the pace of the chemical process: it just tells us what there is in the beginning and at the end of the transformation. I.e. it speaks only of net events. (Moreover r is ambigous; it uses class symbols, such as 'A' and 'B', to refer to members of such classes.) In any case, this is how the above reaction schema can be translated into the state space language: the system is initially in the composition state

$$\mathbb{F}_c(t_i) = \langle C_\mathrm{A}, C_\mathrm{B}, 0, 0\rangle$$

and finally in the composition state

$$\mathbb{F}_c(t_f) = \langle 0, 0, C_\mathrm{C}, C_\mathrm{D}\rangle$$

so that the size of the net change is

$$r = \mathbb{F}_c(t_i) - \mathbb{F}_c(t_f) = \langle C_\mathrm{A}, C_\mathrm{B}, -C_\mathrm{C}, -C_\mathrm{D}\rangle.$$

Example The overall photosynthesis reaction (which has a great many intermediate steps) is

$$6CO_2 + 6H_2O \rightarrow C_6H_{12}O_6 + 6O_2.$$

Setting

$$\mathbb{F}_c = \langle C_{CO_2}, C_{H_2O}, C_{C_6H_{12}O_6}, C_{O_2}\rangle,$$

we get

$$r = \langle 6, 6, 0, 0\rangle - \langle 0, 0, 1, 6\rangle = \langle 6, 6, -1, -6\rangle.$$

And the converse reaction, which (barring differences in intermediate steps) is that of respiration, is of course

$$-r = \langle -6, -6, 1, 6\rangle,$$

a systems theoretic rewrite of

$$C_6H_{12}O_6 + 6O_2 \rightarrow 6CO_2 + 6H_2O.$$

We compress the foregoing into the following semantic assumption. Let $\mathbb{F}_c$ be a composition state function for a chemical system σ between the initial time t_i and the final time t_f (relative to a certain reference frame). Then

(i) the *overall* or *net reaction* in σ between t_i and t_f is representable by the difference between the initial and final values of $\mathbb{F}_c$:

$$r = \mathbb{F}_c(t_i) - \mathbb{F}_c(t_f);$$

(ii) while the positive components of r represent the *reactants*, the negative ones represent the *reaction products*;

(iii) the *inverse* (or *converse*) of a reaction r is the reaction $-r$;

(iv) the *null reaction* is $\mathbb{O} = \langle 0, 0, \ldots, 0 \rangle$.

As we know from elementary chemistry, reactions can add to form further reactions: see Figure 2.4. For example, the enzymatic reaction sequence (or pathway) represented by the system of reaction schemata

$$\begin{array}{l} A + C \rightarrow AC \\ AC + B \rightarrow ACB \\ ACB \rightarrow AB + C \end{array}$$

adds up to the overall reaction

$$A + B \rightarrow AB.$$

In our state space representation we form the composition state function

$$\mathbb{F}_c = \langle C_A, C_B, C_C, C_{AB}, C_{AC}, C_{ACB} \rangle,$$

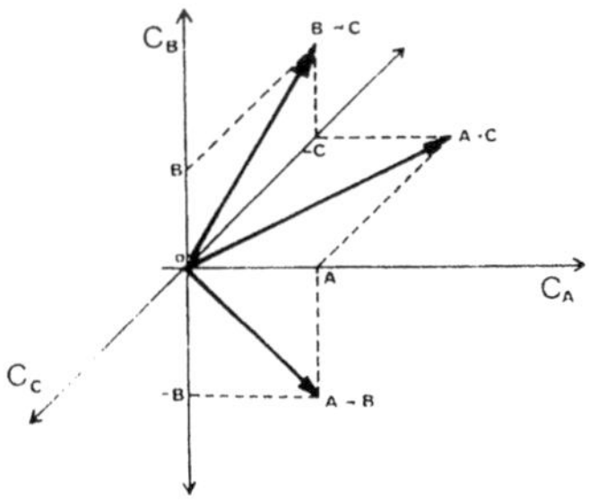

Fig. 2.4. The reactions A → B and B → C add up to the overall reaction A → C. Each reaction is represented by an arrow in the cartesian three-space of the concentrations (or the numbers of atoms or molecules).

so that the three successive reactions are

$$r_1 = \langle 1, 0, 1, 0, 0, 0\rangle - \langle 0, 0, 0, 0, 1, 0\rangle = \langle 1, 0, 1, 0, -1, 0\rangle$$
$$r_2 = \langle 0, 1, 0, 0, 1, 0\rangle - \langle 0, 0, 0, 0, 0, 1\rangle = \langle 0, 1, 0, 0, 1, -1\rangle$$
$$r_3 = \langle 0, 0, 0, 0, 0, 1\rangle - \langle 0, 0, 1, 1, 0, 0\rangle = \langle 0, 0, -1, -1, 0, 1\rangle.$$

The overall reaction is formed by adding the coordinates term by term:

$$r = r_1 + r_2 + r_3 = \langle 1, 1, 0, -1, 0, 0\rangle,$$

which, in effect, represents ⌜A + B ⇀ AB⌝. That is, the catalyzer C and the intermediate products cancel out, so that only the consumed reactants and the newly formed chemicals occur in the overall reaction. If the net result had been a return to the initial state of the system, we would have had the null overall reaction.

We summarize the preceding into

DEFINITION 2.10 Let the m-tuples

$$r_1 = \langle a_1, b_1, \ldots, m_1\rangle, \qquad r_2 = \langle a_2, b_2, \ldots, m_2\rangle$$

represent each one reaction in a given chemical system. Then the *sum* of the two reactions in the same system is the third reaction represented by

$$r = r_1 + r_2 = \langle a_1 + a_2, b_1 + b_2, \ldots, m_1 + m_2\rangle.$$

Since in principle every reaction has an inverse, any two reactions in a chemical system can add up to form a third, and the null reaction adds to any reaction leaving it unchanged, we have (Aris, 1965)

THEOREM 2.1 Let R be the set of all chemical reactions that can occur in a given chemical system, $+$ the binary operation of reaction addition, $-$ the unary operation of reaction inversion, and $\mathbb{O}$ the null reaction. Then the structure $\mathscr{R} = \langle R, +, -, \mathbb{O}\rangle$ is a group.

Now, whereas some reactions are concurrent, others are successive. The notion of reaction succession is obvious:

DEFINITION 2.11 Let $\langle R, +, -, \mathbb{O}\rangle$ be the group of possible reactions in a chemical system, and call

$$r_i = \mathbb{F}_c(t_i) - \mathbb{F}_c(t_f), \qquad r_j = \mathbb{F}_c(t_i') - \mathbb{F}_c(t_f')$$

two members of R. Then r_i *precedes* r_j in time iff r_j sets on after r_i. I.e.

$$r_i < r_j =_{df} t_i' \geq t_f.$$

This concept forms part of a notion of particular interest to the chemistry of living systems, namely that of pathway. A pathway is a set of reactions such that the product of one occurs as the reactant in the next. More precisely, we make

DEFINITION 2.12 Let $\langle R, +, -, \mathbb{O} \rangle$ be the group of possible reactions in a chemical system, and call

$$R_n = \langle r_i \in R \mid i \in \mathbb{N} \rangle$$

a list of n members of R. Then R_n is a *pathway* or *reaction sequence* iff, for any value of i occurring in R_n,

(i) r_i precedes r_{i+1} in time, i.e. $r_i \prec r_{i+1}$, and

(ii) some reaction products of r_i are reactants of r_{i+1}.

In closing this subsection, it will be fair to warn the reader that there are alternative ways of representing chemical compounds and chemical reactions, in particular those proposed by Aris (1965); Sellers (1967, 1970, 1971); Krohn, Langer and Rhodes (1967); and Feinberg (1972). All these alternative representations accept the usual chemical notation, which conflates individual chemicals (e.g. molecules) with their species, writing for example 'H_2O' for both the water molecule and the chemical species water. Another ambiguity of the usual chemical notation is that, when "reading" (interpreting) a reaction schema such as ⌜A + B → C + D⌝, we are asked to interpret the first '+' as symbolizing the combination of the reactants, and the second '+' as the mixture of the reaction products. For these reasons the standard chemical notation is a paradigm of double talk – of Platonism and materialism. Therefore we had to go our own, even if mathematically less sophisticated, way.

1.5. *Chemical Control*

Every system is controlled by its environment; in particular, the milieu of a chemical system is a variable source and sink of chemicals and energy. If the environment is "fit", the chemical system will come into being and subsist for a while. On the other hand a "hostile" environment, e.g. one that supplies either too little or too much energy, or else one that is not wet enough, or far too oxydating (like Mars), will preclude certain reactions or even all of them. Therefore the existence of a chemical system is a sufficient condition for that of a "congenial" environment, i.e. one compatible with the reactions going on in the system.

Certain chemical systems have internal controls in addition to the external or environmental ones. That is, some of either the reactants or the reaction products of a chemical system contribute to controlling the pace of the reactions, often to the point of starting them or bringing them to a halt. In some cases a reaction is accelerated by the action of a catalyzer (e.g. an enzymatic protein). In particular the reaction product itself may catalyze the reaction (*autocatalysis*). In other cases the reaction product, as it builds up, inhibits the further progress of the reaction until the latter stops even if the supply of reactants is not exhausted (*product inhibition*). Or the catalyzer itself may at some point be broken down by some other enzyme – as is the case with a protein when no longer "needed". And a number of processes of these various kinds may occur sequentially (in particular cyclically) or even at the same time in the same system, e.g. a living cell (cf. Eigen, 1971; Lehninger, 1975).

We are particularly interested in chemical systems equipped with controls, because the living cells and its organelles are of that kind. Hence we define explicitly:

DEFINITION 2.13 A chemical system with composition state function $\mathbb{F}_c = \langle F_1, F_2, \ldots, F_m \rangle$(in some representation) is *self-controlled* iff some of the components of $\mathbb{F}_c$ depend exclusively upon other components of it (rather than on any components of the total $\mathbb{F}$ representing properties of the environment of the system).

Consider for example an enzymatic reaction in an aqueous solution – the standard biochemical process. Let E denote an enzyme acting upon a substrate S to form an intermediate compound ES (or enzyme-substrate complex), which eventually dissociates into a product P with liberation of E. Moreover suppose that this is a product inhibition reaction, i.e. one in which the product gets in the way of the reaction because it binds with the enzyme, thus inhibiting the latter from acting upon the substrate. In conventional chemical notation, we have

$$\mathrm{E + S} \underset{k_1'}{\overset{k_1}{\rightleftharpoons}} \mathrm{ES} \underset{k_2'}{\overset{k_2}{\rightleftharpoons}} \mathrm{EP} \underset{k_3'}{\overset{k_3}{\rightleftharpoons}} \mathrm{E + P}$$

$$t_1 \qquad\qquad t_2 \qquad\qquad t_3 \qquad\qquad t_4$$

where the last two reaction constants are comparable, i.e. $k_3 \approx k_3'$. We leave the reader the task of representing the six reactions in our state space

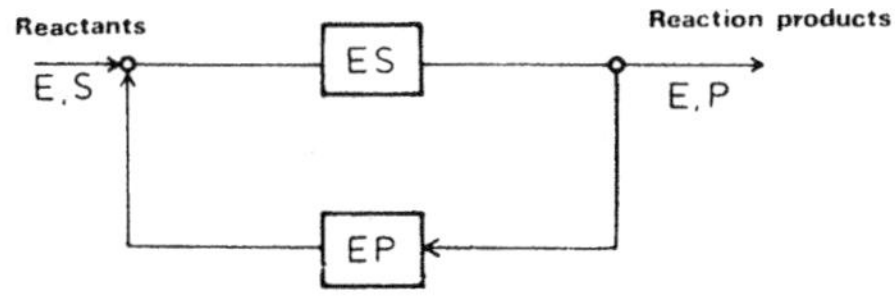

Fig. 2.5. A self-controlled chemical system of the negative feedback type. The reaction product P combines with the catalyzer E forming the 'ntermediate compound EP and thus slows down the left-to-right reactions.

notation. At this point we are only interested in emphasizing that the composition of our system is variable and that it is a self-controlled system and moreover one that is self-accelerated during certain periods and self-decelerated during others. The system can be diagrammed like an ordinary controlled system: see Figure 2.5.

Control systems of this kind fall under the general theory of control. However, chemical self-controlled systems do not accord with the now popular paradigms of control systems, namely the balancing wheel, the thermostat, and the self-controlled missile. Indeed in the latter cases the control is effected by a detachable subsystem. Moreover the whole system has fixed components and its behavior is usually assumed to be linear. Being detachable from the controlled system, a controlling system of this kind can be studied separately, not only as a component of the total self-controlled system. On the other hand if either the enzyme or the reaction product were removed from our chemical system represented by Figure 2.5, the system would cease to be chemical: it would become a physical system. Second, by definition of a chemical system, its composition is not constant but varies in the course of time. Thirdly, most of the chemical kinetics equations – in particular the ones for the formation rate of ES from the reactants – are nonlinear. These three peculiarities of chemical self-controlled systems seem to have prevented most scientists and engineers from recognizing them as systems, hence from using the systems theoretic approach to chemical problems.

2. Biochemical System

2.1. *Biomolecule*

We shall call *biomolecule* any macromolecule that, like a protein or a nucleic acid, is essential to some typically biological process such as cell division, morphogenesis, or development. And we shall call *biochemical* any system, natural or artificial, where biomolecules are formed or broken down. There are myriads of kinds of processes of this type, and some of them culminate in the formation of subsystems of live cells, such as ribosomes: see Figure 2.6.

The key biomolecules are the proteins and the nucleic acids: as Szent–György used to say, while the former carry on the business of life the latter are the guardians of life. Both are polymers (Definition 2.4(iii)). Whereas the building blocks (monomers) of the latter are nucleotides of mainly 8 different kinds, proteins are formed by amino acids, of which 20 species are commonly found in bioproteins. The biomolecules are then macromolecules, and they are constituted by so many units that they can attain remarkable sizes and weights. Because they are so large they are quite sensitive to environmental disturbances. Hence their stability requires fairly constant physical environmental conditions. To a first approximation some such molecules can be modeled as very long chains that can fold-in forming closely packed particles. (Cf. Vol'kenshtein, 1970; Lehninger, 1975; Watson, 1976.)

We shall borrow from biochemistry the notions of nucleotide and of amino acid. We shall also borrow the postulate that, whereas the nucleic acids are formed from eight nucleotide species, proteins involve 20 kinds of amino acids. (We shall disregard proteins other than those occurring in living beings.) More precisely, we incorporate the following biochemical axiom in our ontology:

POSTULATE 2.3. The family M of molecular species contains eight species of mononucleotides and 20 species of amino acids.

Each of the members of the mononucleotide family is composed of phosphoric acid, a sugar, and a nitrogenous base. There are two genera of nucleotides: the ribonucleotides and the deoxyribonucleotides. These genera differ by the kind of sugar: the former contain ribose whereas the latter contain deoxyribose. (A RNA molecule is a polymer of blocks of the ribo type, and a DNA molecule is a string of units of the deoxyribo kind.) In turn, there are four species of ribonucleotides and another four of deoxyribonucleotides.

The nitrogenous bases of greatest biological interest are adenine (A), guanine (G), cytosine (C), uracil (U) and thymine (T). The first four combine with ribose and phosphoric acid to form ribonucleotides or constituents of RNA:

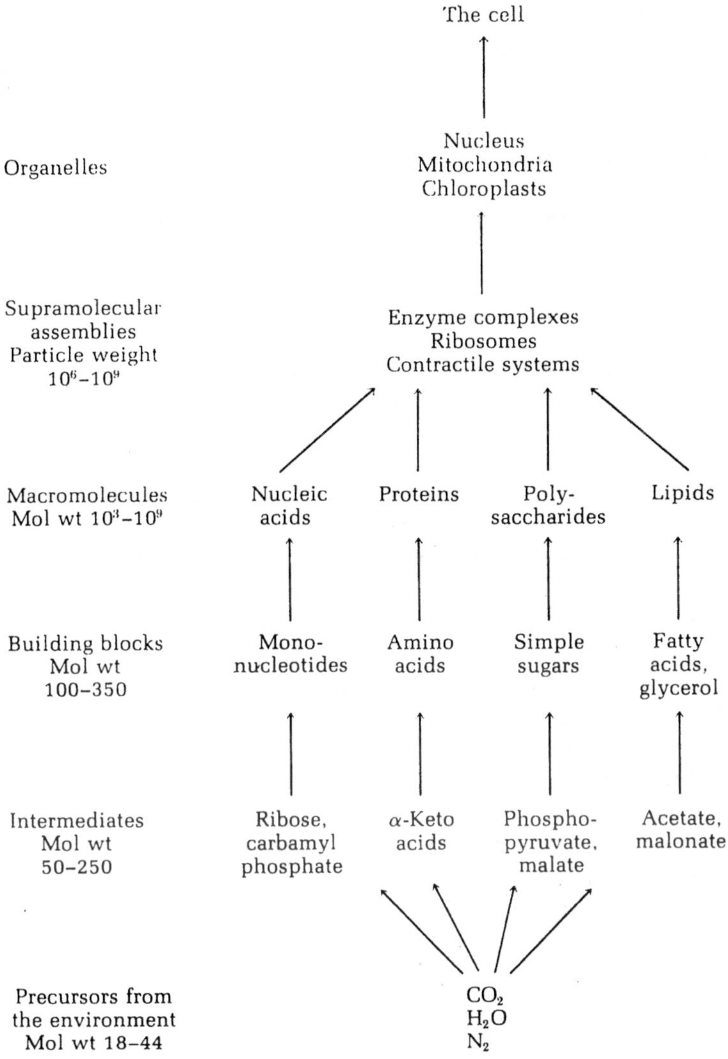

Fig. 2.6. Build-up processes that start with simple molecules in the environment and end up in living cells. 'Mol wt' is short for 'molecular weight'. From Lehninger (1975).

$$a = \text{AMP}, \qquad g = \text{GMP}, \qquad c = \text{CMP}, \qquad u = \text{UMP}.$$

('AMP' is short for 'Adenosine 5′-phosphoric acid'. Similarly with the other three.) On the other hand the nitrogenous bases A, G, C and T combine with deoxyribose and phosphoric acid to form deoxyribonucleotides or constituents of DNA:

$$a' = \text{dAMP}, \qquad g' = \text{dGMP}, \qquad c' = \text{dCMP}, \qquad t' = \text{dTMP}.$$

In short, there are two sets of units (monomers) from which the polynucleotides or nucleic acids are formed:

$$\text{Ribonucleotides } R = a \cup g \cup c \cup u$$
$$\text{Deoxyribonucleotides } D = a' \cup g' \cup c' \cup t'.$$

We proceed to characterize the nucleic acids and in the first place the members of the DNA family. We start with the building blocks in the set $D = a' \cup g' \cup c' \cup t'$ and form the set of all possible strings or sequences of members of D, with the sole restriction that the composition of each compound be equal to the union of the compositions of its constitutents or precursors. I.e. if x and y are in D and combine to form $z = xy$, then $\mathscr{C}_A(z) = \mathscr{C}_A(x) \cup \mathscr{C}_A(y)$, where '$\mathscr{C}_A(u)$' stands for the atomic composition of u. (Cf. Chap. 1, Sec.1.2.) The new set D^* formed from members of D contains all the conceivable combinations, with repetitions and permutations, of deoxyribonucleotides. In particular D^* contains all the polynucleotides forming the DNA molecules as well as monotonous strings such as $x^n = xxx...x$ with $x \in a'$. We shall restrict our attention to those molecules that contain all four nucleotides, i.e. members of a', g', c', and t'.

Call D_4^* the subset of D^* formed by the strings containing *all four* deoxyribonucleotides in all conceivable numbers and locations, such as $wxyz$, $zxwyx$, $yywxxyzzz$, where $w \in a'$, $x \in g'$, $y \in c'$, and $z \in t'$. The set D_4^* contains all the components of all the DNA molecules we need to consider – both the actual and the merely possible ones. The structure of D_4^* is that of a semigroup. More precisely, $\mathscr{D}_4 = \langle D_4^*, \cdot \rangle$ is a subsemigroup of the free semigroup $\mathscr{D} = \langle D^*, \cdot \rangle$ generated by D.

DNA molecules are double strands of things of kind D_4^*. These double strands are coupled in such a way that every A base in one of the strands is paired off to a T base in the other, and likewise every C base is paired off to a G base in the other. Hence we make

POSTULATE 2.4 Let $\mathscr{D} = \langle D^*, \cdot \rangle$ be the free semigroup generated by the set $D = a' \cup g' \cup c' \cup t'$ of deoxyribonucleotides under the operation $\cdot$ of chemical combination, subject to the condition that, for any x and y in D^*, if $z = xy$, then $\mathscr{C}_D(z) = \mathscr{C}_D(x) \cup \mathscr{C}_D(y)$. Further, call $\mathscr{D}_4 = \langle D_4^*, \cdot \rangle$ the subsemigroup formed by those sequences that contain molecules of all four kinds in D. There are systems composed of two chains of molecules in D_4^*, whose structure includes the bijection $p : D \to D$ such that for any $x_A \in a'$, $x_G \in g'$, $x_C \in c'$, and $x_T \in t'$,

$$p(x_A) = x_T,\ p(x_C) = x_G,\ p(x_G) = x_C,\ p(x_T) = x_A.$$

DEFINITION 2.14 The systems characterized by Postulate 2.4 are called *DNA molecules.*

Remark We are assuming that an arbitrary string of molecules satisfying Postulate 2.4 is not just a formal string but a chemically possible compound. The justification for this hypothesis is that there are no known restrictions on the stringing of nucleotides nor, for that matter, on carbon (or even boron) compounds in general. In short, using the notation of Sec. 1.2, $D_4^{**} = D_4^*$.

We proceed similarly with the RNA molecules, which are constituted by mononucleotides of the species a, g, c, and u:

POSTULATE 2.5 Let $\mathscr{R} = \langle R^*, \cdot \rangle$ be the free semigroup generated by the set $R = a \cup g \cup c \cup u$ of ribonucleotides, and call $\mathscr{R}_4 = \langle R_4^*, \cdot \rangle$ the subsemigroup formed by those sequences that contain molecules of all four kinds in R. Every member of R_4 is a possible *RNA molecule.*

The union of all kinds of DNA and RNA molecules is the totality of *nucleic acids*:

$$N = D_4^* \cup R_4^*.$$

The case of proteins is similar. Here the building blocks are the amino acids or rather some of them, such as glycine or NH_2CH_2COOH. More precisely, the *protein amino acids* form the genus

$$AA = \text{Ala} \cup \text{Arg} \cup \text{Asn} \cup \text{Asp} \cup \text{Cys} \cup \text{Gln} \cup \text{Glu} \cup \text{Gly} \cup \text{His} \cup \text{Ile} \cup \text{Leu} \cup \text{Lys} \cup \text{Met} \cup \text{Phe} \cup \text{Pro} \cup \text{Ser} \cup \text{Thr} \cup \text{Trp} \cup \text{Tyr} \cup \text{Val}.$$

This set has the semigroup property under the operation $\cdot$ of chemical combination. Also: the structure $\mathcal{AA} = \langle AA^*, \cdot \rangle$, where AA^* is the set of all possible strings or sequences of members of AA, is the *free semigroup generated by* AA. The members of this group are called *polypeptides* and the longer among them *proteins.* More often than not a protein contains at least one member of every species included in AA. However, there are important exceptions to this rule. Therefore we cannot restrict our attention to the subsemigroup formed by those sequences of amino acids that contain representatives of all 20 kinds of protein amino acids.

We summarize the foregoing in

POSTULATE 2.6 Let $\mathcal{AA} = \langle AA^*, \cdot \rangle$ be the semigroup generated by the set AA of protein amino acids. Every member of AA^* is a possible *polypeptide.*

DEFINITION 2.15 A *protein* is a polypeptide composed of at least 100 amino acids. Symbol for the set of proteins: $\mathcal{P}$.

It is estimated that there are at least 10,000 kinds of proteins in the cells of contemporary organisms. Roughly half of them are enzymes, i.e. reaction controllers.

Postulates 2.3 to 2.6 entail

COROLLARY 2.1 (i) Differences among kinds of DNA molecules consist in differences in composition or in the order (sequence) of the deoxyribonucleotides (members of D);

(ii) differences among kinds of RNA molecules consist in differences in composition or in the order (sequence) of the ribonucleotides (members of R);

(iii) differences among kinds of protein molecules consist in differences in composition or in the order (sequence) of the 20 protein amino acids (members of AA).

Molecules with the same composition but a different structure – in particular polymers differing only in the relative position of the component monomers along the chain – are called *isomers.* A polymer constituted by n distinct monomers has $1.2. \ldots n$ isomers.

So much for the composition and structure of nucleic acids and proteins. Let us now turn to their formation and breakdown.

2.2. *DNA Replication, and RNA and Protein Synthesis*

Not only are the nucleic acids the stuff of heredity: they are assumed to control whatever goes on in the cell, from protein synthesis to cell division. We must therefore take a closer look at them, particularly at their mode of emergence and the way they control the synthesis of proteins. However, we are forced to simplify, even to the point of disregarding the differences among the three kinds of RNA (ribosomal RNA, mRNA and tRNA).

The typical DNA molecule is composed by two polynucleotide chains that twist about each other forming what has become the most popular microsystem – the double helix. This has the well known property of self-duplication. This process involves the gradual detachment of the two strands, each of which serves as a mold or template for the formation of a complementary strand out of the molecules that pullulate in the rich cellular background. The attachment of these molecules is activated by enzymes. The overall process is captured by

POSTULATE 2.7 The duplication of a DNA molecule is a process whereby the molecule joins with molecules from its D-environment E to form two DNA molecules. In other words, the structure of a DNA molecule contains a function

$$\rho\colon D_4^* \times 2^E \longrightarrow D_4^* \times D_4^*$$

such that $\rho(d, e) = \langle d', d''\rangle$, with $d, d', d'' \in D_4^*$, $e \in 2^E$, and $\mathscr{C}_D(d)$, $\mathscr{C}_D(d')$, $\mathscr{C}_D(d'') \in 2^E$.

As mentioned above, DNA molecules control their own reproduction, i.e. are their own templates. But they are not the only molecular systems acting as templates: RNA molecules template for the synthesis of proteins. We must therefore elucidate the notion of a template. See Figure 2.7.

DEFINITION 2.16 Let $\tau \in T$ be a concrete system of kind T and let E be a set of things in the environment of τ. Assume that τ pre-exists a second system σ of kind Σ. Further, call i_T the identity function on T, and $\mathscr{C}_A\colon \Sigma \to 2^E$ the atomic composition function that assigns each (assembled) system σ its atomic composition $\mathscr{C}_A(\sigma) = e \in 2^E$. Then f is a *molding function* for σ, and τ a *template* for σ (or a σ-*template*) iff

(i) $f = i_T \times \mathscr{C}_A\colon T \times \Sigma \to T \times 2^E$ such that

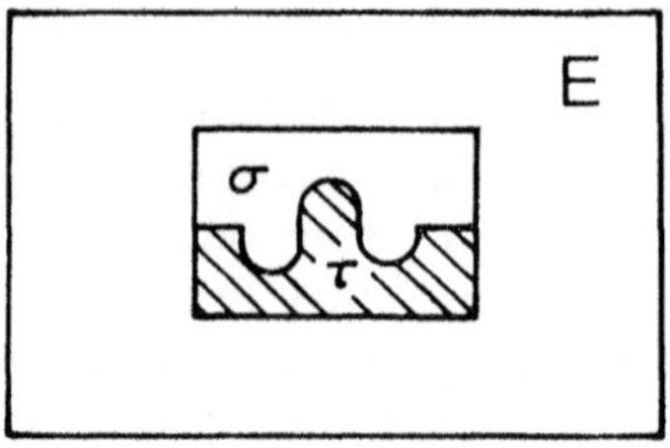

Fig. 2.7. System τ is a σ-template, i.e. τ provides a sort of mold for the assembly (e.g. synthesis) of σ out of environmental entities.

$$f(\tau, \sigma) = \langle \tau, e \rangle \quad \text{for } \tau \in T, \sigma \in \Sigma, \mathscr{C}_A(\sigma) = e \in 2^E;$$

(ii) the boundaries of τ and σ match in shape.

Molecular templates are, of course, molecules with the capacity to act as templates for the formation of further molecules. DNA and RNA are among them: DNA are autotemplates as well as RNA-templates, and RNA are in turn protein-templates. This is part of the well confirmed hypothesis misnamed *central dogma* of molecular biology, summarized in the heuristic diagrams of Figure 2.8, where the solid arrows represent the "information transfer" occurring in all cells, whereas the dotted arrows represent the "flow of sequence information" occurring in some cells.

We assume then, to begin with,

POSTULATE 2.8 RNA synthesis is the process whereby a DNA molecule

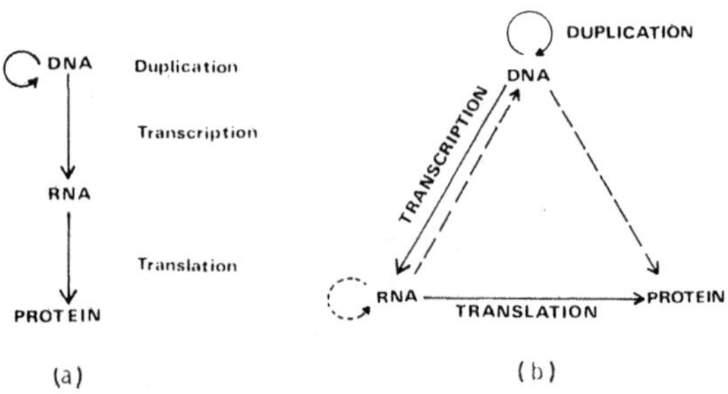

Fig. 2.8. Two versions of the "central dogma". (*a*) The standard version (e.g. Watson, 1976). (*b*) An alternative version (Crick, 1970). For a translation of the metaphorical expression 'information flow' into the nonmetaphorical language of quantum chemistry, wait a few decades.

(template) joins with nucleotides from its environment resulting in the same molecule (template) together with an RNA molecule. I.e. the structure of a DNA molecule contains a function

$$s_R = i_{D_4^*} \times \mathscr{C}_A: D_4^* \times R_4^* \longrightarrow D_4^* \times 2^E,$$

where $i_{D_4^*}$ is the identity function on D_4^* ,and $\mathscr{C}_A$ the composition function $\mathscr{C}_A: R_4^* \to 2^E$, such that

$$s_R(d, r) = \langle d, e\rangle \quad \text{for } d \in D_4^*, r \in R_4^* \; \mathscr{C}_A(r) = e \in 2^E.$$

DNA duplication and RNA synthesis are of course chemical processes, for in both cases a batch of reactants (the set e of precursors "swimming" in the environmental solution) combine into the reaction product. Moreover the entire system is a self-controlled chemosystem, since both processes are controlled by DNA molecules. However, the DNA molecules act not as ordinary chemical controls (e.g. enzymes) but as templates: theirs is template chemistry. (The same holds for the control exerted by RNA molecules in protein synthesis.) This is why the processes summarized by Postulates 2.7 and 2.8 are not represented by ordinary reaction equations, and why we had to introduce the functions ρ (DNA replication) and s_R (RNA synthesis). But this is precisely one of the peculiarities of biochemosystems in contrast to plain chemosystems.

Another peculiarity of biochemosystems is their capacity to synthesize biomolecules and, in particular, proteins. As will be recalled from Definition 2.14, a protein is a compound of amino acids and, more particularly, a member of the set AA^* of strings or sequences containing (protein) amino acids of all kinds. Again, the synthesis of proteins is controlled by RNA molecules acting as templates. More precisely, we introduce

POSTULATE 2.9 Protein synthesis is a process whereby an RNA molecule (template) joins with amino acids from its environment E, to yield the same molecule together with one protein. I.e. the structure of RNA contains the function

$$s_P = i_{R_4^*} \times \mathscr{C}_A: R_4^* \times AA^* \longrightarrow R_4^* \times 2^E,$$

where $\mathscr{C}_A: AA^* \to 2^E$, such that

$$s_P(r, p) = \langle r, e\rangle,$$

where $r \in R_4^*$, $p \in \mathscr{P} \subset AA^*$, $\mathscr{C}_A(p) = e \in 2^E$.

This axiom describes only the overall process: it does not tell us which segments of the RNA molecule correspond to which amino acids in the sequence characterizing the given protein. The fine structure – though by no means the mechanism – of this matching is given by the misnamed *genetic code*. This "code" is nothing but the detailed correspondence between the nucleotides in the RNA template and the amino acids in the protein synthesized on it. Let us describe this correspondence.

Let $R = a \cup g \cup c \cup u$ be the set of ribonucleotides forming the RNA molecules in R_4^* (recall Postulate 2.5). The latter is the set of strings (or sequences) containing ribonucleotides of each of the four nitrogenous bases, and therefore all the possible triplets of such, e.g. *uca*, *uuc*, and *gcg*, enter into the composition of members of R_4^*. Call R_3^* the set of $4^3 = 64$ such triplets, called *codons*. See Figure 2.9.

Take next the set AA of 20 kinds of protein amino acids, i.e. the protein building blocks. The full genetic "code" is the function that assigns each codon (triplet of nucleotides in an RNA molecule) a given amino acid in a protein. (Cf. Watson, 1976.)

We reformulate the preceding as a convention and an axiom:

DEFINITION 2.17 Let R_3^* be the set of 64 triplets of nucleotides in $R = a \cup g \cup c \cup u$ such that, for every $x \in R_3^*$, there exists a segment $y \in R_4^*$ of an RNA molecule for which $x \sqsubset y$ (x is a part of y). Every member of R_3^* is a *codon*.

POSTULATE 2.10 The codon-amino acid (or RNA-protein) correspondence is the partial function $\pi: R_3^* \to AA$ that

(i) assigns every codon, other than the starting and the terminating codons, one amino acid in a protein. I.e. with the noted exceptions, if $r \in R_3^*$, then $\pi(r) = a \in AA$, where $a \sqsubset p$ and $p \in \mathscr{P}$ is some protein;

(ii) is order preserving: i.e. if $r, r' \in R_3^*$, and r precedes r' in a given codon sequence, then $\pi(r)$ precedes $\pi(r')$.

In the literature a different function f occurs implicitly, namely one assigning each amino acid one codon, i.e. $f: AA \to R_3^*$, where '$f(a) = r'$, for

RNA strand	*aaugcuuaaaccgggauu...*
Triplets	(*aau*)(*gcu*)(*uaa*)(*acc*)(*ggg*)(*auu*)...

Fig. 2.9. The sequence of nucleotides in the RNA molecule, when acting as a template for protein synthesis, is organized into triplets (codons). The kind and the position of each amino acid in the protein are determined by the composition and the position of the corresponding codon.

$a \in AA$ and $r \in R_3^*$, is read '*r codes for a*'. However, it has been found experimentally that different codons can "code for" one and the same amino acid. (I.e. the "code" is degenerate or many-one, hence not a proper code allowing one to break any ciphered molecular message.) In other words, no such function f from AA to R_3^* exists, so that the converse "translation" of protein structure into RNA structures is impossible.

A genuine code is a certain correspondence between two sets of *artificial* signs and, more particularly, *languages* (cf. Birkhoff and Bartee, Ch. 8). The coding and decoding functions of errorless codes are one to one, not many to one like π. Therefore the use of the linguistic concept of code in chemistry is metaphorical. The same holds for the expressions 'the templates instruct (or code for) protein synthesis' and 'the genetic instructions contained in a gene'. That these are analogies is shown by the following facts. First, a genuine information system is one composed of a transmitter, a receiver, and an information channel between them; no such components are apparent in a chemical system. Second, the computations of the quantity of information allegedly locked in a chunk of genetic material (i.e. a segment of a DNA molecule) are phoney – so much so that they occur in no biological law statements and everyone gives his own favorite arbitrary estimate. For these reasons the information-theoretic language employed in molecular biology is a heuristic prop that has now exhausted its heuristic power. Moreover it is a stumbling block on the road to transforming the "central dogma" into a hypothetical-deductive system explaining the synthesis and the duplication of nucleic acids, as well as the synthesis of proteins, exclusively in physical and chemical terms instead of describing them in analogical terms such as 'informational molecule', 'information flow', 'transcription', and 'translation'.

Note finally that, like the previous mappings introduced in this section – namely ρ (DNA replication), s_R (RNA synthesis) and s_P (protein synthesis) – π is not defined by Postulate 2.10, which characterizes a whole class of such functions. However, unlike ρ, s_R and s_P, π is known in detail: indeed it is given by the table that summarizes the so-called genetic code. What is still unknown is the origin of the "code". Nor is it known whether nucleic acids have always been "used" to synthesize proteins.

2.3. *Biochemical System*

A biochemical system may be characterized as a system wherein biomolecules are synthesized or broken down. A ribosome and an insulin synthesis reactor are biochemical systems. On the other hand a cellular membrane is a plain chemosystem because it is not involved in the production or disintegration of biomolecules.

Our explicit characterization is

DEFINITION 2.18 A system σ is a *biochemical system* (or a *biochemosystem*) iff σ is a self-controlled chemical system such that

(i) the composition of σ includes proteins or nucleic acids;

(ii) the environment of σ contains all the precursors of the components of σ and moreover is compatible with the processes of synthesis of proteins or of nucleic acids;

(iii) the components of σ exchange things and energy with the environment (i.e. σ is semi-open) and moreover some of them participate in the reactions happening in σ;

(iv) the structure of σ contains the protein synthesis (or the breakdown), the DNA replication, or the RNA synthesis function.

The reactions in a chemical system can eventually come to a halt because of exhaustion of reactants or as a result of product inhibition. In contrast, a biochemosystem is capable – environment willing – of maintaining its reactions or switching over to alternative reactions by predating on its environment and self-assembling its catalyzers (enzymes), and eventually also its templates. In fact, processes of two kinds are constantly going on and intertwining in a biochemical system: (*a*) the building up or self-assembly of complex molecules from simpler precursors (*anabolism*), and (*b*) the breaking down or dismantling of complex chemicals (*catabolism*). Furthermore these two processes are interdependent: the break down processes require enzymes that had to be synthesized to begin with, and such syntheses are often dependent upon further enzymes occurring as intermediates in breakdown processes. Moreover the anabolic processes consume energy released chiefly by catabolic ones. In short, biochemical systems are characteristically metabolic. We refine this notion by introducing

DEFINITION 2.19 If σ is a biochemical system, then

(i) the *intermediary metabolism* of σ is the set of all the enzymatic reactions involving only components of σ;

(ii) the *metabolism* of σ is the set of all the chemical reactions involving components of σ and the environment of σ.

Certain cell components are metabolizing biochemosystems. However, it is assumed that the precursors of the first organisms, on the Earth and probably elsewhere, were extracellular biochemical systems, and that the first cells emerged by self-assembly from, or the merger of, some such biochemical systems. And it is sometimes conjectured that the immediate precursors of cells may have been systems containing both proteins and nucleic acids. Such systems, though not possessing all the properties of contemporary cells, would have their main emergent properties, namely the abilities to metabolize, self-reproduce, and mutate. (Cf. Küppers, 1975.) However, the problem of the origin of life will be taken up in the next chapter.

2.4. *Concluding Remarks*

In this chapter we have distinguished three kinds of system that are usually conflated: physical, chemical, and biochemical. See Figure 2.10.

Chemical systems are distinguished from plain physical systems by the fact that some of their components are continually engaged in reactions in the course of which some chemicals are consumed while others are produced. Chemical systems are thus paradigms of qualitatively changing things. Not that qualitative change fails to occur in physical systems: the emission and absorption of light are qualitative changes, and so are nuclear fission and fusion as well as changes in the state of aggregation. However, many a physical system can be rendered immune to qualitative changes for a while. In contrast, a chemical system is by definition one whose composition changes relentlessly, so that it eventually becomes

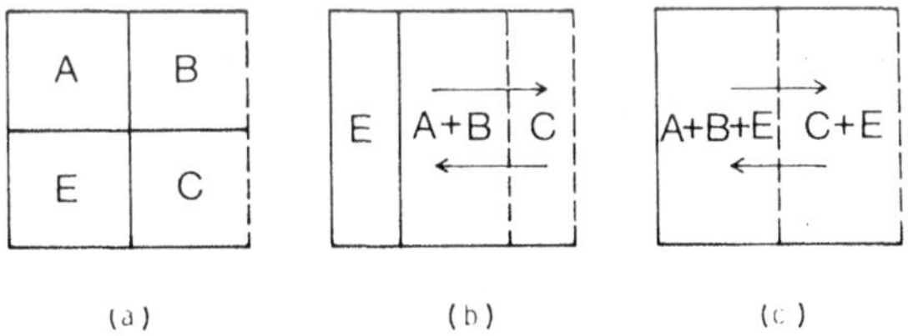

Fig. 2.10. Three kinds of system: (*a*) physical system with four non-reacting components; (*b*) chemical system with one nonreactive component (E); (*c*) biochemical system: all components react, and one of them (E) is an enzyme or proteinic catalyzer. Whereas physical and chemical systems may be almost completely closed (full line) or open only in some respects, biochemical systems must be semiopen (dashed line), for they prey on their environment.

something qualitatively different from what it was in the beginning. That such changes can be explained, at least in principle, with the help of physical theories – though not with them alone – does not turn them into physical processes. We are here concerned with ontological, not epistemological matters: with objective level distinctions not with the understanding of the higher levels with the help of knowledge concerning the lower ones.

Biochemical systems are the seats of even deeper and more sustained qualitative changes. They engage not just in any old chemical reactions but in reactions resulting in the self-assembly or the self-dismantling of huge molecules (often millions of times bigger than the inorganic ones). Moreover some such reactions involve templates, objects that do not occur in ordinary chemical reactions. Finally biochemical systems have a sustained activity thanks to their enzymes, and they can help themselves to environmental morsels. Because of these properties biochemical systems come rather close to biosystems.

The differences among physical, chemical, and biochemical systems are not just differences in complexity: they are momentous qualitative differences, so great in fact that we are justified in speaking of different levels: the physical and chemical levels, and perhaps also the biochemical level. Of course we must not forget that all biochemical systems are chemical systems, which in turn are composed of physical systems: otherwise we won't succeed in explaining biochemistry with the help of chemistry, or chemistry with the help of physics. Nevertheless we must not exaggerate similarities at the expense of differences, particularly if we hope to understand one more qualitative jump, namely the one from biochemical systems to biosystems. Let us take this leap.

CHAPTER 3

LIFE

Life has long been a mystery on which mystery mongers have thrived. It is no longer such: biologists are beginning to understand what makes organisms tick, and physicians to manipulate the control systems responsible for that ticking – which, incidentally, is not like a clock's. This understanding could not be forthcoming as long as organisms were studied either on their own level exclusively (holistic approach) or as physical systems devoid of emergent properties (levelism), and in either case apart from their history. The theory of evolution and molecular biology are to be credited for having transformed the mystery of life into the problem of life: its origin and maintenance, its evolution and extinction.

We have come to understand that many of the peculiarities of organisms have molecular roots. Thus cell division, a typically biological event, is the outcome of a chemical process. Moreover we understand now that other biological properties and processes – such as those of emerging by way of self-assembly of nonliving units, and of being subjected to natural selection – are properties shared by all systems. (Recall Ch. 1, Sec. 3.2.) In other words, we have come to realize that organisms are systems composed of chemical systems, and moreover that they have evolved from lower level things. This suggests the plan of this chapter: we start out with biochemical systems (studied in Ch. 2), endow them with certain emergent properties to obtain biosystems, and examine some peculiarities of biological adaptation and evolution.

The upshot of our study will be a merger of ontological pluralism (or emergentism) with moderate epistemological reductionism. Indeed we shall take the view that ontological monism – whether mechanistic or spiritualistic – is false because it denies the emergence of novel properties and laws as one moves from physical systems through chemosystems to biosystems. On the other hand the program of explaining the higher (more complex and newer) levels by the lower ones – i.e. epistemological reductionism – has proved fertile and is in keeping with contemporary biology. Indeed the latter – particularly molecular biology and evolutionary biology – is explaining life without explaining it away, i.e. without leveling down (let alone up), and is doing it with the help of chemistry, though not in terms of it exclusively.

1. From Chemism to Life

1.1. *Self-Organization of Biochemical Systems*

According to contemporary biology (*a*) modern organisms descend ultimately from primitive biosystems (protocells), and (*b*) the latter in turn emerged from biochemical systems, which evolved out of biomolecules. (See Oparin (1968), Miller and Orgel (1974), and Oró, Miller, Ponnamperuma and Young (Eds.) (1974).) In turn, all of the biomolecules are carbon compounds, all of which contain also hydrogen and oxygen, and most of them also nitrogen. This is no accident: of all the 100-odd elements, hydrogen, carbon, nitrogen and oxygen constitute the most stable molecules and by far the greatest variety of them. Boron too has the capacity to form large stable molecules, but it is far less common than the other four elements, which exist almost everywhere in the explored part of the universe.

Moreover, the physical conditions necessary for the synthesis of biomolecules – such as plenty of solar energy and humidity, as well as moderate temperatures and pressures – have existed on our planet for about four billion years. Besides, it is reasonable to conjecture that the primitive atmosphere was reducing rather than oxydizing, hence lenient rather than hostile to the self-assembly of biomolecules and systems composed thereof. Under such conditions it is no wonder that biomolecules should have formed and eventually merged to constitute larger systems. In short, ours is "the best of all possible environments for life" (Henderson, 1913, p. 273) – or at least one among a number of environments fit for the origin and maintenance of life. This is not to say that the emergence of life was inevitable: scientists maintain only that it was possible (i.e. lawful rather than miraculous). Moreover, since the conditions for the emergence of life are rather stringent, they are also probably rather exceptional, so that the likelihood that they have occurred elsewhere in exactly the same manner must be vanishingly small. Therefore, although it is very likely that there are extraterrestrial organisms, it is unlikely that they, or at least the more complex among them, much resemble those we know (Jacob, 1977).

Until recently it was assumed to have taken aeons for biochemical systems to form from hydrogen, carbon, oxygen, and nitrogen. And this because it was taken for granted that such self-assemblies could occur only by the simultaneous chance encounters of all the atoms composing a biochemical system. Indeed, such events are so unlikely that they can be ruled out. Although chance is ever present, particularly in chemical dissociations, the chemical reactions of the buildup and substitution kinds are brought about by chemical bonds against the disorganizing random motions of the atoms and molecules in the surroundings. (Surely there is randomness in the very collisions ensuing in chemical compounds, but the latter are unstable, hence short lived, unless the atoms are held together by interatomic bonds. So, chance within the system works against chance in the environment. And even the dissociations caused by random thermal motions may result in reaction products that are in turn reactants forming eventually more stable systems.)

Nowadays we know that chemical and biochemical systems cannot help but self-assemble under the action of bonds of various kinds. Moreover we know that such self-assembly processes are more likely to occur in stages than at one stroke. In particular, the formation of primitive proteins may have proceeded in two stages: amino acid synthesis followed by polymerization. Therefore the first biochemical systems, and even the first organisms, may have formed on our planet and elsewhere as soon as the requisite conditions were met. This explains the short time span between the origin of rocks and the appearance of the first bacteria and blue-green algae.

As soon as biomolecules assembled they could associate with others forming larger molecules or transmolecular systems. About the simplest systems of the latter kind are the colloidal particles. For example, a protein molecule in an aqueous solution attracts a number of water molecules forming a colloidal hydrophilic complex. And in turn a number of such systems can coalesce to form a coacervate: see Figure 3.1. In either case the resulting system is a distinct thing, separate from its liquid medium and with a structure of its own (Oparin, 1968). Presumably nucleic acids can form similar systems. In particular a DNA molecule placed in a suit-

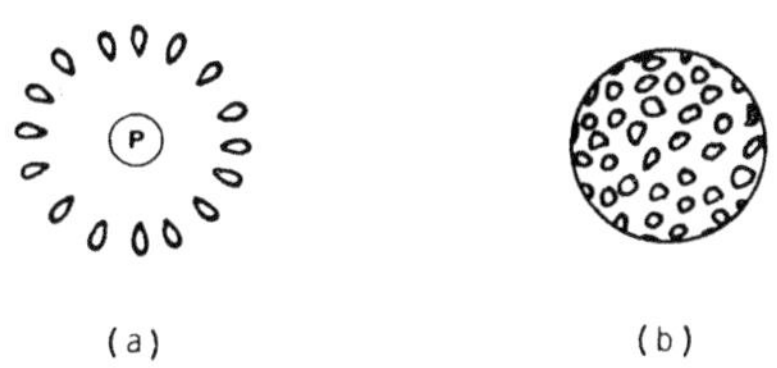

Fig. 3.1. The self-assembly of colloidal particles around a protein molecule (P) in an aqueous solution (*a*), and the coalescence of colloidal particles into a coacervate (*b*).

able solution of nucleotides and nucleic acids could perhaps start synthesizing RNA molecules, which in turn would control the synthesis of proteins as well as of further DNA molecules. Furthermore a protecting double layer of lipid molecules could self-assemble around such a DNA-RNA-protein system. Such a semipermeable membrane would increase the concentration of biomolecules, hence the rate of biochemical reactions, and would keep out disrupting (e.g. dissociating) agents. New reactions would thus be possible, ending up in the formation of new kinds of biopolymers and even of transmolecular systems such as vesicles and organelles of various kinds.

Given biomolecules in a suitable medium, the formation of organelles such as chloroplasts and ribosomes is far from miraculous. Indeed a ribosome, for example, is not an impossibly complicated system: it consists of 58 macromolecules, three of them RNA molecules, and the rest proteins of various kinds. What the ribosome does – its function – is, chiefly, to synthesize proteins from precursors in its environment.

The next step is, of course, the self-assembly of a few such transmolecular systems to form a cell, however primitive – hence inefficient, ergo ill-adapted. The rest is history – evolutionary history driven by mutation and selection. In short, the process of emergence of life may have looked like Figure 3.2.

This diagram condenses what is common to various rival hypotheses concerning the precise self-assembly of primitive cells. (See Florkin (1960), Oparin (1968), Fox and Dose (1972), Miller and Orgel, (1974), Oró, Miller, Ponnamperuma and Young (Eds.) (1974).) There is rapidly accumulating

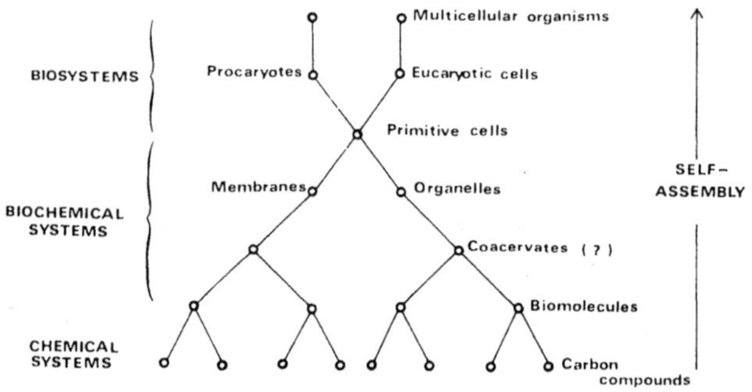

Fig. 3.2. The step-wise self-assembly of organisms from carbon compounds.

evidence that most of the buildup processes depicted in Figure 3.2 have been going on on our planet, and at least the first few steps elsewhere as well. In particular, the analysis of Precambrian rocks, of meteorites, and recently of some celestial objects, has been yielding amino acids, fatty acids, and even nucleic acids.

We compress and generalize the above into

POSTULATE 3.1 Let $\Sigma = \langle \Sigma_i \mid 1 \leq i \leq n \rangle$ be a sequence of really (nomologically) possible species of system, such that the systems of kind Σ_i are composed of systems of kind Σ_{i-1}. Then every actual system belongs in the sequence Σ and has self-assembled from systems in the immediately preceding term of the sequence.

1.2. *Definition of a Biosystem*

We have reached the point where we can list the properties of a system we believe to be jointly necessary and sufficient for it to qualify as a living thing. We assume that an organism is a system such that

(i) its *composition* includes proteins (both structural and functional, in particular enzymatic, the latter enabling it to exploit its environment) as well as nucleic acids (which make for its reproducibility and the likeness of its offspring);

(ii) its *environment* includes the precursors of all its components (and thus enables the system to self-assemble most if not all of its biomolecules);

(iii) its *structure* includes the abilities to metabolize, to self-repair, and to reproduce.

We shall spell out this hypothesis into an axiom and a convention. Here is the former:

POSTULATE 3.2 There are systems of a kind B such that, for every member b of B,

(i) b is composed of chemical and biochemical (hence metabolizing) subsystems;

(ii) b incorporates some of the biomolecules it synthesizes (rather than releasing them to its environment);

(iii) the possible activities of b include (*a*) the rearrangement, assembly, and dismantling of components; (*b*) the replacement of nonfunctional components and the removal or neutralization of some chemicals; (*c*) the capture and storing of free energy (in ATP molecules, glycogen, and fat) for future consumption;

(iv) *b* can adjust to some environmental changes;

(v) *b* is a component of some supersystem (organism) or else of some population of systems of the same kind;

(vi) some of the subsystems of *b* are capable of reproducing parts of *b*;

(vii) some of the subsystems of *b* control some of the processes occurring in *b* in such a way that a fairly constant inner environment (*milieu intérieur*) is maintained in the system;

(viii) all of the control systems in *b* are interconnected by signals (diffusion of chemicals, propagating chemical reactions, electrical signals, etc.) and so constitute a signal or information network;

(ix) one of the control systems of *b* (its genetic system or genome) is composed of nucleic acid molecules, is unique to *b*, and controls the development and the reproduction of *b*;

(x) the descendants of *b* are similar to *b* but may possess some new traits of their own (by mutation or gene recombination);

(xi) *b* competes in some respects with other systems of the same species as well as of other species, and cooperates with the same or others in some other respects;

(xii) *b* descends proximately or remotely from biochemical systems devoid of some of the properties listed above, and its nucleic acid molecules keep traces (hence some record) of the ancestry of *b*;

(xiii) *b* lasts, as a member of *B*, for a limited time lapse.

The name of the *B*'s should come as no surprise:

DEFINITION 3.1 The systems *B* referred to by Postulate 3.2 are called *biosystems* or *living things.*

Surely biosystems have many more properties than those we have just listed, but these seem to be necessary and sufficient to distinguish terrestrial biosystems from other things. In particular, clause (ii) exhibits one of the pecularities of biosystems as opposed to nonliving systems. In fact it is possible to assemble in the laboratory systems capable of synthesizing biomolecules that do not go into forming cell components such as organelles and membranes, let alone whole cells. To put it in metaphorical and therefore unscientific terms: a characteristic of metabolism in biosystems is that, far from being either "indifferent" or "self-serving", it "serves" the organism as a whole. (No teleonomy here: those systems that do not metabolize this way do not make the grade: they are either nonliving or sick.) Likewise clause (iii): a nonliving (e.g. biochemical) system need not be self-cleaning and self-repairing but could accumulate reaction-inhibiting chemicals (e.g. toxic substances) that would eventually bring some or even all of its reactions to a halt.

Note that in clause (xi) we have included not only competition, the heart of Darwinism, but also cooperation. Of course in the vast majority of cases cooperation is not deliberate, i.e. it does not consist in solidarity, but it is not less real for that. Without the unwitting cooperation of plants, animals, and other organisms, the terrestrial atmosphere might not be very different from what it was in the beginning, i.e. about five billion years ago: for one thing it might still be reducing rather than oxydizing. Note also a peculiarity of heredity in biosystems (clause (xii)). Like other complex systems, such as stars, mountains, and artifacts, biosystems betray their past. But, unlike nonliving systems, biosystems concentrate their archives in a tiny part of themselves, namely their genome.

Not all of the functions (properties, processes and activities) we attribute to an organism are actually carried out by it throughout its life history. Thus metabolism and cellular division may be temporarily suspended, as is the case with spores and dormant seeds. Some of the properties of biosystems are then dispositions, or potential properties that may actualize under favorable environmental circumstances. (For the notion of potentiality see Vol. 3, Ch. 4.)

We have not included certain properties that are obviously possessed by biosystems – and therefore often listed as definitory – because they are also common to all biochemical systems and even to some physical systems. Among them are (*a*) having a more or less definite boundary separating the system from its environment, and (*b*) being subject to natural selection. That the former property is shared by many a macrophysical nonliving system, is clear. As for the latter property, Postulate 1.6 attributes it to all systems.

Finally note that, according to Definition 3.1, chromosomes are not alive because they do not metabolize, and ribosomes are not alive because they do not replicate. Likewise viri are not alive because they do not function at all outside some host cell – so much so that independent viri are often crystals. (Only the host cell-virus system is alive.) Nor do robots, however sophisticated, qualify as biosystems, if only because they are made of mechanical and electrical components instead of biochemical ones. A couple of properties, no matter how important, do not suffice to characterize a biosystem; the property system consisting of the 13 properties listed in Postulate 3.1 is necessary and sufficient – until further notice.

1.3. *Cell and Organism, Biospecies and Biopopulation*

To begin with let us take a closer look at a feature of biosystems that is of particular philosophical interest, namely their cellular composition. To identify the smallest unit of living matter, as well as that of life, we shall use the notions of biosystem (Definition 3.1) and of system component (Definition 1.2):

DEFINITION 3.2 (i) A *cell* is any biosystem such that none of its components is a biosystem. In obvious symbols,

$$x \in C =_{df} x \in B \ \& \ \forall y \, (y \in \mathscr{C}(x) \Rightarrow y \notin B);$$

(ii) the *cellular composition* of a biosystem is the set of cells that compose it:

$$x \in B \Rightarrow \mathscr{C}_C(x) =_{df} \mathscr{C}(x) \cap C, \quad \text{where} \quad \mathscr{C}(x) = \{y \in \Theta \mid y \sqsubset x\};$$

(iii) an *organism* is a cell or a multicellular biosystem which is not a proper subsystem of a biosystem:

$$x \in O =_{df} x \in C \vee x \in B \ \& \ (\mathscr{C}_C(x) \subset C \ \& \ \neg(\exists y) \quad (y \in B \ \& \ x \prec y)$$

Remark 1 The intersection of C and O is the set of unicellular organisms. *Remark 2* Useful synonyms for C and O are *cellular level* and *organismic level* respectively. *Remark 3* According to the above definition, the living part of a coral reef may be regarded as an organism. *Remark 4* A bacterial culture, though a system, is not a biosystem; in fact it is a population. *Remark 5* It has been held that organisms could be construed either as "classes of cells' or as "parts of biological species" (Hull, 1974, p. 48). Both disjuncts are false. Classes, in particular biospecies, are sets, hence concepts, whereas organisms are things and, more particularly, concrete systems. A thing can be a member ($\in$) of a certain set and a part ($\sqsubset$) of another thing: it cannot be a part of a set and a member of a thing, and this by definition of the membership and the part-whole relations.

We can now define the concept of a biospecies. We shall construct it with the help of the notion of a natural kind (Vol. 3, Ch. 3, Sec. 3.5) as determined by a bunch of natural laws, and the concept of descent (Definition 1.17):

DEFINITION 3.3 A species is a *biospecies* iff
(i) it is a natural kind (rather than an arbitrary collection);
(ii) all of its members are organisms (present, past, or future);
(iii) it descends from some other natural kind (biotic or prebiotic).

Remark 1 This definition avoids the difficulties that beset some of the hazy notions of a biospecies encountered in the biological and biophilosophical literature. In particular we are not requiring reproductive isolation, for this is irrelevant to organisms that do not reproduce sexually, and it founders in the cases of plant and animal hybrids that fail to be sterile. *Remark 2* We do not confuse a biospecies with a biosystem, as even some eminent biologists have done. (E.g. Dobzhansky, 1970, p. 354: "A species is composed of individuals as an individual is composed of cells, or as a termite or ant colony is composed of fertile and sterile members.") Like any other species, a biospecies is a set, hence a concept. Only, it is not an arbitrary set but one defined by a definite collection of laws. And a biospecies is not an unstructured set but one partially ordered by the relation of descent. But a species is not a concrete system, because (*a*) its members do not all share the same environment, and (*b*) its past and future members are not coupled to its present members. *Remark 3* Because of clause (iii) of Definition 3.3, the organisms that may eventually be synthesized in the laboratory would belong not to biospecies but rather to artificial biospecies.

Finally let us deal with three bioentities that are neither individual organisms nor sets of such:

DEFINITION 3.4 A system is
(i) a *biopopulation* iff it is composed of individuals of the same biospecies;
(ii) a *community* or *ecosystem* iff it is composed of several interacting populations of organisms belonging to different biospecies;
(iii) a *biosphere* iff it is the largest system in which a given biosystem participates.

Remark 1 It is usual to include, in the definitions of biopopulation and ecosystem, the sharing of a territory or habitat, such as a swamp or a tree canopy. This is unnecessary in our case because we have defined those entities as concrete systems: they would not be such unless there were bonds amongst its components, and such bonds are possible only if the individuals concerned are not far apart. *Remark 2* Nor is it necessary to specify the planet in defining the general notion of a biosphere: the definition applies to any planet and it does not hold for the set of ecosystems on all possible planets because such a set is not a concrete system.

1.4. *Biolevel*

We have distinguished, more or less explicitly, six different biotic levels, namely those of cell, living component of an organism, organism, population, ecosystem, and biosphere. We may as well display them for reference purposes:

B_1 = *cell level* = the set of all cells
B_2 = *organ level* = the set of all organs
B_3 = *organismic level* = the set of all organisms
B_4 = *population level* = the set of all populations
B_5 = *ecosystem level* = the set of all ecosystems
B_6 = *biosphere level* = the set of all biospheres.

The family $\beta = \{B_1, B_2, B_3, B_4, B_5, B_6\}$ illustrates the concept of a level structure (Definition 1.8). So, the set β together with the level precedence relation deserves a name of its own:

DEFINITION 3.4 The set β of biotic levels together with the relation $<$ of level precedence, i.e. $\mathscr{B} = \langle \beta, < \rangle$, is the *biolevel structure.*

$\mathscr{B}$ has also been called the *scala naturae* and the *hierarchy* of life. These are misnomers: a ladder proper is supposed to lead somewhere, and a hierarchy proper involves a dominance or subordination relation, while $\mathscr{B}$ does not.

Now, the biotic levels we have included in β are not the only logically possible ones. In particular, one can imagine the existence of subcellular organisms; in fact it has occasionally been claimed that such biosystems exist. We shall assume that whatever is alive or composed of living beings is in β:

POSTULATE 3.3 Every biosystem and every system composed of biosystems belongs in some level of the biolevel structure $\mathscr{B} = \langle \beta, < \rangle$.

This version of the "hierarchical organization" of life is static: it talks about levels and their order but not about their origin. However, most biologists nowadays would claim that biotic levels are so many stages of an evolutionary process: that every biolevel has emerged spontaneously from the preceding (prebiotic or biotic) level. Actually we need no separate axiom to incorporate and refine this assumption, as it is a mere specification of Postulate 3.1:

COROLLARY 3.1 Every concrete system belonging to a given biolevel has self-assembled from things at the preceding level.

In turn, an immediate consequence of this corollary is that every system at a given biolevel is preceded in time by its components or precursors, Hence level precedence and temporal precedence, though not cointensive, are coextensive. The biolevel structure is no longer static: it has become part of an ontology which is not only pluralistic but also evolutionary. Biolevels are not static layers that happen to be piled atop one another. They succeed each other in time (metaphorically speaking) and they do so by virtue of a definite and pervasive mechanism, namely self-assembly. And they are not rungs in a hierarchy leading from lowly atom through middling man to the Supreme Being: levels are but stages in a natural evolutionary process that may have occurred and indeed may be occurring at various places and epochs in the history of the universe, though not twice in the same manner.

1.5. *Concluding Remarks*

We have characterized biosystems as systems composed of chemical and biochemical subsystems. Hence while some of the properties of biosystems can be traced back to their biochemical constituents, others are emergent or peculiar to the former. The most obvious property of this kind is of course that of being alive – a derivative property not a root one. Thus the chloroplasts and genes of a plant cell are essential components of the latter but are not alive. Nor are biopopulations and communities alive. Thus a forest teeming with plants, fungi, animals, and bacteria is not alive.

It follows that the class of biosystems is not included in that of biochemical systems, and that of biopopulations is not included in that of biosystems. What we do assert instead is that, just as biochemical systems compose biosystems, so the latter compose populations. Therefore we reject ontological reductionism, according to which biosystems are just a kind (subset) of biochemical systems, and biopopulations in turn a kind (subset) of biosystems. We assert instead that the components of members of the biopopulational level are members of the biosystem level, which are in turn composed of members of the biochemical level. This emergentist thesis is compatible with the program of explaining the higher in terms of the lower or, more precisely, the system in terms of its components (and the interactions among them). We shall come back to these matters in Sec. 4. (See also Bunge, 1977f.)

2. Biofunction

2.1 *Health and Death*

Systems, in particular organisms, can cope with their environment with varying degrees of success: recall from Definition 1.14 that the selection pressure exerted by the environment on a population of systems of some kind is a number lying between 0 (total success) and 1 (utter failure). But that concept of degree of adaptation was collective, nonspecific, and phenomenological: it does not carry over to an individual biosystem and it is not physiological. We need a separate concept of degree of adaptation of a given biosystem, not necessarily an organism. Here is one:

DEFINITION 3.5 Let b be a biosystem in an environment e. Then b is *healthy or normal in e* iff b, when in e, possesses all of the properties (can discharge all of the functions) listed in Postulate 3.1.

Remark 1 This is a physiological concept of health or normality, not a statistical one. It is possible for an entire population of healthy organisms to be wiped out by an environmental catastrophe, such as a severe drought. And it is also possible for most of the individuals of a biopopulation to be sick – a case of statistical normality and physiological abnormality. *Remark 2* Sickness is sometimes regarded as a form of adaptation. This is so in the case of infectious diseases, when the organism synthesizes antibodies that "fight" the invaders. But it is not so in other cases: there is nothing adaptive about arthritis, the obstruction of the bile canal, an infarct, stroke, or stress. Quite the contrary, these are glaring cases of failure to adapt. *Remark 3* Our definition is not limited to whole multicellular organisms: it may apply to cells or organs of a multicellular organism. Nevertheless in such cases it is preferable to restrict oneself to the specific functions discharged by such subsystems. More on this in Sec. 2.2.

The concept of health elucidated by the previous definition is qualitative. It is desirable to have also a quantitative concept of (degree of) health (or sickness) as well as of total health (or health in all respects). One obvious candidate is this: The degree of sickness of a biosystem in a given environment, in a given respect, at a given time, is its relative imbalance (or deviation from the physiological norm) in that respect at that time. Assuming optimistically that such normal values can always be found, we propose

DEFINITION 3.6 Let F_i: $B \times E \times T \to \mathbb{R}$ be the ith component of the state function of biosystems of kind B in environments of type E, and $\bar{F}_i$ the corresponding (physiological) normal value of F_i. Further, let $w = \{w_i \in [0,1] \mid 1 \leq i \leq N\}$ be a set of positive reals that add up to unity, and such that w_i measures the weight or relative importance (for the whole biosystem) of the property represented by the function F_i. (Cf. Vol. 3, ch. 2, Sec. 4.2, Definition 2.15 of property weight.) Then

(i) the *degree of sickness* of the biosystem $b \in B$ in the environment $e \in E$, in the ith respect and at time $t \in T$, is

$$s_i(b, e, t) = \left| F_i(b, e, t) - \bar{F}_i \right| / \left| F_i(b, e, t) + \bar{F}_i \right|;$$

(ii) the *degree of health* of the biosystem b in the environment e, in the ith respect and at time t, is the complement of $s_i(b, e, t)$ to unity, i.e.

$$h_i(b, e, t) = 1 - s_i(b, e, t);$$

(iii) the *total degree of sickness* and the *total degree of health* of organism b in e at t are

$$S(b, e, t) = \sum_{i=1}^{N} w_i s_i(b, e, t) \quad \text{and} \quad H(b, e, t) = 1 - S(b, e, t)$$

respectively.

The values of s_i and h_i are real numbers in the unit interval. And each is the same whether $F_i(t)$ equals $n\bar{F}_i$ or $(1/n)\bar{F}_i$, with n an arbitrary integer.

The conceptual interest of the above formula resides in its making sense only within a theoretical model of biosystems of the species of interest. Its methodological interest lies in its reminding us that scientific medicine is just a chapter of biology (including psychology in the case of man).

Finally we come to

DEFINITION 3.7 A thing is *dead* iff, having been a biosystem, it has ceased to perform all of the functions listed in Postulate 3.1.

Remark 1 We could not define death as maximal degree of sickness, or as minimal degree of health, because the concepts of sickness and health refer to living things (Definition 3.6). *Remark 2* Surely not all of the vital functions stop at the same time: death is a process rather than a point event. However, we could not define dying as the gradual cessation of biofunctions, because recovery is not unusual, whereas dying is irreversible. *Remark 3* Because all biosystems emerge from biochemical systems or even from whole cells rather than from scratch (i.e. from physical entities), the process of dying is not the inverse of the birth process. See Figure 3.3 *Remark 4* Life and death are not entities but states of, or processes in, biosystems. Death is the process of breakdown of a biosystem into an aggregate of chemicals. It starts when the biochemical reactions that maintain a steady state – namely the metabolic processes – come to a halt. In particular, no new enzymes necessary for the synthesis of biomolecules are produced and, in general, breakdown processes start to prevail over buildup processes. *Remark 5* That health and sickness, life and death are not entities but states of entities, is obvious from a systems theoretic viewpoint but not from alternative points of view. Indeed, sickness and death have been treated as entities, not only by primitive men but also by some medical and philosophical schools. Having reified and even personified death, it was possible to regard life either as a fight against Death or as the transit towards it. The anxieties elicited by such primitive views (contained in contemporary existentialism) are allayed by the naturalistic doctrine proposed by Epicurus, according to which dying is a process of dismantling of an organism.

2.2. *Function and Value*

Every cell, no matter how simple or primitive it may look at first sight, is composed of a number of biochemical systems of enormous complexity

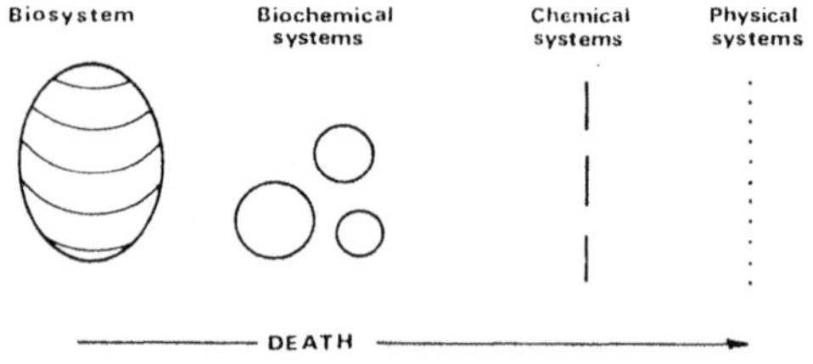

Fig. 3.3. The successive collapses of the state space of a dying biosystem: a gradual loss of properties and of systemicity.

which are linked to one another, and succeed one another, in various ways. If eucaryotic (nucleated), a cell consists of a nucleus and a cytoplasm. The former is composed of DNA and RNA threads, and the latter of subsystems such as ribosomes, mitochondria, chloroplasts (in plants), and cilia – none of which are alive.

Each of the nonliving subsystems of a cell performs certain activities, usually called 'functions', such as storing energy, synthesizing proteins, breaking down excess biomolecules, detecting imbalances, or transmitting signals. Such subsystems of the cell are therefore called 'functional units'. In multicellular organisms (metazoans) this division of labor often carries over to entire cells or even cell groups such as tissues and organs. In either case each subsystem of a biosystem, whether or not the former is alive, discharges certain functions, i.e. does certain things, that are peculiar to it, and others which are done also by subsystems of other kinds. For example, neurons not only generate and transmit signals but also synthesize ATP, nucleic acids, and proteins,

We elucidate these notions by means of the following definition, which employs the notion of a process elucidated in Ch. 1, Sec. 1.7:

DEFINITION 3.8 Let b be an organism, and $a \prec b$ a subsystem of b, of kind $[a]$. Further, call $\pi(a)$ the totality of processes or activities involving a during a certain period. Then

(i) any subset of $\pi(a)$ that includes any of the processes listed in Postulate 3.1 is a *biological function*;

(ii) the *specific biological functions* of a are those performed by a and its likes but not by any other subsystems of b. I.e.

$$\pi_s(a) = \pi(a) - \bigcup_{x \prec b} \pi(x), \quad \text{with} \quad a \neq x \prec b \;\&\; x \notin [a].$$

The biological functions of a subsystem of an organism are valuable or disvaluable to the organism as a whole. For example, the specific functions of the liver are to synthesize certain substances (e.g. albumin and cholesterol) and to store others (e.g. glycogen and iron). No mammal can live without a liver: the latter's value to the organism is maximal, though of course not greater than the values of the heart and the brain. On the other hand a mammal can live without a gall bladder although this organ markets (stores, concentrates, and delivers) the bile produced by the liver, and the bile is useful for digesting fats and other substances. That is, the value of the gall bladder, though positive, is far smaller than that of the liver. Finally, some subsystems of an organism perform disvaluable biological functions: think of tumors.

The above suggests the following measure of the biological value of a subsystem a of an organims b in an environment e: The value of a to b in e equals the degree of health of b in e when equipped with a, minus the degree of health of b in e when deprived of a. More precisely, with the help of Definition 3.6 we lay down

DEFINITION 3.9 Let $H_A\colon B \times E \times T \to \mathbb{R}$ be the health function for organisms of kind B, in environments of type E, when subsystems of kind A are acting in the B's (i.e. when $a \prec b$ and $\pi_s(a) \neq \varnothing$ for $a \in A$ and $b \in B$). Similarly, call $H_{\bar{A}}\colon B \times E \times T \to \mathbb{R}$ the corresponding health function when the A's are not acting in the B's (i.e $\neg(a \prec b)$ or $\pi_s(a) = \varnothing$). Then the *value of A for B in E* is the function

$V\colon A \times B \times E \times T \to \mathbb{R}$ such that, for any $a \in A$, $b \in B$, $e \in E$ and $t \in T$,

$$V(a, b, e, t) = H_A(b, e, t) - H_{\bar{A}}(b, e, t),$$

which is the *value of subsystem a for organism b in environment e at time t*.

Since health ranges between 0 and 1, biological values are comprised between -1 and 1. In particular

a is *maximally valuable* to b in e at t iff $H_A(b, e, t) = 1$ and $H_{\bar{A}}(b, e, t) = 0$;
a is *worthless* to b in e at t iff $H_A(b, e, t) = H_{\bar{A}}(b, e, t)$;
a is *maximally disvaluable* to b in e at t iff $H_A(b, e, t) = 0$ and $H_{\bar{A}}(b, e, t) = 1$.

Thus in humans the heart is maximally valuable, the normal appendix worthless, and a malignant tumor maximally disvaluable. And the values of other systems, such as the reproductive system, the thymus gland and the tonsils, vary with age.

The preceding definition allows one to elucidate another concept, this time of particular interest to surgeons and bioengineers:

DEFINITION 3.10 Two subsystems a and a' of a given biosystem b are *functionally equivalent* iff

(i) a and a' perform the same specific functions, i.e. $\pi_s(a) = \pi_s(a')$, and

(ii) a and a' have the same value during the time interval of interest, i.e. $V(a, b, e, t) = V(a', b, e, t')$ for t and t' in the given interval.

Remark 1 Some biofunctions, such as metabolism and cell division, are common to all organisms. Others are species-specific. For example photosynthesis is peculiar to plants, and the synthesis of enzymes that occurs in the digestion of meat is peculiar to carnivores. The grouping of organisms by biofunction clusters provides just as natural and fundamental a classification as evolutionary taxonomy; only, it is out of fashion. *Remark 2* The more numerous the functions an organism can perform, the better equipped it is to cope with environmental challenges. Multicellular organism that are highly differentiated will therefore be selected by rich and changing environments. Also, because they have spare cells, metazoans are covered against the accidental loss of some cells, so they have better chances of survival than organisms without such spares. (Recall the discussion of reliability in Ch. 1, Sec. 4.1.) These two factors help explain the prosperity of metazoans on our planet. In short, the terrestrial environment has been fit for complex and versatile organisms endowed with a high degree of redundancy. *Remark 3* Alternative definitions of biological value are of course thinkable. One alternative worth looking into is to equate biovalue with efficiency. Another is teleological: Valuable to an organism (or a species) X is whatever contributes to X's attaining its goals. We cannot adopt this alternative because we are not assigning goal-seeking activities to all organisms. Goals will be introduced in the next chapter, with reference to organism endowed with complex nervous systems. *Remark 4* As evolution proceeds, the value of some subsystems increases (preadaptation) and that of others decreases, sometimes to the point of vanishing altogether. In the latter case the subsystem may disappear (as in the loss of the ability to synthesize certain vitamins), or it may be kept functionless for hundreds of thousands of years. Thus it is possible, nay quite likely, that man has a number of fossil (functionless) proteins, genes, entire cells, and even cellular systems – i.e. subsystems that have long ceased to serve any biofunction. But this is very hard to prove. *Remark 5* It is just as likely that certain hereditary pathological conditions were valuable in the past or may yet become valuable. This is the case with sickle-hemoglobin, which is associated with resistance to malaria. Likewise certain congenital malformations, such as flat feet and poor sight, have spared the lives of many men in modern wars. (War selects the unfit.)

2.3. *Biocontrol*

All systems are under environmental control. And biochemical systems are also under internal – in particular enzymatic – control. (Recall Definition 2.17.) Biosystems are then controlled by their environment (in particular by the availability of precursors of biomolecules) and by their biochemical subsystems. In particular, enzymes and nucleic acids exert internal controls. Besides, multicellular organisms have additional control systems, such as the control of growth and size by specific hormones. In short, all organisms are subject to dual controls: internal and environmental, or self and external.

Naturally, the self-control of a biofunction is called *biocontrol.* Most biocontrols are of the feedback type. (See Appendix A, Sec. 1.3.) In such cases, if the output of a subsystem falls below (above) a certain normal value, the control system increases (decreases) the input in such a way that the imbalance is gradually redressed. (The normal output level is sometimes misleadingly called the set of *goal states* or *Sollwerte.* But there is no evidence of goal directed behavior except for certain features of the behavior of animals equipped with an advanced brain: see Ch. 4.)

When all the controls of an organism work properly, the organism attains or maintains a steady state or *homeostasis*: it stays in or near such a state regardless of external perturbations, provided these are bounded. However, homeostasis is usually a global result of frantic molecular and activity. Moreover there are a number of nonhomeostatic states in the life history of every organism, namely all those leading to and from homeostasis – such as the processes of birth and death, of sickness (or injury) and recovery from it. In these cases one speaks of *homeorhesis.*

Biocontrol starts at the molecular level. Thus DNA controls the synthesis of RNA, which in turn controls protein synthesis. (See Ch. 2, Sec. 2.1.) But DNA itself, far from being an unmoved prime mover, is under severe control. In fact, according to the Jacob-Monod hypothesis, genes are turned on or expressed only when such control is lifted. (It is likely that, at least in multicellular organisms, many genes are repressed or inactive most of the time and, like firemen, become activated only during emergencies.) In other words, only when the control is off the gene is free to participate in the synthesis of whatever protein(s) it may. And this is not the whole story: some enzymatic proteins in turn control RNA synthesis. (Furthermore, catalytic cycles formed by RNA molecules and enzymes may well exist: see Eigen (1971).) In short, genetic control is not ultimate and hierarchical. This is a point of philosophical interest in view of the fashionable claim that life is at the mercy of a bunch of selfish and smart genes.

In addition to the enzymatic and genetic controls, which regulate directly the thousands of biochemical reactions going on in the organism, there are highly specialized biocontrol systems, such as those that control acidity, temperature, growth, and motion. These are unicellular or multicellular systems, every one of which can be analyzed into a *sensor* and an *effector* (or *corrector*). The sensor evaluates both the input and the output, and "informs" the effector whether they are adequate. (Actually there is no need for the information concept at this level: all the sensor does is to trigger the effector as soon as it detects an imbalance.)

In a sense every concrete thing can "sense" or detect some other things. Thus an atom can detect light (i.e. react to it) as well as the proximity of other atoms. However, biocontrols are *specific*: some detect gravity (or rather deviations from the vertical); others light intensity; still others light frequency, or acidity differences, or temperature differences, and so on. Furtermore, whereas physical and chemical detectors usually detect absolute values or intensities, all biosensors detect *imbalances*, i.e. differences between actual and normal ("goal") values. This they achieve by being components of negative feedback loops. In short, biocontrols are specific and they correct imbalances.

Furthermore, unlike a bunch of physical detectors, the biocontrols in an organism constitute a network: they interact and, in the healthy organism, they do so synergically, i.e. harmoniously rather than independently from one another. (Such interaction may involve so little energy, and yet may have such important effects, that it is usually called 'informational'. See Appendix A, Sec. 1.4.) For example, the various internal secretion glands in a vertebrate are interconnected by hormones carried by the bloodstream, and the various parts of the nervous system keep in touch ("talk to each other") by synaptic contacts as well as by hormonal "messengers". In short, the control systems in an organism constitute signal (or informational) networks. Actually every organism, however primitive, is equipped with two interconnected signal networks: an internal network constituted by its internal biocontrols, and an interfacial network constituted by biocontrols that sense environmental changes. These two networks are coupled and between them they control all the functions of the organism.

This, then, is one of the peculiarities of organisms: namely that they have multiple control systems grouped into tightly integrated *signal networks*. Another peculiarity is that such networks function well – amazingly well in most cases. 'Well' does not mean either at full blast or else with maximal economy, but instead in whatever manner is most valuable to the organism as a whole. (Recall Definition 3.9 of biovalue.) For example, the rate at which any given protein is synthesized is neither minimal (maximal economy of resources) nor maximal (waste) but somewhere in between – above equilibrium and short of waste. This near optimality is peculiar to biosystems.

Unlike biology, physics and chemistry are dominated by extremal principles such as Hamilton's, i.e. basic law statements according to which, of all the conceivable processes of a given kind, only those are actualized that either minimize or maximize a certain magnitude, e.g. the action. (Cf. Appendix B, Sec. 1.4.) These principles "operate" of course for the physical components of a biosystem, but these components intertwine in such a manner that certain properties of value to the organism as a whole are near-optimized rather than either minimized or maximized. I.e. such near optima are not determined locally by each separate system but result from the interaction among the various subsystems of the organism. In fact since the various components of the control network of the whole organism are interdependent, the near optimization of certain local variables will hamper that of others. So, what would be best for each subsystem is unattainable. But the next best is attainable, and this is a compromise among a number of optima. This compromise may be called *overall* (or organismic) *near optimality* – not full optimality because no organism is perfect. (A perfect biospecies would not be subject to evolution.) Only the biological organization of *successful* species *tends* to optimality.

An organism may be said to possess an efficient biocontrol network if it is capable of redressing whatever imbalances may occur – i.e. if it can detect and correct them. If not, the organism becomes sick and may die. More precisely, every malfunction of a specific biocontrol system is a specific sickness. (The converse is false. For example infectious diseases, if overcome, are so by the proper functioning of the immune response system.) And if several biocontrol systems malfunction, or do not function synergically, a nonspecific (generic) sickness supervenes. Natural death is probably the final state in a process of generic (nonspecific) sickness.

Biocontrol has given rise to the belief that organisms are goal-seeking: that they act in everything so as to ensure their own preservation or at least that of their species. First the theory of evolution through natural selection, and in recent times the theory of control, should have corrected that illusion. In the first place self-control is a property of all biochemical systems – by virtue of their enzymatic activity – whether or not they are alive. And if system integrity were defined as equilibrium (steady state), then every biochemical system would be seen to behave *as if* it reached for its own preservation. Secondly, biocontrol is not always efficient: it falls short of optimality when the organism becomes sick and it fails utterly at death. Thirdly, biocontrol can fail, even in healthy organisms, when certain imbalances grow beyond bounds during catastrophes such as floods and droughts. (After all, the great majority of terrestrial biospecies have become extinct.) Fourthly, biocontrol is occasionally defective. For example, only a small part of the RNA synthesized in the cell nucleus migrates to the surrounding cytoplasm to control protein synthesis: most of it is broken down into nucleotides without ever being "used". Another example: many genes are never "expressed" or activated. Fifthly, genetic control is not so strict as to prevent mutations, most of which are deleterious. In any case, we need not invoke goal-directedness to explain why biocontrol is normally beneficial for the organism as a whole: the organisms not endowed with efficient control networks succumb before reproducing, and that's all.

2.4. *Development*

The history of any system can be represented as the trajectory of its representative point in a suitable state space for the system. (Cf. Ch. 1, Sec. 2.2.) In particular, the *life history* of an organism can be represented as a line in its state space. The part of the life history comprised between the beginning (birth) and the attainment of the adult state of the organism is called its *development*. (See Bonner (1974).) The development process of an organism has three main strands, each of which is taken care of by a bunch of components of the state function representing its possible states. These strands or aspects of development are:

(i) *growth*, or multiplication of cells of certain kinds (at uneven rates for different subsystems of the organism);

(ii) *morphogenesis*, or formation of new subsystems;

(iii) *differentiation*, or specialization of cells.

All three aspects of development proceed under a triple control: enzymatic, genetic, and environmental. The environment supplies or withholds the wherewithall, the enzymes make the transformation of environmental precursors possible, and the genes guide the entire process – provided their controls permit them to do so, as we saw in the previous subsection. (Some molecular biologists tend to play down the control effected by the environment. However, the latter can be dramatic, as illustrated by the sporulation of moulds that occurs in a poor environment.) One way of characterizing the salient features of the development process is by means of

DEFINITION 3.11 If *b* is a biosystem, then

(i) *b grows* iff (*a*) some DNA molecules in *b* replicate, and (*b*) the rates of RNA and protein syntheses in *b* are greater than the corresponding breakdown rates;

(ii) *b* is involved in *morphogenesis* iff some processes in *b* consist in the formation of new subcellular systems (*intracellular morphogenesis*), or new cells (*cellular morphogenesis*), or new organs (*organogenesis*);

(iii) *b differentiates* iff *b* is a multicellular organism such that (*a*) the number of specific functions of *b* increases as the number of cells increases, and (*b*) different specific functions are performed by different cells or multicellular systems.

Of these three aspects of development, morphogenesis has always seemed the most mysterious and the one most clearly indicative of design and purpose. To begin with, the very word 'morphogenesis' suggests a process whereby the formless (e.g. an egg) acquires a definite form (shape or spatial pattern). While this is so at the physical and chemical levels, as in the cases of the generation of waves in a homogeneous fluid, or the growth of crystals in solutions, it is not so at the biolevel: here morphogenetic processes are those where new forms emerge out of old. (For examples and models of self-organization processes at various levels, see Glansdorff and Prigogine (1971), Gierer and Meinhardt (1972), Glass (1973), Tyson (1976), and Nicolis and Prigogine (1977).) Particularly in embryology every developmental phase is structured rather than amorphous. Moreover, shape change is not the primary or driving feature of development but an outcome of a profound process such as the diffusion of the products of an enzymatic reaction, or the multiplication of cells under the control of morphogens – enzymes, hormones, and genes. (Classical morphology was restricted to describing the net outcome of such processes: it neither attempted nor could explain them by hypothesizing biochemical and cellular mechanisms. Likewise the pretentious work of

Thom (1972), with its total disregard for morphogens, physical constraints, natural selection, chance, and history, does not go beyond classical morphology except for the addition of a mathematics of dubious utility.)

Morphogenesis may be regarded as a particular case of self-assembly of molecules, organelles, or cells, as the case may be. Like other self-assembly processes, the morphogenetic ones are guided by both internal forces and environmental resources and circumstances. The formation of an organelle, a cell, or an organ is controlled by genes and other morphogens in conjunction with the *milieu intérieur* and the environment. The DNA and RNA molecules, the hormones and other morphogens control the chemical reactions and the diffusion processes, as well as the assembly and dismantling of the various units, but nothing gets done unless the environment supplies the materials and the energy.

A self-organization process can start spontaneously in an initially homogeneous system of reactants and morphogens. Thus a local chance fluctuation in their relative concentration can diffuse throughout the medium, ending up in a differentiated system (Turing, 1952). However, reaction-diffusion processes are not the only morphogenetic mechanisms, nor are they typically biological. Alternative mechanisms are: (*a*) sorting processes (grouping of functionally similar units, as in the case of polyribosomes), (*b*) cooperative interactions among molecules or cells of different kinds, (*c*) packing, folding, and other mechanical effects of accretion and external pressure; (*d*) enzymatic syntheses of morphogens proceeding until a new entity is formed – and inhibited thereafter.

Let us take a quick look at processes of the first kind, i.e. sorting processes (Steinberg, 1963; Mostow (Ed.), 1975). Let C be a set of cells of two kinds, A and B. Assume that all these cells are motile and that like cells adhere to one another more strongly than do unlike cells. Moreover, suppose that the cells of type B adhere to one another less strongly than those of type A. In sum, assume that the adhesion works (or energies) compare like this:

$$W_A > W_B \geq W_{AB},$$

where W_A measures the adhesion of cells of kind A, W_B that of cells B, and W_{AB} that of unlike cells. One can imagine, and indeed prove rigorously, that the cells clump together and sort out all by themselves. In particular, if $W_{AB} = 0$, the sorting will be complete: two separate tissues will ensue. See Figure 3.4.

Although the details of such self-assembly processes are far from being

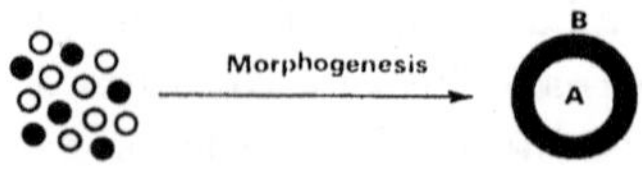

Fig. 3.4. The formation of an onion-shaped organ constituted by two concentric spheres composed of cells of different kinds, A and B, according to Steinberg (1963).

known, there is no excuse for regarding them as unknowable. Morphogenetic processes look mysterious only if one adopts either a vitalistic view or a primitive version of mechanism – one according to which matter is a mere aggregate of inert particles. But biology has outgrown both extremes: it has no use for supramaterial entities or forces, and it has learned that molecules, particularly carbon compounds, can stick together in numerous ways, forming systems of great complexity and often also of great beauty.

One outcome of morphogenesis in multicellular organisms is differentiation. The degree of differentiation of a system may be taken to equal the total number of specific functions performed by its various subystems. Now, for a complex system to keep together, its various subsystems must be well coordinated. In other words, while systemicity in general requires the existence of bonds among the system components, *organic systemicity* requires that these bonds be well coordinated, so that nearly every system component be valuable to the system as a whole. In particular, not only the right reactants must be present under favorable circumstances to their reactions, but the latter must proceed at the proper pace and in the proper sequence; and no component must block the operation of the other components – unless such blocking be of value to the organism as a whole. All this is compressed into the well worn platitude that "The greater the differentiation of an organism, the greater its coordination and integration". (Needless to say, it is not the case that coordination is a goal of development, but rather that the organisms that fail to achieve efficient coordination perish.)

2.5. *Reproduction and Heredity*

One of the properties of organisms is that they are capable of reproducing. (Sexually reproducing organisms do not reproduce unless they mate, but most of their cells divide. And neurons do not normally divide but other cells, linked to them, do.) Moreover organisms reproduce "true to type" i.e. their offspring belong to the same species. (Speciation takes many generations because it is a result of cumulative mutations and successive selections.)

The root of reproduction is the chemical process of DNA replication. (See Ch. 2, Sec. 2.1.) This explains the invariant aspect of heredity, particularly in asexually reproducing organisms. Sexually reproducing organisms are far more variable because of the recombination (nearly random shuffling) of parental genes. In either case the genetic system, or genome – which is composed of DNA molecules – is "in charge" of heredity. Organisms are not pulled by goals but pushed by the genetic memory of the past – pardon the metaphor.

The genetic system, at one time thought to be immutable, is known to be subject to mutation. The mutation mechanism is molecular. Consider the DNA molecules in a cell. As we know, these molecules can replicate. Now, during the replication process accidents ("mistakes") of two kinds can occur: (*a*) the daughter molecular strings may differ from the parent string in the order or sequence of the nucleotides (i.e. the daughter DNA molecules are isomers of their parents), or (*b*) one or more nucleotides may be either deleted or inserted (substitution). And, because DNA controls protein synthesis, such "mistakes" are likely to ensue in the synthesis of proteins of kinds different from those contained in the parental cells(s) – i.e. the mutant DNA can "code for" (control the synthesis of) new proteins. And the latter make it possible for the offspring to interact in new – not necessarily more efficient – ways with its environment. The end result is a mutant organism – one that may or may not be viable. See Figure 3.5.

Mutations are not frequent but they are bound to occur either in large populations or along many generations. And the number of possible mutations, hence of possible new proteins, and therefore of possible organisms more or less different from their ancestors, is staggering. (Speak of the poor "possible worlds" imagined by philosophical logicians!) Mutation is indeed an inexhaustible source of biotic novelty. And because mutation occurs by chance and irrespective of its value or disvalue to the organisms, chance has been said to be the sole source of newness in the biosphere (Monod, 1970, p. 127). But this is an exaggeration. The chemical affinities at the root of the nucleotide sequence are hardly chancy, and the selective action of the environment has causal components as well as random ones. (See Sec. 3.)

The large number of possible mutations explains the proteinic uniqueness of each organism. Equivalently: it accounts for many individual differences within any given species – or what is misleadingly called "indi-

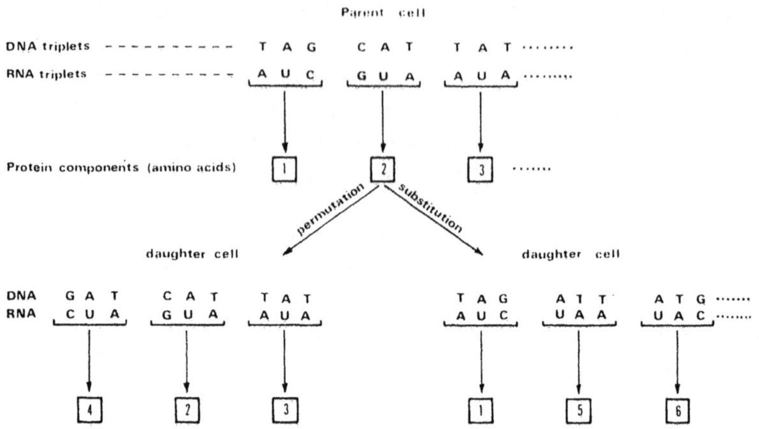

Fig. 3.5. Two kinds of mutation: by permutation and by deletion (or insertion) of nucleotides.

vidual variability". It also explains the large number of biospecies. And the limited environmental resources, with the accompanying competition for them, explains why biopopulations do not grow exponentially (except for limited periods). Finally, changes in environmental conditions – particularly important changes such as glaciations and continental drifts – explain why (*a*) most biospecies that have been on Earth have become extinct, and (*b*) the number of species can jump up or down during certain brief periods. (One such revolutionary period was the beginning of the Cambrian, when many species of multicellular organisms emerged.)

In view of the fashionable claim that whatever an organism is and does is programmed in its genome, it may be helpful to insist that what is inherited is a bit of genetic material carrying certain potentialities. These potentialities may or may not be actualized in the course of the organisms' life, according to the possibilities offered by the environment. (For the concepts of potentiality and actuality see Vol. 3, Ch. 4.) We may as well be more precise and adopt

DEFINITION 3.12 If b is an organism, then

(i) the *genome* of b = the potentiality of b at birth;

(ii) the *phenome* of b = the set of molar (global) properties of b in adult state.

While geneticists focus on the genome, others focus on the phenome. Both visions are myopic, for only a suitable genetic equipment allows the organism to cope with its environment, and only a suitable environment enables the genetic traits to "express" themselves (i.e. to become activated). Thus, while every organism inherits its genome, it evolves its own phenome in a given environment; the phenome depends then upon both the genome and the environment. Given the variability of both, evolution should come as no surprise. But evolution will concern us in Sec. 3, after we look at the philosophical background of the contemporary doctrines on development and heredity.

2.6. *Development, Heredity, and the Metaphysical Tradition*

Contemporary studies of development and heredity have been influenced, mostly unwittingly, by ancient metaphysical ideas on the relations between substance and property, on the one hand, and between potency and actuality on the other. In turn, the results of such studies should be allowed to influence our general ideas on the substance-property and the potency-actuality pairs.

Consider first the organ-function relation as a particular case of the substance-property relation. Every biofunction is the function of some organelle or some organ, but the converse is false: i.e. there may be organs that have no biofunction (the vestigial or "fossil" organs), and others that perform specific functions only at certain stages of development. In the latter case the organ sometimes precedes the functioning organ – or, as the misleading idiom has it, "organ precedes function". For example the eye, the reproductive system, and large parts of the brain of a cat are formed before they function fully.

The temporal precedence of the nonfunctional organ over the fully functional one would seem to refute the Platonic doctrine that form (or essence or idea) precedes existence. Not so, for the Platonist could rejoin that the form (function) is inserted into the organ when needed. What does refute the Platonic doctrine of forms is the well confirmed generalization that there is no form (function) without organ and, in general, no property without substance. (Cf. Vol. 3, Ch.2, Sec. 5.2.) The controversy between the substantialists ("organ determines function") and the functionalists ("function determines organ") is a particular case of the age-old metaphysical controversy over substance and property. This controversy dissolves upon realizing that neither determines the other. All that happens in the case of organs that do not perform their specific function during the entire life of an organism, is that their nonfunctional states precede in time the functional ones.

A second point where developmental biology and the theory of heredity make contact with metaphysics is over the question of preformation *vs.* emergence. There have been two classes of solution to the problem of explaining the development of an organism from egg through adulthood to final breakdown. One is to assume that all of the properties and patterns that appear in the course of development are just the unpacking or unfolding (*development*) of what was contained in potency (*preformed*) in the egg. Or, to put it into modern parlance, the assumption is that the genome contains a full set of "instructions" steering the organism through the successive stages of its development. This is the *preformationist* hypothesis held by Aristotle and Leibniz, and resurrected by some contemporary scientists who believe that biology is fully reducible to molecular biology. Preformationism denies real novelty, in particular the emergence of new properties that accompanies self-organization (Ch. 1, Sec. 3.2); and it minimizes or even ignores the role of the environment.

The partisans of *epigenesis* hold, on the contrary, that development is an epigenetic or creative process in which novelty occurs at every stage. They maintain that the organism starts at a low state of organization and acquires new components and properties, as well as loses others as it develops. Epigeneticism is of course the hypothesis underlying all of classical embryology as well as the contemporary studies on the origin of life.

The molecular biology revolution has all but eclipsed epigeneticism. What is fashionable nowadays is genetic preformationism. Yet fashion need not concide with truth. The claim that development is just the deployment or implementation of a blueprint contained in the genome is no more illuminating or correct than holding that it is guided by an immaterial *élan vital* (Bergson, 1907) – or that it consists ultimately in the embodiment of Platonic self-existing geometric shapes (Thom, 1972). Surely development can be neither purely preformative nor purely epigenetic, if only because the environment supplies the raw materials for it and weeds out the maladapted organisms, as well as offers chances to the preadapted ones.

It seems that each view, either preformationism or epigeneticism, has alternatively accelerated and blocked the development of biology. Preformationism may have stimulated the purely descriptive phase of embryology and the beginnings of molecular biology, whereas epigeneticism may have stimulated experimental embryology (particularly investigations into the mechanisms of morphogenesis) and ecology. But neither view explains anything because each is just a programmatic hypothesis or a framework for discussing development and heredity. Indeed, far from being full-fledged theories, preformationism and epigeneticism are ontological hypotheses summarizing entire bodies of ideas, where sound data and conjectures are inextricably mixed up with metaphysical fossils.

It seems quite possible that the next breakthrough in the fields of development and heredity will come from a sort of merger of the two frameworks, one harmonizing two lines of research: (*a*) the search for specific development mechanisms (e.g. the mechanical constraints each cell is subjected to, the synthesis and action of morphogenes, and the interactions with the nutrient medium), and (*b*) the genetic "programming" of development possibilities – potentialities that are "encoded" in the genome in the sense that the latter contains, not "instructions", but some of the key reactants and templates required for the occurrence of certain biochemical reactions. But let us not attempt to write the history of the future.

3. Evolution

3.1. *Adaptation*

A salient feature of our biosphere is its rich variety. There are at least three million biospecies (Dobzhansky, 1970), and no two strictly identical individuals in any of them. (Even so-called identical twins are somewhat different.) This variety is a result of four billion years of evolution. And the evolutionary process is in turn the net effect of a combination of genic changes (mutations and recombinations) with environmental selection. The process seems to be continuing: there is no reason to believe that it stopped the moment man became aware of it. In fact the extinction of a number of species has been certified during our own lifetime, and it is likely that new species are forming right now. In short, bioevolution is revolutionary – or creative (Bergson, 1907), or emergent (Lloyd Morgan, 1933).

Anthropomorphically speaking, evolution is a trial and error, or success and failure, story. Success is called *adaptation*, and failure *maladaptation*. A biopopulation of well adapted individuals thrives or expands, whereas one of maladapted individuals declines. This looks obvious but ceases to be so as soon as one stops and examines the very term 'adaptation': he then realizes how tricky it is (Williams, 1966). In fact, the word 'adaptation' is ambiguous, as it designates at least three different concepts, namely the following:

A_1 = *suitability* of a subsystem (organelle, organ, etc.) to a function, or high biovalue of the subsystem for the entire organism;

A_2 = *adjustment* of the organism to its environment;

A_3 = *fertility* of a biopopulation.

That these are indeed different concepts, is realized upon displaying their structure. A_1 is a relation between subsystems and organisms, A_2 a relation between organisms and environments, and A_3 an intrinsic property of organisms. A methodological examination confirms the difference. In fact the way to find out whether something has the property A_1 is to conduct certain biochemical or physiological experiments or observations in a constant environment. By varying environmental factors we can ascertain whether an organism possesses property A_2. And the criterion for A_3 is the count of offspring.

In sum, we must be reconciled with the fact that 'adaptation' is an ambiguous term. This seems obvious, yet it is overlooked by those who claim that only population genetics uses "the correct" concept, namely A_3. And it is equally overlooked by those who claim that the Darwin–Spencer principle of the *survival of the fittest* is tautological. Indeed, survivorship or fertility (A_3) is not the same as fitness or adjustment (A_2) but an indicator of the latter. The logical relations among the three concepts are these: $A_3 \Rightarrow A_2 \Rightarrow A_1$, i.e. subsystem biovalue is necessary for adjustment, which is in turn necessary for fertility.

It goes without saying that adaptation of any kind is time-dependent. I.e., A_1, A_2 and A_3 vary in time, either because of internal changes, or environmental changes, or both. The case of sudden environmental changes is particularly interesting, for they are bound to favor some biopopulations while hurting others. In the first case one speaks of *preadaptation*, in the second of *premaladaptation*. In keeping with the above distinction among three adaptation concepts, we distinguish three preadaptation types:

PA_1 switch of function(s) to other(s) of greater biovalue;

PA_2 improved adjustment to the environment;

PA_3 greater fertility.

As with the case of adaptation, the first is necessary for the second, which is in turn necessary for the third, i.e. $PA_3 \Rightarrow PA_2 \Rightarrow PA_1$.

Neither of the three kinds of adaptation or preadaptation calls for teleological considerations. The adapted or preadapted organisms may survive and reproduce, the maladapted ones don't, and that is the end of the story. To invoke adaptive *plans* at all evolutionary stages (Holland, 1975) is to indulge in anthropomorphism. There is not a shred of evidence for such master plans except in the case of man, and no serious biological theory involves them. Instead of imagining adaptive plans we should hypothesize and check adaptation mechanisms.

3.2. *Adaptation Mechanism*

Random mutation and recombination give rise to a great variety of unequally adapted organisms. The environment selects the best adapted amongst them and thus steers evolutionary change. Hence the environment is just as creative as the genetic material. But, whereas the latter varies randomly, the environment exerts a directional action. (Directional, not goal-directed – i.e. changing in a given direction but not directed towards a goal, let alone by a goal.) That, in a nutshell, is what the modern theory of evolution is all about.

A classical experiment confirming this theory is the following (Waddington, 1959). Larvae of fruit flies (*Drosophila*) are exposed to diverse environmental changes, such as excess salt, or heat, or ether vapour. If fed salted food, many of the flies die. The survivors happen to exhibit an enlargement of certain papillae conjectured to be involved in the regulation of salt concentration in the body. (The hypothesis is that this is a preadaptation, i.e. that those individuals that happen to be born with the more efficient salt controls, i.e. which are preadapted to the new circumstances, survive.) The 20th generation is definitely more resistant to salted food than the unselected or control group: roughly twice as many selected flies survive when given food with 5 percent salt added. The new character, i.e. the ability to cope with abnormally high salt concentrations, does not appear at one stroke but is instead the result of a gradual accumulation brought about by genic changes in conjunction with environmental pressures.

No master plan guides the gradual adaptation of organisms to their environment – until the latter changes unfavorably and maladaptation sets in, and eventually the population is decimated or even wiped out. Indeed, adaptation can be explained as the outcome of a random process powered by genic variation (mutation or recombination) and guided by the environment. To clarify this point consider a single property P, such as the size of the salt control system of *Drosophila* in Waddington's experiment. (Surely such a restriction to a single property is artificial, since every property is lawfully related to some other property of the same system. However, paying attention to a selected property does not entail denying the existence of other properties.) Make the following assumptions about P:

(i) P has a symmetric (e.g. bell-shaped) distribution in each population of the given lineage;

(ii) the successive generations do not overlap – so that the evolution of the population, as far as P is concerned, can be represented as a denumerable sequence of symmetric distributions;

(iii) there is an optimal value P_f of the given property, i.e. the individuals that come closest to possessing this value of P have the greatest survival chance. For the sake of definiteness assume that P_f lies somewhere to the right of the value P_i of the center of the initial distribution: see Figure. 3.6.

Variation and selection ensure that a gradual shift of the distribution in P to the right will occur as long as the optimal value P_f has not been attained. Indeed only those organisms with a P value close to the optimal P_f will be fit and they will tend to pass that characteristic on to their progeny – not because they want to secure the preservation of their species but because of the blind molecular process at the root of reproduction. As adaptation increases, survivorship increases and spread – i.e. individual variability – decreases. Where the vitalist sees a master plan or "teleonomic project", the evolutionist sees only a random process guided by natural selection. (Cf. Levins, 1968, Ch. 2; Dobzhansky, 1974.)

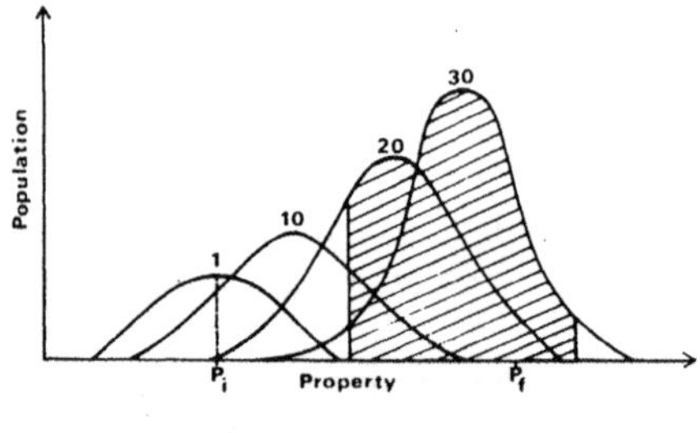

Fig. 3.6. Distribution of P-ness along 30 generations of organisms of a given lineage in a constant environment. The shaded areas represent fractions of survivols. In the beginning a tiny fraction of the total are adapted. At the end survivorship has increased enormously but variability has decreased. Imaginary and exaggerated.

3.3. *Evolution and Coevolution: basic ideas*

We have been referring to evolution without defining the concept. We do not have to look very far: the concept of bioevolution is just a specification of the general concept of history of a system (Ch. 1, Sec. 2.2) to biopopulations (Definition 3.4). That is,

DEFINITION 3.13 The *evolution* of a biopopulation is the history of the latter. I.e. if x is a biopopulation, then

$$\text{The evolution of } x = h(x).$$

What evolve, then, are biopopulations regarded as concrete systems spread out in space and time, rather than either individuals or species. (Individuals develop and species can succeed one another.) If we focus on qualitative changes and disregard intermediate steps, we can construe an evolutionary line as a succession of species or lineage (Definition 1.17).

Before Darwin proposed his theory of evolution by natural selection, bioevolution either was denied or was conjectured to have proceeded in a different way – e.g. by individual adaptation and the inheritance of acquired characters (Lamarck). We adopt of course the core of the Darwinian theory:

POSTULATE 3.4 All populations evolve, and the evolution of each is a process whereby spontaneous (genic) changes ensue in unequally adapted individuals, amongst which the environment selects the best adapted ones.

Selection is called *natural* if it is spontaneous, and *artificial* if controlled by man.

Because biopopulations are not isolated from each other but interact with other biopopulations forming ecosystems, the evolution of each must affect that of others. This process has a name of its own:

DEFINITION 3.14 Two or more populations *coevolve* iff the evolution of each affects that of the other I.e. if x and y are biopopulations, then x and y coevolve iff $h(x \mid y) \neq h(x)$ and $h(y \mid x) \neq h(y)$.

But we just said that every change occurring in any single biopopulation is bound to affect the biopopulations with which it interacts – in particular cooperates or competes. We obtain thus, albeit informally, a result important enough to be dignified with the name

THEOREM 3.1 Every biopopulation coevolves with every other biopopulation in the same ecosystem.

This is why it has been held that the unit of evolution (hence the referent of the theory of evolution) is not so much the monospecific population as the entire ecosystem (Dunbar, 1960). And why not the entire biosphere? These heresies are common sense in our systemic ontology, for any change in one system component is bound to affect every component to which it is coupled. (The only way to escape Theorem 3.1 is to deny that there are ecosystems and so to fly in the face of evidence.) The moral for the building of mathematical models of evolution is clear: Start by modeling the evolution of one biopopulation; proceed to model the coevolution of the (sub)system composed of the given biopopulation and its closest partner(s) or rival(s) – e.g. rock fish and the algae they eat; and aim at modeling the whole ecosystem of interest. Easier said than done!

In view of Definition 3.13, every evolutionary process can be represented as the trajectory of the representative point of the state of a biopopulation (or ecosystem) in its state space. However, it is not at all obvious how to choose the components of the state function spanning that state space. (Cf. Lewontin, 1974.) One possibility is to choose the gene frequencies (actually probabilities) as the coordinates or components of the state function. This is the course adopted by population genetics. It has not been too fruitful because selection operates on phenomes not genomes. (What the environment tolerates or wipes out are whole organisms, not just genomes.) The other extreme is of course to choose only molar or phenotypic variables as components of the state function. But this would leave the primary source of individual variability in the dark. And neither of the above strategies makes room for environmental variables. So we are led to a third approach, namely taking the census counts of the various subpopulations of a given biopopulation in the course of time, and viewing them as the net effects of genic variability and environmental selection. This will be done in the next subsection – with apologies to the biological profession for encroaching on their territory.

3.4. *A Schema of the Evolutionary Process*

We propose to sketch a simple model of evolutionary processes (Bunge 1978a). Consider a biopopulation of organisms such as bacteria of some kind, that reproduce by binary division. Assume that each individual lives one time unit, at the end of which it divides simultaneously with all the other members of the population. Assume also that mutants appear at each generation and that they can be grouped into distinct varieties. Call V_i the ith variety, where $1 \le i \le n$, and count the population of each variety at each generation. Further, call N_t^i the population of variety V_i at time (generation) t. If the fraction of each generation that dies without leaving offspring is $d_i < 1$, then the population of variety V_i at time $t + 1$ is

$$N_{t+1}^i = 2(1 - d_i)\, m_{ii}\, N_t^i, \tag{3.1}$$

where m_{ii} is the probability that division be not accompanied by mutation, and the factor 2 accounts for the division of each organism into two new individuals.

The previous formula assumes that all of the offspring of generation t belong to the same variety. In point of fact the number of individuals of variety V_i at time $t + 1$ will differ from the above by two terms: the fraction leaving the given variety and the fraction entering it, in both cases by mutation. Call m_{ij} the probability per unit time of a mutation from variety V_i to variety V_j. These probabilities are subject to the normalization condition

$$\sum_j m_{ij} = 1 \quad \text{for every } 1 \le i \le n. \tag{3.2}$$

The average number of defectors from variety V_i between times t and $t + 1$ will be

$$D^i = 2 \sum_{j \ne i} m_{ij}(1 - d_i)N_t^i. \tag{3.3}$$

And the converts to variety V_i in the same time interval number

$$C^i = 2 \sum_{j \ne i} m_{ji}(1 - d_j)N_t^j, \tag{3.4}$$

for they will be drawn from all varieties but V_i. The factor 2 in D^i and C^i shows that we are making the additional hypothesis that mutants come in pairs – i.e. that if an organism is mutant so is its sister.

Collecting (3.1), (3.3) and (3.4) we obtain the net average population of variety V_i at time $t - 1$:

$$\begin{aligned} N_{t+1}^i &= 2m_{ii}(1 - d_i)N_t^i - D^i + C^i \\ &= 2\Big[m_{ii}(1 - d_i) - \sum_{j \ne i} m_{ij}(1 - d_i)\Big]N_t^i \\ &\quad + 2 \sum_{j \ne i} m_{ji}(1 - d_j)N_t^j. \end{aligned} \tag{3.5}$$

Because of the normalization condition (3.2) of the mutation probabilities, the total probability of mutation from V_i is

$$\sum_{j \neq i} m_{ij} = \sum_{j} m_{ij} - m_{ii} = 1 - m_{ii}. \tag{3.6}$$

Introducing this value into (3.5) we are finally left with

$$N^i_{t+1} = 2(2m_{ii} - 1)(1 - d_i)N^i_t + 2\sum_{j \neq 1} m_{ji}(1 - d_j)N^j_t. \tag{3.7}$$

This system of n linear difference equations summarizes our model of evolution. The solution to these equations can be obtained as follows. Set

$$\begin{aligned} a &= \|a_{ij}\| = \|\, 2(2m_{ii} - 1)\,(1 - d_i)\delta_{ij}\| \\ b &= \|b_{ij}\| = \|\, 2m_{ij}(1 - d_i)\,(1 - \delta_{ij})\|, \end{aligned} \qquad N_t = \left\| \begin{matrix} N^1_t \\ N^2_t \\ \vdots \\ N^n_t \end{matrix} \right\|, \tag{3.8}$$

where $\delta_{ij} = 1$ iff $i = j$ and 0 otherwise. (The transpose of N_t is the value of the state function of the system at t.) The system (3.8) can be condensed into the matrix difference equation

$$N_{t+1} = (a + b^T)N_t, \quad \text{where} \quad \|b_{ij}\|^T = \|b_{ji}\|. \tag{3.9}$$

Iterating (3.9) we find the general equation of evolution

$$N_{t+k} = (a + b^T)^k\, N_t, \quad \text{where} \quad k \in \mathbb{N}. \tag{3.10}$$

The solution to this equation is

$$N_t = (a + b^T)^t\, N_0, \tag{3.11}$$

where N_0 is the column matrix of the initial populations of the n varieties being considered.

Suppose that all of the entries of the evolution matrix $a + b^T$ are positive, and that initially a single variety is present. Then all the other varieties can emerge simultaneously at the next generation, and the mutation-and-selection process proceeds relentlessly. Every one of the n varieties prospers once emerged, and the prosperity of each is conducive to that of all. (The only qualitative change in this process can be induced by a sudden environmental change causing some death rates to jump to unity.) But in general the process will not be explosive: while some entries of the evolution matrix may be positive, others may be zero or negative. An overall negative matrix corresponds of course to extinction.

As long as the death rates d_i and the mutation probabilities m_{ij} are not specified, the model is a phenomenological skeleton. Let us proceed to flesh it out.

3.5. *The Underlying Mechanism*

Let us now specify the architecture of our evolving organisms as well as their mutability (the m_{ij}) and viability (the d_i). We assume that the n varieties of our biopopulation differ by the composition or arrangement of a fixed number of modules or basic units that can be repeated. For the sake of simplicity we suppose that there are just two kinds of module, which we call 0 and 1. Thus if each entity is composed of four such modules arranged differently, we obtain a total of $n = 2^4 = 16$ varieties, namely 0000, 1000, 0100, 0010, 0001, 1100, 0110, 0011, 1010, 0101, 1001, 1110, 0111, 1011, 1101, and 1111. The general formula for a variety whose members are composed of m modules is $abc\ldots m$, where each letter is a binary digit. So much for architecture.

Every change in at least one of the m binary digits occurring in the formula for a variety is a mutation. Thus a mutant will differ from its parent organism either by the total number of 0's (or 1's) or by the order of 0's or 1's. Thus in the above example the transitions 0000 → 1000, 0100 → 0110, and 0111 → 1111 are mutations and, at the same time, changes in variety. Of course not all mutations are equally probable. We shall determine certain relations among mutation probabilities with the following considerations on viability.

Enter the environment. We lump all of the environmental factors into a single real valued function e with values in the $[-1, 1]$ real interval. And we assume that the probability that an organism belonging to variety V_i survives up to and including division is symmetrical about the origin. More precisely, we assume that the death rates d_i in (3.1), far from being constant, depend upon the environmental variable e in the following way:

$$1 - d_i = \begin{cases} v_i(1 - e^2) & \text{for } -1 \leq e \leq 1 \\ 1 \text{ otherwise,} & \end{cases} \tag{3.12}$$

where $0 \leq v_i \leq 1$ for each variety V_i.

Since v_i is characteristic of each variety and is environment-independent, we call it the *adaptedness* of organisms of the variety V_i. The factor $(1 - e^2)$ represents the selection pressure effected by the environment. A substantial increase or decrease in e will have a considerable impact on the fraction $1 - d_i$ of a given population surviving until the next division. In particular, if a catastrophe occurs ($e = \pm 1$), no survivors are left, i.e. $d_i = 1$ for all i. On the other hand if the environmental conditions are optimal ($e = 0$), then the fraction v_i of individuals of variety V_i survive long enough to divide. The variable e is, in sum, the causal or nonstochastic component of evolution.

We introduce now an assumption that will allow us at least to rank the viability values. Our hypothesis is that the more varied the composition of an organism the better its chances of survival. Thus in the case $m = 2$, where we have only the varieties 00, 01, 10, and 11, we set

$$\begin{aligned} v_1 = v_4 &\equiv v \quad \text{for 00 and 11} \\ v_2 = v_3 &= \alpha v \quad \text{with } \alpha > 1 \text{ for 01 and 10.} \end{aligned} \tag{3.13}$$

Finally we relate viability to mutability, namely thus: we postulate that a mutation is the more probable the less viable its product is. (This assumption is in keeping with the well known generalization that most mutants are lethal.) More precisely, we assume

$$m_{ij} \geq m_{ik} \quad \text{if and only if } v_j \leq v_k. \tag{3.14}$$

We can now build the matrix of the mutation probabilities in the simple case where $m = 2$, whence $n = 4$. In this case the viabilities are those indicated in (3.14). Hence the reproduction chances are

$$\begin{aligned} 1 - d_1 = 1 - d_4 &= v(1 - e^2) \\ 1 - d_2 = 1 - d_3 &= \alpha v(1 - e^2). \end{aligned} \tag{3.15}$$

The hypothesis (3.14), that most mutations are deleterious, together with the assumption (3.13) on the adaptedness values, entails

$$m_{41} = m_{14}, \qquad m_{32} = m_{23} = m \tag{3.16}$$

(where this m is not to be confused with the number of kinds of modules per organism, which we have taken to be 2), and

$$\begin{aligned} m_{21} &= \beta_1 m_{12}, \qquad m_{31} = \beta_1 m_{13} \\ m_{24} &= \beta_2 m_{42}, \qquad m_{34} = \beta_2 m_{43}, \end{aligned} \tag{3.17}$$

with $\beta_1, \beta_2 > 1$. We simplify by setting $\beta_1 = \beta_2 = \beta$, and so are left with

$$\begin{aligned} m_{21} &= \beta m_{12}, \qquad m_{31} = \beta m_{13} \\ m_{24} &= \beta m_{42}, \qquad m_{34} = \beta m_{43}. \end{aligned} \tag{3.18}$$

Since module permutations are far less radical than changes in composition, we make the further simplifying assumptions that

$$m_{12} = m_{13}, \qquad m_{24} = m_{34}. \tag{3.19}$$

Similary, we posit that the variety 01 is just as stable as 10, and that 00 and 11 are equally stable:

$$m_{22} = m_{33}, \qquad m_{11} = m_{44}. \tag{3.20}$$

Our final assumption is that the mutations between the extreme varieties 00 and 11 proceed via the intermediate species 01 and 10, i.e.

$$\begin{aligned} m_{14} &= m_{12} \cdot m_{24} + m_{13} \cdot m_{34} \\ m_{41} &= m_{42} \cdot m_{21} + m_{43} \cdot m_{31}. \end{aligned} \tag{3.21}$$

A first consequence of the above assumptions is

$$\begin{aligned} m_{21} &= m_{31}, \qquad m_{42} = m_{43} \\ m_{14} &= 2m_{12} \cdot m_{24} = m_{41} = 2m_{21} \cdot m_{42}. \end{aligned} \tag{3.22}$$

Finally we use the normalization condition (3.6) which, together with the previous equalities, yields the mutability matrix in terms of m_{11}, m_{22} and m:

$$\|m_{ij}\| = \begin{Vmatrix} m_{11} & \dfrac{(1 - m_{11})}{(3 - m_{22} - m)} & \dfrac{(1 - m_{11})}{(3 - m_{22} - m)} & \dfrac{(1 - m_{11}) \cdot (1 - m_{22} - m)}{3 - m_{22} - m} \\ \frac{1}{2}(1 - m_{22} - m) & m_{22} & m & \frac{1}{2}(1 - m_{22} - m) \\ \frac{1}{2}(1 - m_{22} - m) & m & m_{22} & \frac{1}{2}(1 - m_{22} - m) \\ \dfrac{(1 - m_{11}) \cdot (1 - m_{22} - m)}{3 - m_{22} - m} & \dfrac{(1 - m_{11})}{(3 - m_{22} - m)} & \dfrac{1 - m_{11}}{3 - m_{22} - m} & m_{11} \end{Vmatrix} \tag{3.23}$$

We proceed now to trace the evolution of our four-variety biopopulation. To compute their respective numbers at successive generations we must know what their initial numbers were. Suppose a single variety is initially present, say V_1 or 00. Call $N_0^1 = N$ its initial population. Then by (3.7) one generation later the average numbers are

$$
\begin{aligned}
N_1^1 &= 2(2m_{11} - 1)(1 - d_1)N \\
N_1^2 &= \frac{2(1 - m_{11})}{3 - m_{22} - m}(1 - d_1)N \\
N_1^3 &= N_1^2 \\
N_1^4 &= (1 - m_{22} - m)N_1^2.
\end{aligned}
\tag{3.24}
$$

If $m_{11} \leq \frac{1}{2}$ or $d_1 = 1$, the first variety becomes extinct forthwith. And if $m_{11} = 1$ while $d < 1$, V_1 prospers without change. The *necessary and sufficient condition* for the evolutionary process to occur is then

$$\tfrac{1}{2} < m_{11} < 1 \quad \text{and} \quad 0 \leq d_1 < 1. \tag{3.25}$$

If this condition obtains then the lower bound of the population of varieties V_2 and V_3 at first generation is

$$\tfrac{2}{3}(1 - m_{11}) \cdot (1 - d_1)N.$$

Since this number equals at least 2, we infer that the lower bound of the stability of the parent variety is given by

$$m_{11} \geq 3/(1 - d_1)N. \tag{3.26}$$

Because the upper bound of m_{11} is 1, we infer that for a change of variety to occur it is necessary that

$$N > 3/1 - d_1. \tag{3.27}$$

For a low death rate of the parent variety, as few as four members of it could give rise to the differentiation process.

The above conditions concern only varieties V_2 and V_3. For variety V_4 to emerge a more severe condition must be met, as we gather from the last of equations (3.24), namely

$$(3.25)\ \&\ (m_{22} + m < 1) \quad \text{or, equivalently,} \quad (3.25)\ \&\ (m_{21} + m_{24} > 0).$$

Finally we write the general evolutionary equations (3.7) for an arbitrary generation (t value). Using the mutation matrix (3.23), we obtain the following system of first order linear difference equations:

$$
\begin{aligned}
N_{t+1}^1 = {} & 2(2m_{11} - 1)(1 - d_1)N_t^1 + (1 - m_{22} - m)(1 - d_2)N_t^2 \\
& + (1 - m_{22} - m)(1 - d_3)N_t^3 \\
& + \frac{2(1 - m_{11})(1 - m_{22} - m)}{3 - m_{22} - m}(1 - d_4)N_t^4
\end{aligned}
$$

$$\begin{aligned}
N^2_{t+1} &= 2(2m_{22} - 1)(1 - d_2)N^2_t \\
&\quad + \frac{2(1 - m_{11})}{3 - m_{22} - m}(1 - d_1)N^1_t + 2m(1 - d_3)N^3_t \\
&\quad + \frac{2(1 - m_{11})}{3 - m_{22} - m}(1 - d_4)N^4_t \\
N^3_{t+1} &= 2(2m_{33} - 1)(1 - d_3)N^3_t + \frac{2(1 - m_{11})}{3 - m_{22} - m}(1 - d_1)N^1_t \\
&\quad + 2m(1 - d_2)N^2_t + \frac{2(1 - m_{11})}{3 - m_{22} - m}(1 - d_4)N^4_t \\
N^4_{t+1} &= 2(2m_{44} - 1)(1 - d_4)N^4_t \\
&\quad + \frac{2(1 - m_{11})(1 - m_{22} - m)}{3 - m_{22} - m}(1 - d_1)N^1_t \\
&\quad + (1 - m_{22} - m)(1 - d_2)N^2_t + (1 - m_{22} - m)(1 - d_3)N^3_t.
\end{aligned}$$

This is the kind of linear algebra problem that computers love. But its solution is of no philosophical interest. What are of philosophical interest are the following points. First, the general evolution equations (3.7) can be turned, from a mere descriptive black box, into a mechanismic model representing the mutation-selection process – a model capable of making precise predictions about a certain evolutionary line. Second, this is a peculiarity not of our problem but of the synthetic theory of evolution, which is not a predictive theory unless it is adjoined precise data and hypotheses concerning the genetic makeup and the environment of the organisms concerned. (We shall come back to this problem in Vol. 6.)

3.6. *Bioprogress*

Evolutionary biologists agree that all modern complex organisms descend from primitive organisms: recall Figure 3.2. So, there would seem to be no doubt that evolution has been progressive. Not so fast. First of all, the following kinds of evolutionary process are usually distinguished:

- Evolution
  - Stasigenesis (e.g. horseshoe crab and shark)
  - Evolution proper
    - Regressive (e.g. endoparasites, cave fish)
    - Progressive
      - Increasing specialization (e.g. fly, horse)
      - Increasing generalization (e.g. primates)

Some lineages, then, have advanced over certain periods while others have not. The second problem is that there is no agreement on the very notion of bioprogress – the more so since progress through increasing specialization is sure to lead to stagnation or extinction.

A number of alternative definitions of bioprogress have been proposed (cf. Williams, 1966, Bowler, 1976). Here are some: bioprogress is increasing histological differentiation, morphological complication, survivorship, adaptability, freedom from the environment, or amount of genetic information. All of these proposals presuppose that progress is global rather than partial (in particular respects), and none of them lends itself easily to quantitation. In particular, the information-theoretic definition is pseudo-quantitative, as there is no unambiguous and objective way of measuring or computing the "quantity of information stored in an organism" (Lwoff, 1962; Apter and Wolpert, 1965). After all, biosystems are not identical with their genomes, and the latter are more than just information processing devices.

We shall take it that bioprogress consists in increasing biovalue, and shall proceed to refine the distinction (Ayala, 1974) between uniform (relentless) progress, on the one hand, and overall (net, average) progress on the other. Our elucidation will make use of the concept of system progeny (Definition 1. 15(v)) and Definition 3.9 of biovalue. Since the former was framed for systems in general, we shall first adapt it to biopopulations; and this adaptation will involve the notion of a variable collection, i.e. one with changing membership (Vol. 3, Definition 5.13). Let us then start with

DEFINITION 3.15 Let $\mathscr{P} = \{p_t \mid t \in T\}$ be a family of biopopulations of a certain kind, indexed by time (T). For any $p_0, p_t \in \mathscr{P}$, where $t > 0$,

(i) p_t *descends* from p_0, or $p_0 \prec p_t$, iff every member of p_0 descends from some member(s) of p_0;

(ii) the *progeny* of p_0 is the family of biopopulations that descend from p_0:

$$P(p_0) = \{p_t \in \mathscr{P} \mid p_0 \prec p_t\}.$$

We now add the idea that, if x and y are successive populations, x constitutes an improvement over y in a certain respect z if x performs better than y in doing z. More precisely, we propose

DEFINITION 3.16. Let p_0 be a biopopulation of organisms of a certain kind K, and call $P(p_0)$ the progeny of p_0. Further, call Σ a certain kind of subsystem (e.g. organ) of every member of p_t in $P(p_0)$, and assume that all the subsystems of that kind perform (roughly) the same function(s). Then

(i) the K's have *advanced on the whole* with regard to Σ between p_0 and p_t, for $t \gg 0$, iff the value of subsystem of kind Σ for the members of p_t is significantly greater than its value for the members of p_0;

(ii) the K's have *advanced uniformly* along $P(p_0)$ with regard to Σ between p_0 and p_t, for $t \gg 0$, iff the value of subsystem of kind Σ for the members of every population p_u in $P(p_0)$, between p_0 and p_t, is greater than its value for the immediate ancestors of the p_u's.

Whereas the first clause elucidates the notion of net or average progress, the second clarifies that of uniform progress. Uniform progress implies net progress but not conversely: i.e. there may be temporary setbacks in a progressive lineage,

Finally we make

DEFINITION 3.17 Under the nomenclature and assumptions of Definition 3.16, the K's have *progressed globally* between p_0 and p_t iff the value of every function of the members of p_t is greater than the value of the corresponding function for the members of p_0.

Remark 1 The above definitions do not discriminate between progress in a roughly constant environment, therefore at the cost of severe selection, and the accidental improvement of environmental conditions (or preadaptation). *Remark 2* Our definitions involve the concept of biovalue, hence that of goodness. Indeed, x is *good* for $y =_{df}$ the value of x for y is positive. However, these concepts are biological in our system, because biovalue was defined in strictly biological terms (Definition 3.9). Instead of injecting an anthropomorphic concept of value into biology, we have tried to derive an objective concept of value from the biological concept of health of a biosystem,

So much for definitions. Now is the time for commitment. Is there progress in evolution and, if so, is it universal and global or local and partial? According to some philosophers and biologists, from Aristotle to Lamarck, bioevolution is gradual and slow progress toward perfection. This hypothesis presupposes that the environment has remained constant for over 4 billion years, or that it has changed for the better of every species. This presupposition has been refuted by the evidence for continental drifts, glaciations, desertifications, earthquakes, and the depletion of a number of resources. While some such changes have been favorable to some species (those preadapted to the new environments), and even to the emergence of new species, they have been fatal to others. In any case most species have become extinct instead of persisting on the road of improvement.

Another presupposition of the hypothesis of universal and global progress is that there exists a fixed goal – an ideal or paradigm – towards which successive generations move. This hypothesis would make sense in a Platonic biology, where the archetype is perfect, every real organism being a sort of blurred carbon copy of the former. But of course there is no such thing as Platonic biology, although there are vestiges of Platonism in biotaxonomy. The very notion of an immaterial object – such as the Idea of the Perfect Mosquito – guiding the evolution of concrete things is inconsistent with the ontology of modern science. (Cf. Vol. 3, Introduction and Ch. 2.) Besides, since the environment changes rather erratically and sometimes dramatically, if there were a target then it should be moving rather than fixed, and so evolutionary processes zig-zag rather than go in a definite direction.

In sum, the biosphere does not obey the so-called "law of progress". Progress, when it *does* occur, is a trend not a law, and one occurring in *some* lineages, in *some* respects, and for *some* time. In the biosphere, just as in the sociosphere, there are not only progressives (e.g. mammals) but also conservatives (e.g. ants) and reactionaries (e.g. lice). Besides, whereas many progressives have become extinct, some conservatives are still around: drastic environmental changes and fierce competition can wreck the most brilliant careers.

For all the above reasons, the most we can assert is

POSTULATE 3.5 Every biospecies advances in some respects during some periods and in certain environments.

That is, bioprogress is often real but always partial, relative, and temporary. It is *partial* because an improvement in some respect(s) may constitute a drawback in other(s); e.g. an increase in size may confer greater strength but it also calls for greater energy consumption and less mobility. Progress is *relative* to the environment – moreover it can be active or passive: earned by mutation and selection, or won in the lottery of environmental cataclysms. And progress is *temporary* or limited to definite periods rather than eternal.

4. Concluding remarks

The traditional biometaphysics are vitalism and mechanism. The strength of the former lies in its ability to ask questions, some of them pertinent, that the latter fails to answer. Whereas classical vitalism has hardly evolved, mechanism has. Two kinds of mechanism are popular today: (a) *physicochemicalism*, or the thesis that organisms are just physicochemical systems, and (b) *machinism*, or the thesis that organisms are machine-like systems, hence fashioned after a design or plan, and acting in a goal-seeking manner.

Physicochemicalism is the heir to traditional mechanism, and machinism a hybrid of mechanism and vitalism. Whereas the model of the first is the chemical reactor, that of the second is the computer. Neither of these biometaphysics is consistent with the whole of biology, although physicochemicalism has been fruitful where machinism has been sterile. Physicochemicalism is incompatible with biology because it denies emergence and levels, and machinism because it takes for granted and posits plans and goals where there are none.

The alternative we propose is *biosystemism*, or the application of systemism (Ch. 1, Sec. 4.2) to the study of life. Biosystemism recognizes the *bios* as an emergent level rooted to the chemical one. Therefore it accepts the lay difference between living and nonliving – denied, e.g. by Pirie (1960), Kendrew (1966) and Keosian (1968) – as well as the molecular-biological account of the self-assembly of biosystems from biochemical precursors, and the theory of evolution by genic changes and natural selection.

We are faced then with four different biometaphysics, each with its accompanying problem circle, method, and goal: see Table 3.1. The oldest and most persistent of them is vitalism, which is also the most popular biophilosophy. Actually there are two kinds of vitalism: ontological and methodological. The former states that organisms are distinguished by some immaterial principle, the latter that they are so complex and variable, that they cannot be tackled with the ordinary methods of science. Ontological vitalism has been pronounced dead long ago, even by biologists who continually lapse into it – such as Monod (1970) and Jacob (1970) when fantasizing about the *project téléonomique* of every organism. On the other hand the claim that organisms cannot be studied by the ordinary scientific method but call for a special method (i.e. methodological vitalism) is still very much alive except among molecular biologists (Bunge, 1973a, Ch.3).

As for machinism, its distinctive thesis is that organisms are like machines in their having a design or plan and a purpose. However, two kinds of machinism must be distinguished: verbal and serious. Verbal machinism restricts itself to repeating uncritically the organism-machine

TABLE 3.1

Four approaches to the study of life

	Vitalism	Mechanism: Machinism	Mechanism: Physico-chemicalism	Biosystemism
Ontological Framework	Life transcends matter and is guided by an immaterial principle	Organisms are machine-like: they have plans and goals.	Organisms are physico-chemical systems.	Organisms are *sui generis* systems composed of biochemical systems and possessing emergent properties and laws.
Problem Circle	'What is it for?' questions.	'What is it for?' and 'How does it work?' questions.	'How is it composed?' and 'What are its properties?'	All biological problems, in particular regarding subsystems, their self-assembly, and evolution.
Method	Description of organism as a whole.	Physico-chemical methods plus building of analogs.		The whole array of particular methods of the various chapters of natural science.
Goals	Intuitive understanding of the whole	Technology-wise explanation and forecast.	Physicochemical explanation and prediction.	Biological explanation and prediction.

analogy and to warning that biology may learn more from engineering that from physics and chemistry (Polanyi, 1968). This kind of machinism can be discarded by noting that biology does not have a single law statement, let alone theory, containing the concepts of plan (or design) and purpose. (Even the most serious attempt to incorporate the concept of goal-directedness into biology, namely that of Sommerhoff (1974), fails to mathematize correctly that concept as well as to exhibit examples of laws and theories where the concept occurs.)

Serious contemporary machinism is the heir to Descartes and La Mettrie and culminated with the theory of self-reproducing automata (von Neumann, 1966). It is rather popular with the computer science crowd, as can be seen by scanning the *J. Computer and System Sciences*. This approach has stimulated, in particular, biological applications of the theories of automata and information. However, three decades of work along this line have yielded precious little; in particular, it has not contributed to the molecular biology revolution or to evolutionary biology. No wonder, for machinism skips the chemical level and ignores history: it thinks in terms of ready made systems rather than of enzymatic reactions or natural selection.

Machinism combines the negative aspect of mechanism (downward reductionism) whith the negative aspect of vitalism (upward reductionism). Its sole virtue is its clarity. Surely, machinism has stimulated the design and construction of ingenious mock organisms, such as Grey Walter's mechanical tortoise. However, since such artifacts are avowedly very different from the real thing they simulate, they do not teach us much about the latter. The only use of biosimulates is as toys or as slaves. In short, machinism is mistaken because it ignores that organisms are composed of chemical systems, rather than electromagnetic ones, and assigns them a property very few animals exhibit, namely that of seeking goals, and another that none have, namely optimal design or perfection.

As for physicochemicalism, it has undoubtedly been a fertile biophilosophy. In fact it has stimulated the birth of entire branches of science, such as biophysics, biochemistry, molecular biology, and bioengineering. In particular molecular biology might not have been born but for the hypothesis that chemical (in particular enzymatic) reactions constitute the root if not the marrow of life. In any case this conjecture sparked off, among other things, the chemical theory of heredity, which was the initial phase of molecular biology. (On the other hand physicalism, which tried to skip the chemical level, failed to contribute anything to the developments that culminated in the discovery of the double helix and the genetic "code": see Olby (1974). This is not to deny that physics, in particular X-ray crystallography, was indispensable. But it is one thing to use physical techniques and another to conceive of the organism as a physical system.)

Molecular biology, largely an offspring of the physicochemicalist branch of biometaphysics, has gone beyond digging up the chemical roots of life: it has shown that a number of properties that used to be regarded as typically biological – such as metabolism and self-reproduction – are actually biochemical. However, in their eagerness to rid biology of myth and explain it in terms of chemistry, molecular biologists have often stressed composition at the expense of organization or structure, as well as heredity to the detriment of environment. Consequently they have often overlooked the multi-level structure of the organism – organelle, cell, organ, multicellular organism. For this reason, although molecular biology is indispensable for explaining evolution, it is not sufficient: phenotypes and environments are just as important as genotypes. After

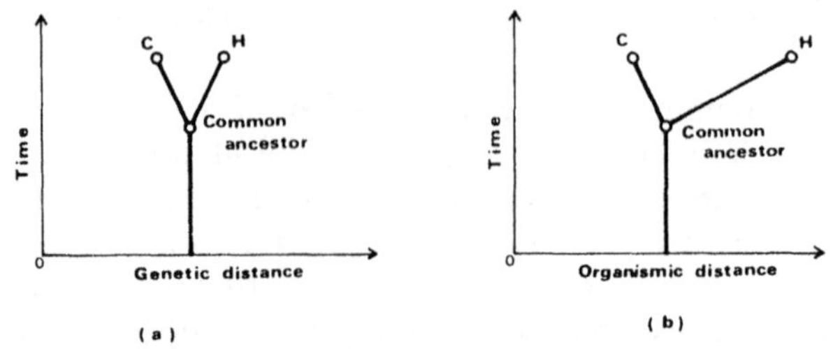

Fig. 3.7. Humans (H) and chimpanzees (C) are genetically about as different as two subspecies of mice (*a*). They differ mainly at the organ level (*b*), particularly in respect of the nervous system, hence of behavior. See King and Wilson (1975).

all, man and chimpanzee are hardly distinguishable at the biochemical level: see Figure 3.7.

To conclude. Vitalism is untenable but it has bequeathed us two important true principles:

V1 Life is emergent: it transcends the physico-chemical level.

V2 The biosphere has a level ("hierarchical") structure going from cell to organ to organism to population to ecosystem.

Physicochemicalism denies these principles but contributes another two important true hypotheses:

M1 Life emerged by the self-assembly of macromolecules.

M2 Some physical and chemical laws continue to operate in organisms.

Biosystemism adopts all four principles and adds its own:

S1 Organisms are systems composed of biochemical systems and possessing emergent properties.

S2 Every level in the biosphere has its own laws: laws of cells, of organisms, of populations, and of ecosystems.

S3 The units of biological study are the organism-in-its environment and its various subsystems (molecules, cells, organs) and supersystems (population, ecosystem, biosphere).

So much for the ontology of biosystemism. As for its methodology, it is epistemic reductionism, which is perfectly compatible with ontological pluralism (Bunge, 1977f), and boils down to

RULE 3.1. Each bioentity must be studied on its own level and should be explained with the help of its adjoining levels.

Thus an organ should not only be studied on its own level but should also be analyzed both downward (in terms of its subsystems) and upward (in terms of the entire organism in its environment). This holds particularly for the human brain, which is not only a neural system but also a subsystem of a component of social systems. But this is the story to be told next.

CHAPTER 4

MIND

The present chapter deals with one of the oldest, most intriguing, and most perverse of all the problems lying in the intersection of philosophy and science, namely the so-called mind-body problem. This is the system of questions about the nature of mind and its relations to matter.

Some eminent thinkers – notably Spencer (1862), E. du Bois-Reymond (1872), and Popper and Eccles (1977) – have pronounced the problem to be insoluble. I submit that part of the problem is that it is usually formulated in inadequate terms – namely those of ordinary language. These are inadequate not only because ordinary language is notoriously poor and inexact, to the point of being of little use in investigating the problem of matter, but also because it begs the question. Indeed European languages are loaded with a preconceived solution to the problem of mind, namely psychophysical dualism, or the doctrine that mind is something apart from matter. The very expression 'mind-body problem' suggests that mind and body are distinct entities on the same footing – i.e. things. We must then look to science for inspiration.

In science we do not speak of the motion-body problem, or of the reaction-substance problem, or of the digestion-digestive tract problem, or of the mobility-society problem. We speak instead of the motion *of* bodies, the reaction *of* chemicals, the digestive functions *of* the digestive tract, and the mobility *of* a society. We do not reify properties, states, or events – except when it comes to the properties, states, and events of the nervous system. We shall close this gap, that keeps the study of mind a scientific anomaly in the grip of myth, by replacing ordinary language by state space language, which is mathematically precise and is shared by science and scientific ontology.

Another difficulty besetting the so-called mind-body problem is the extreme youth of neuroscience: recall that Ramón y Cajal's neuronal hypothesis is less than one century old. Surely there were important sporadic discoveries in earlier times, starting with Hippocrates' hypothesis that the brain – rather than the heart or the liver – is the organ of emotion and thought. However, it is only in recent years that a concerted attack at all levels has been launched (Worden *et al.* (Eds.), 1975).

A third difficulty is that, far from being a strictly scientific problem, ours is also a philosophical one. In fact it is a traditional metaphysical problem and the one that has aroused the greatest passion – hence also the greatest caution. Moreover its very formulation presupposes a number of notions that are far from clear in traditional metaphysics, in particular those of substance, property, state, event, emergence, and level of organization. Thus any argument as to whether or not there is a mental substance, or whether there are mental states that fail to be brain states, calls for a prior elucidation of the general concepts of substance and state – an elucidation that most philosophers do not care to perform.

Fourth and last, another reason for the backward state of research into the so-called mind-body problem is that it belongs not only to science and to philosophy but also to ideology. Indeed all religions and some political ideologies have a vested interest in it and some even claim exclusive ownership. So, far from being interested in the investigation of the problem, they are anxious to have us accept their own ready-made solution.

In short, there have been many obstacles in the way of a scientific study of the problem. Fortunately some of these obstacles are disappearing rather quickly, and the science of mind – i.e. psychology – is making great strides. Three formerly separate currents are converging on it: neuroscience, psychology, and philosophy. Indeed, over the past few decades neurophysiologists and neurologists have begun to study the mental functions of neural systems, physiological psychologists have started to disclose the neural mechanisms "mediating" behavior, and philosophers have begun to analyze some of the key concepts involved in the question, and to overhaul the ancient doctrine that mental functions are a kind of bodily functions. There is still a long way to go before maturity is attained, but the science of mind has finally made a start.

This chapter, like the others in this book, will offer only a general framework – but, of course, one consistent with the basic ideas of system, property, law, and change prevailing in science and expounded earlier in the work.

1. Central nervous system

1.1. *Philosophical Background*

Perceiving, feeling, remembering, imagining, willing, and thinking are usually said to be mental states or processes. Since in science and in our ontology there are no states or processes in themselves, but only states *of* some entity and processes *in* some thing, we must ask *what* minds – i.e. what is the thing that perceives, feels, remembers, imagines, wills, and thinks. This is the very core of the so-called mind-body problem, i.e. the identification of the subject or referent of the mentalist predicates.

Those who hope the mind-body problem to be soluble have proposed two solutions to it: that what does the minding (perceiving, feeling, thinking, etc.) is the mind (or soul or spirit), and that it is the brain. According to the former the mind is an immaterial entity wherein all mental states and processes occur: feelings, memories, ideas, and the like would be in or of the mind, which would in turn be an entity separate from the body. According to the second answer the mind is not a separate entity but a set of brain functions: perceiving, imagining, thinking, and the like would be brain processes. The former thesis is called *psychophysical dualism*, the latter *materialism* or *naturalism*.

However, there are a number of varieties of dualism and of materialism. (Cf. Vesey, 1964; Smythies, 1965; Armstrong, 1968; O'Connor, 1969; Borst, 1970; Popper and Eccles, 1977; Margolis, 1978.). Our main task in this chapter is not to discuss them but to formulate a version of materialism consistent with the principles laid down in the previous chapters and in Vol. 3 of this work. This is *emergentist materialism*, or the doctrine according to which (*a*) all mental states, events and processes are states of, or events and processes in, the brains of vertebrates; (*b*) these states, events and processes are emergent relative to those of the cellular components of the brain, and (*c*) the so-called psychophysical (or psychosomatic) relations are interactions between different subsystems of the brain, or among them and other components of the organism (Bunge, 1977e). This kind of materialism is monistic with respect to substance and pluralistic with respect to properties.

Here are some reasons for favoring emergentist materialism:

(i) because it eschews the ghostly mental substance without thereby denying mental facts, emergentist materialism is *compatible with the scientific approach*;

(ii) emergentist materialism – as we hope to show – is *free from the fuzziness that characterizes dualism* with its talk of an immaterial mind and of "correlations" between it and the body;

(iii) unlike dualism, emergentist materialism is *consistent with the general concepts of state and event that can be gleaned from all the sciences.* (On the other hand according to dualism mental states would be the *only* states that fail to be states of some thing, and mental events the *only* events that fail to be changes of state of some thing – this being why dualism agrees more with theology than with science.)

(iv) Unlike dualism, emergentist materialism *fosters interaction between psychology and the other sciences*, in particular neuroscience, and this precisely because it regards mental events as special biological events;

(v) unlike dualism, which digs an unbridgeable chasm between man and beast, emergentist materialism *accords with evolutionary biology*, which – by uncovering the gradual development of mental faculties along certain lineages – refutes the complacent superstition that only Man is endowed with Mind;

(vi) unlike dualism, which postulates an unchanging mind, emergentist materialism *agrees with developmental psychology* and neurophysiology, which exhibit the gradual maturation of the brain and behavior.

The only drawback to emergentist materialism is its so far not having been formulated in a precise and detailed way. Indeed, emergentist materialism is little more than a programmatic hypothesis with enormous heuristic power. We shall try and remedy this shortcoming in the course of this chapter, by building a general theory of the mind consistent not only with our ontological principles but also with current neuroscience and psychology.

1.2. *Neural Units*

We shall deal throughout this chapter with systems of a special kind, namely organisms endowed with nervous systems. And we shall pay special attention to animals endowed with a central nervous system, or CNS for short. Nervous systems come in a number of degrees of complexity, from primitive worms with half a dozen neurons, to insects with about 100,000, to human with nearly a hundred billion (i.e. 10^{11}). The corresponding behavior levels range from automatic to highly creative. And the varieties of inner life from nil to extremely rich. Since the lower forms of nervous system functioning and animal behavior are generally recognized to be explainable along strictly biological lines, we shall concentrate on the higher functions – i.e. on those which, according to popular belief, require the presence of a soul, spirit, or mind inaccessible to science.

The following levels may be distinguished in a CNS:

Subcellular: neuron's membranes, synaptic boutons, dendrites.

Cellular: neurons and glial cells.

Neural microsystems: multineuronal systems (fixed or itinerant assemblies of thousands or millions of neurons in primates).

Neural macrosystems: systems of thousands of neural microsystems. In primates the following sublevels are distinguished:

Uncommitted cortex, somethetic system, visual and hearing systems, temperature regulating system, etc.;
cerebral hemispheres;
brain;
central nervous system (CNS), or brain *cum* spinal cord;
neuroendocrine system (NES), or nervous system *cum* endocrine glands (e.g. pituitary and adrenals).

Every mental process presumably engages millions of neurons and occurs not only at the neural system level but also at the cellular and subcellular levels. Moreover, given the strong coupling between the CNS and the rest of the body, every mental act is bound to engage also some non-nervous tissue. E.g., voluntary movement is an activity not only of certain neural systems but also of certain muscles and endocrine glands. Thus the various neural systems (micro and macro) can be distinguished from one another but they cannot be detached from each other or from the rest of the body, They do the minding but not in isolation from other subsystems of the body.

While some neural systems have permanent components others don't. Thus the visual system has a constant composition, i.e. is composed of anatomically identifiable subsystems – the eyes, the optic nerves, the optic tracts, the lateral geniculate nuclei, and the visual cortex. Other neural systems may not have constant components and so may not be carved out by the surgeon: far from having a constant location, they may be itinerant neuron assemblies formed just for the occasion (Craik, 1966; Bindra, 1976). See Figure 4.1.

We shall call a *psychon* any neural unit capable of discharging mental functions of some kind. There are three main views concerning the nature and size of psychons: neuronism, holism, and systemism. According to neuronism single neurons may possess certain mental abilities – e.g. that of issuing commands. There is no evidence for this view. Holists, on the other hand, assume that nothing short of the whole brain can mind: they speak of 'mass action' and favor holographic theories of memory and other mental functions. Although there is plenty of evidence for the strong coupling among a number of neural systems, there is also evidence for localization—of, e.g., pleasure and speech.

So we are left with *psychosystemism*, or the hypothesis that the brain,

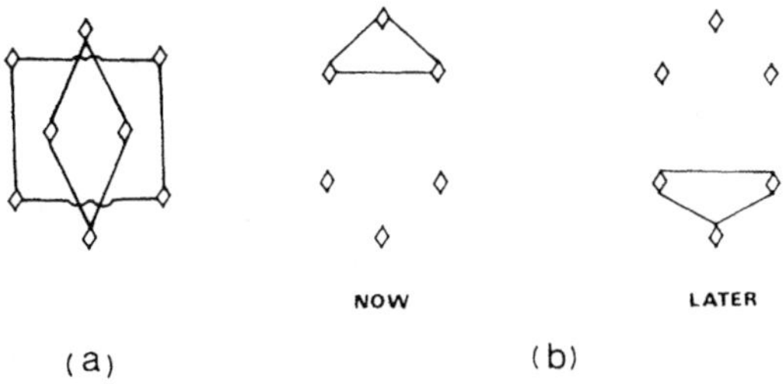

Fig. 4.1. (*a*) Two neural microsystems occupying roughly the same spatial region yet functionally detached: they 'subserve' different functions. (*b*) Psychons can be itinerant even if the components stay in place, because their connections change in time.

far from being either a heap of self-sufficient units or a homogeneous jelly, is a *system of specialized subsystems* or organs, some of which are itinerant rather than fixed. There is multiple evidence for psychosystemism, so we shall adopt it. Besides, psychosystemism accords with our systemist world view and furthermore it suggests the most fruitful research strategy, namely that of investigating the brain as a whole as well as all of its subsystems, their mutual couplings, and their links with non-neural subsystems, particularly the endocrine system.

The systemic approach to the CNS is necessary but insufficient: we must also take development and evolution into account. Dualism regards the mind as an invariable entity, holding that each person keeps his or her "personal identity" throughout life. However, the empirical evidence is against this thesis: the newborn human has hardly a mind, after a few months has learned a number of tricks, and continues to learn until senility sets in. In short the CNS of vertebrates is plastic. Moreover it has evolved from modest beginnings. The believers in the myth of the immaterial mind must ignore the evidence that the mind has evolved along with other bodily functions. So, we shall ignore the dualistic myth of personal identity: we shall hold instead that the only constant thing about John Smith is his name.

1.3. *Initial Definitions*

We start by making

DEFINITION 4.1 An *animal* is an organism that
(i) feeds on other biosystems;

(ii) can move as a whole toward or away from some stimuli;

(iii) can explore its surroundings in search of food, water, or stimulation, or to avoid stimuli that might impair its health,

We shall next define a nervous system as an information biosystem in an animal. We can and must do without the concept of information in biochemistry and in cellular biology, not however when dealing with organisms endowed with endocrine or, *a fortiori*, neuroendocrine systems. Here the environmental and the internal inputs not only are nonspecific energy transfers but sometimes also generate specific signals conveying definite information about events. Indeed there is a definite code – a naturally evolved one of course – whereby stimuli of a certain kind (e.g. thermal) are "translated" (transduced) into neural pulses in the organism; and these signals in turn activate other components, e.g. muscles or glands. These are signals proper not just neutral energy fluxes: they carry information and they trigger the release of much greater quantities of energy than they transport. The nervous system, in sum, is a genuine information network, for it detects, generates, and transmits information in accordance with definite natural codes. Hence we make

DEFINITION 4.2 A system is a *nervous system* iff it is an information biosystem (i.e. a living signal-detecting-and-processing system) such that

(i) it is composed of (living) cells;

(ii) it is or has been a proper part of a multicellular animal (i.e. the complement of the latter in the immediate environment of the system);

(iii) its structure includes (*a*) the regulation or control of some of the biofunctions of the animal, and (*b*) the detection of internal and environmental events, as well as the transmission of signals triggered by such events.

DEFINITION 4.3. A biosystem is a *neural* (or *neuronal*) system iff it is a subsystem of a nervous system.

DEFINITION 4.4. A biosystem is a *neuron* iff it is a cellular component of a neural system.

Remark 1 The vertebrate nervous system has a third structural property: it interacts directly with every other system of the animal. And the CNSs of higher vertebrates are more than information processors: they are also information generators. More on this in Sec. 4.5. *Remark 2* Neural circuits and ganglions, cell assemblies (Hebb, 1949) and neuron populations

(Freeman, 1973), working constellations (Luria, 1966) and itinerant neural systems or pexgos (Bindra, 1976) are neural systems, and so are the functional columns of cortical neurons (Mountcastle, 1957; Powell and Mountcastle, 1959; Hubel and Wiesel, 1963). *Remark 3* The neural systems capable of perceiving signals, imagining, or willing, are probably composed of millions or billions of neurons. However, even neuron systems composed of a few neurons are bound to possess properties that its components lack, i.e. emergent properties. Therefore neuronal systems are ontologically irreducible entities even if they are (up to a point) epistemically reducible, i.e. explainable in terms of composition and structure. *Remark 4* The preceding remark counters the dualistic argument that, because all cortical neurons look alike (to the untrained eye), they cannot be attributed the diversity of functions imputed to them by the monists. Actually no two neurons are alike. And even if they were, we know that they form systems (in particular, columns) – and, as any child experienced with Lego or Meccano knows, there is no end to the variety of systems that can be formed with a few elements. *Remark 5* Exclusive attention to the individual neuron and its alleged "all or none" firing pattern is misleading for it favors focusing on discontinuous processes and analogies with digital computers. Mental processes involve presumably millions of neurons and therefore continuous or quasi-continuous processes. This allows one to build continuous models of neural systems (Rashevsky, 1972; MacGregor and Lewis, 1977). And it disposes of the so-called "grain objection" to the psychoneural identity hypothesis, namely that whereas brain events are spatially and temporally discrete, we can have experience of continuous expanses of red.

The various neuronal systems differ in their mode of connection. There are several types of interneuronal and intersystem connection: constant or fixed, variable in a regular way, and variable in a random way. In order to discuss neural connectivity we had better define this notion. One way of doing it is as follows. Call $C_t(a, b)$ the strength of the connection from neuron (or neuron assembly) a to neuron (or neuron assembly) b of a given system at time t. In general, $C_t(a, b) \neq C_t(b, a)$. The $a - b$ connection is *excitatory* at t iff $C_t(a, b)$ is positive, and *inhibitory* iff $C_t(a, b)$ is negative. The excitation or inhibition caused by cell a on target cell b equals the output of a times $C_t(a, b)$. And the total input to cell b is obtained by adding all the partial inputs (i.e. by summing over a). The matrix $\|C_t(a, b)\|$ displays the total connectivity of the system at time t. The connectivity of a neural system with 1 million neurons is represented by a 1,000,000 × 1,000,000 matrix (lumping the hundreds of synapses of each cell into a single junction). Since we shall make heavy use of this concept we may as well display it for future reference:

DEFINITION 4.5. Let ν be a neural system and $\mathscr{C}_t(\nu)$ the neuronal composition of ν at time t. Further, call

$$C_t : \mathscr{C}_t(\nu) \times \mathscr{C}_t(\nu) \longrightarrow [-1, 1]$$

the function such that $C_t(a, b)$, for $a, b \in \mathscr{C}_t(\nu)$, is the strength (intensity) of the connection (coupling, link) from neuron a to neuron b at time t. Then the *connectivity* of ν at t is represented by the matrix formed by all the connection values, i.e.

$$\mathbb{C}_t = \|C_t(a, b)\|.$$

DEFINITION 4.6. A connectivity is *constant* iff it does not change once established (i.e. iff $\mathbb{C}_t$ is independent of time). Otherwise it is *variable*.

DEFINITION 4.7. A neuronal system is *plastic* (or *uncommitted*, or *modifiable*, or *self-organizable*) iff its connectivity is variable throughout the animal's life. Otherwise (i.e. if it is constant from birth or from a certain stage in the development of the animal), the system is *committed* (or *wired-in*, or *prewired*, or *preprogrammed*).

DEFINITION 4.8. Every plastic neural system is called a *psychon*.

So much for our preliminary definitions. We are now ready for the first bunch of axioms of our theory of the mind.

1.4. *Basic Assumptions*

Our first hypothesis is rather obvious:

POSTULATE 4.1. All animals with a nervous system have neuronal systems that are committed, and some animals have also neuronal systems that are plastic (uncommitted, self-organizable).

Apparently worms, insects and other lowly animals have only totally (or nearly totally) wired-in neuronal systems. On the other hand the plastic cortex of man is the largest in all animals and a sort of vast reserve army ready to meet emergencies, most of which never happen. Hence whereas the behavioral repertoire of lowly animals is predetermined (programmed), that of man (and many another vertebrate) can evolve in the course of his life. More on behavioral plasticity in Sec. 3.3.

POSTULATE 4.2. The neuronal systems that regulate (control) the internal milieu, as well as all the biofunctions of the newborn animal, are committed (wired-in).

Example The temperature and acidity of the intercellular fluid are regulated mainly by certain wired-in neuronal systems, and so are the breathing and sucking movements of the newborn animal. But they are not the only regulators: also the endocrine system plays an important role in internal control.

POSTULATE 4.3 The plastic (uncommitted) neuronal systems of an animal (i.e. its psychons) are coupled to form a supersystem, namely the *plastic neural supersystem* of the animal. Symbol: P.

POSTULATE 4.4 Every animal endowed with psychons (plastic neuronal systems) is capable of acquiring new biofunctions in the course of its life.

From Postulates 4.1 and 4.4 it follows that some animals are in fact capable of acquiring new biofunctions in the course of their lives. (But of course the emergence of new biofunctions is only an indicator, not a proof, of the existence of plastic neuronal systems.)

Finally we make

DEFINITION 4.9 Every neural function involving a psychon (or plastic neural system) with a regular connectivity (i.e. one that is constant or else varies regularly), is said to be *learned.*

In other words, we assume learning to consist in the formation of new neural systems, i.e. in establishing permanent connections among neurons or facilitating ephemeral (but repeatable) neuron interconnections. Some such connections may be formed accidentally (at random); if valuable, they have a chance of becoming established or of recurring. These ideas go back to Ramón y Cajal and Hebb: they proposed the use-disuse hypothesis, which has been the object of much recent experiment (e.g. Moore, 1976; Rutledge, 1976) and theory (e.g. Malsburg, 1973; Cowan, 1976).

Let us proceed to refine the above ideas with the help of the concepts of state function and state space (Ch. 1, Sec. 2.2), and to introduce the special concept of mental function.

2. Brain States

2.1 *Brain Functions*

We have defined the functions of a system as the set of processes in the system, i.e. what the system does; and the specific functions of the system as those that the system, but no thing of any other kind, is capable of doing (Ch.3, Sec. 2.2, Def. 3.8). Just as there is no system without functions (processes), so there is no function without a system that does the functioning. When a neuroscientist says that function X is *mediated* or *subserved* by "structure" (neuron system) Y, he means that Y *does* X and, more precisely, that X is among the specific activities or processes of Y,

Being a multibillion-neuron system, the mammalian CNS engages in activities of a large variety, from synthesizing proteins to controlling the functions of other organs to performing higher functions such as forming a map of the surroundings. We shall not be concerned with the nonspecific or "household" CNS functions, except to note that, if anything, they are more intense than anywhere else in the body, as judged from the rates of oxygen intake and protein turnover. We shall instead focus on the specific functions of the CNS, which may be grouped into two large classes: control and cognition.

Control can be of internal systems (e.g. digestion), of external moving parts (motor), or of sensory inflow. And cognition can be of external events, of internal activity except the brain's, or of brain activity itself – in which case it is called 'consciousness'. We know from Chapter 3 that control is inherent in every biosystem and, indeed, in every biochemical system. What is peculiar to the control functions of the CNS is that they are ultimate and integrative, i.e. that there is no appealing them and that they concern the whole organism. But we shall not be concerned with the control functions of the CNS because they are nowadays of little philosophical interest. We shall instead deal with the cognitive activities or functions of the CNS.

In the first place there is the cognition of the external world, which can be direct (perceptual) or indirect (intellectual). The former shall concern us in Sec. 3.1, the latter in Sec. 4.3. Next come the various kinds of information that the brain receives from the rest of the body, and integrates and processes. Some of them, like the visual images, are rather specific and localized; others, like elation and depression, are rather nonspecific and nonlocalized. But all of them are functionally related to further bodily events: they are signalled by physiological changes outside the brain and in turn react upon them. Finally comes the brain's cognition of some of its own states, i.e. self-consciousness. We shall deal with it in Sec. 5.1.

We proceed now to explain the state space representation of CNS activity. Consider an arbitrary neuronal system, be it a small neuronal circuit (fixed or itinerant), a sizable subsystem of the CNS, or the entire brain. Like every other system, it can be represented by a state function $\mathbb{F} = \langle F_1, F_2, \ldots, F_n \rangle$. If we are interested not in the precise localization of the system and its components, but only in its overall nonspatial properties, we can assume that $\mathbb{F}$ is only a time-dependent function whose values are n-tuples of real numbers. I.e. we can set $\mathbb{F} : T \to \mathbb{R}^n$, with $T \subseteq \mathbb{R}$, and $\mathbb{R}$ the set of real numbers. (We take the host of the neural system, i.e. the whole animal, as its reference frame.)

Each component F_i of the state function $\mathbb{F}$ can be decomposed into a constant (or nearly constant) part F_i^c and a variable part F_i^v: see Figure 4.2. Obviously, either can be zero during the period of interest. However, the point is that, while the rate of change of F_i^c vanishes at all times (i.e. $\dot{F}_i^c = 0$), that of F_i^v does not. The latter may therefore be taken to represent the activity of the neuronal system in the ith respect. For this reason we make.

DEFINITION 4.10 Let $\mathbb{F} : T \to \mathbb{R}^n$ be a state function for a neural system ν, and let $\mathbb{F} = \mathbb{F}^c + \mathbb{F}^v$, with $\dot{\mathbb{F}}^c = 0$ (zero rate of change) for all $t \in T$. Then

(i) ν is *active* at time t iff $\mathbb{F}^v(t) \neq 0$;

(ii) the *intensity of the activity* of ν over the time lapse $\tau \subset T$ equals the fraction of components of ν active during τ;

(iii) the *state of activity* of ν at time t is $s = \mathbb{F}^v(t)$;

(iv) the (total) *process* (or *function*) ν is engaged in over the time interval $\tau \subset T$ is the set of states of activity of ν:

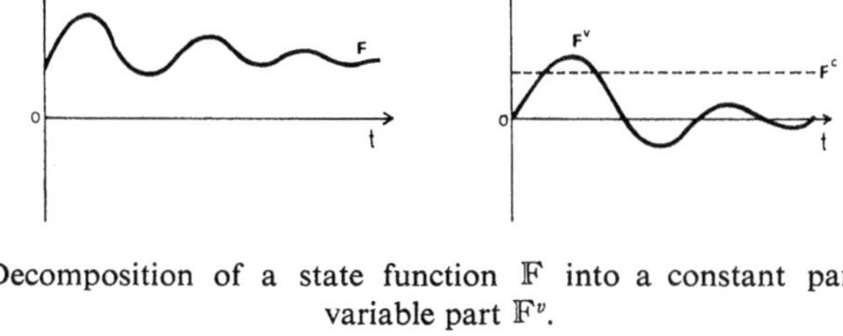

Fig. 4.2. Decomposition of a state function $\mathbb{F}$ into a constant part $\mathbb{F}^c$ and a variable part $\mathbb{F}^v$.

$$\pi(\nu, \tau) = \{\mathbb{F}^{\nu}(t) \mid t \in \tau\}.$$

Whereas single neurons may pass twice through almost the same state of activity, it is most unlikely that a neural system composed of a million neurons will do the same. We may assume that there are no two identical successive states of activity of a medium-sized neural system and, *a fortiori*, of a gross neural system such as, e.g.. the hippocampus. Since mental states will be identified with some such states of activity, it follows that no mental event ever recurs in exactly the same form.

We now introduce the concept of specific activity of a neural system as a mere application (specification) of the concept of specific function of a biosystem (Definition 3.8 (ii)):

DEFINITION 4.11 Let $\pi(\nu, \tau)$ be the total process (or function) of a neural system ν in an animal b during the time interval $\tau \subset T$. The corresponding *specific function* (*activity, process*) of ν during τ is

$$\pi_s(\nu, \tau) = \pi(\nu, \tau) - \bigcup_{\mu \lessdot b} \pi(\mu, \tau) \quad \text{with} \quad \mu \neq \nu.$$

We have postulated that every thing changes spontaneously in some respects, inductively in others (Vol. 3, Postulate 5.11). In particular, neurons are never quiescent, and the human CNS is active even when sensory stimulation stops. We shall therefore adopt the hypothesis that the CNS, and every neural subsystem of it, is constantly active even in the absence of external stimuli. To formulate this principle consider m neurons synapsing on to a given neuron i. The rate of change of the latter's state of activity can be assumed to consist of an autonomous (spontaneous) part and another proportional to the signals of the afferent fibers:

$$\dot{F}_i^{\nu}(t) = A_i(t) + \sum_{j=1}^{m} C_t(j, i)\, F_j^{\nu}(t),$$

where $C_t(j, i)$ is the strength of the $j - i$ connection (cf. Definition 4.5). This system of equations is incomplete. It must be supplemented with a system of equations for the rates of change of the connectivities. A popular hypothesis is

$$C_t(i, j) = c_{ij} \cdot \exp(- |x|/b_{ij}),$$

where the c_{ij} and b_{ij} are real numbers, and $|x|$ the distance between the units i and j. Another interesting candidate would seem to be

$$\dot{C}_t(i, j) = c_{ij}F_i^{\nu}.$$

(For some mathematical models of the dynamics of cortical systems, see Marr (1970), Wilson and Cowan (1973) and Nass and Cooper (1975).)

However, here we are interested only in the general principle:

POSTULATE 4.5 For any neural system ν of an animal, the instantaneous state of activity of ν decomposes additively into two functions: $\mathbb{F}^{\nu} = \mathbb{A} + \mathbb{E}$, where $\mathbb{A}$ does not vanish for all $t \in T$, whereas $\mathbb{E}$ depends upon the actions of other subsystems of the animal upon ν.

DEFINTION 4.12 Let $\mathbb{F}^{\nu} = \mathbb{A} + \mathbb{E}$ be the active part of the state function of a neural system ν. Then $\mathbb{A}(t)$ is the *state of spontaneous activity* of ν at time t, and $\mathbb{E}(t)$ the *state of induced* (or *stimulated*) *activity* of ν at t.

Once we know, or pretend to know, how a neural system works, we may attempt to find out how two or more neural systems combine to form a supersystem. Suppose ν_1 and ν_2 are two neural systems of a given animal, with the same number m of neurons each. Call $\mathbb{F}_1^{\nu}$ and $\mathbb{F}_2^{\nu}$ their respective active state functions. And suppose further that the neurons in ν_1 synapse on to those of ν_2, so that ν_2 has a stimulated or induced activity in addition to its spontaneous activity. The simplest assumption is of course that the induced activity $\mathbb{E}_2$ of ν_2 depends linearly upon $\mathbb{F}_1^{\nu}$, i.e. $\mathbb{E}_2 = C \cdot \mathbb{F}_1^{\nu}$, where C is an $m \times m$ matrix representing the inter-system connectivity. This connectivity is time-dependent. We assume that it strengthens with use and weakens with disuse.

If ν_1 fails to activate ν_2, the value of C will have decreased, after one time interval, to $C(t + 1) = a\ C(t)$, with $0 < a < 1$. After k time intervals, $C(t + k) = a^k\ C(t)$, which approaches zero as time goes by. If on the other hand ν_2 is activated by ν_1, the connectivity will be strengthened in proportion to the simultaneous activity of the two neural systems. I.e. the increment in connectivity will be $\Delta C(t) = b\mathbb{F}_1^{\nu} \cdot \mathbb{F}_2^{\nu}$, where b is an $m \times m$ matrix of real numbers, some positive, others negative. Hence in general the connectivity value at time t will be

$$C(t) = a\ C(t - 1) + b\ \mathbb{F}_1^{\nu}(t) \cdot \mathbb{F}_2^{\nu}(t),$$

so that the state of activity of ν_2 at t will be represented by

$$\mathbb{F}_2^{\nu}(t) = \mathbb{A}_2(t) + aC(t - 1) \cdot \mathbb{F}_1^{\nu}(t) + b(\mathbb{F}_1^{\nu}(t) \cdot \mathbb{F}_2^{\nu}(t)) \cdot \mathbb{F}_1^{\nu}(t).$$

This is sheer speculation, though one in line with current trends in neural modeling. Its philosophical interest lies in showing that it is possible to construe mental states and events as states and events of neural systems, hence to treat them mathematically. More on this anon.

2.2. *Mental States and Processes*

Every fact experienced introspectively as mental is identical with some brain activity: this, in a nutshell, is the neurobiological or materialist hypothesis of the mind. For example, vision consists in the activity of neural systems in the visual system; learning is the formation of new neural connections; and memory is the activation of neural connections.

All mental activity is brain activity but the converse is false. For example some of the activities of the cortical component of the auditory system are mental; those of the committed or wired-in components (e.g. the ear drum and the cochlea) are not mental. Only plastic neural systems are capable of learning and they are the "seats" or "neural correlates" of the mental. More precisely, we assume that the mental is the specific function of certain plastic neuronal systems. Our assumption takes the form of

DEFINITION 4.13 Let b be an animal endowed with a plastic neural system P. Then

(i) b undergoes a *mental process* (or performs a mental function) during the time interval τ iff P has a subsystem ν such that ν is engaged in a specific process during τ;

(ii) every state (or stage) in a mental process of b is a *mental state* of b.

For example, acts of will are presumably specific activities of neuron modules in the forebrain. On the other hand hunger, thirst, fear, rage, and sexual urge are processes in subcortical systems (mainly hypothalamic and limbic), hence are nonmental according to our definition. What is a mental process is the consciousness of any such states – consciousness which is a process in some subsystem of P.

The following are immediate consequences of four previous definition:

COROLLARY 4.1. All and only animals endowed with plastic neural systems are capable of being in mental states (or undergoing mental processes).

COROLLARY 4.2 All mental disorders are neural disorders.

Consequently, the division between neurology and psychiatry, suggested by psychophysical dualism, is unreasonable.

COROLLARY 4.3 Mental functions (processes) cease with the death of the corresponding neural systems.

Examples Brain injuries, strokes, and of course death bring about the cessation of the mental functions normally discharged by the affected neural systems. Dualism does not explain such a "correlation".

COROLLARY 4.4 Mental functions (processes) cannot be directly transferred (i.e. without any physical channels) from one brain to another.

Consequently extrasensory perception is out of the question. (The Montreal magician Henry Morgan suggests that it should be renamed 'extrasensory deception'.)

We come now to the central concept in the philosophy of mind:

DEFINITION 4.14. Let P be the plastic (uncommitted) supersystem of an animal b of species K. Then

(i) the *mind* of b during the period τ is the union of all the mental processes (functions) that components of P engage in during τ:

$$m(b, \tau) = \bigcup_{x \prec P} \pi_s(x, \tau);$$

(ii) the *K-mind*, or *mind of species* K, during period τ, is the union of the minds of its members during τ:

$$M(K, \tau) = \bigcup_{x \in K} m(x, \tau).$$

Since the members of the set called 'mind' are brain functions (processes), it makes no sense to say that the brain is the physical "basis" of the mind. And since the human mind is nothing but the union of all the individual human minds, it makes no sense to speak of the collective mind of mankind as if it were an entity or even a functional system. On the other hand the mind of an individual animal does have functional unity: it is a functional system (Definition 1.9 in Ch. 1, Sec. 1.7), for every process (function) in it is lawfully related to at least one other process (function) in it. The "neural basis" of the unity of the mind, on which many philosophers have insisted, is this: the components of the plastic neural supersystem P, far from being uncoupled, form a system by Postulate 4.3. Hence

COROLLARY 4.5. The mental functions of (processes in) the plastic neural supersystem of an animal are coupled to one another, i.e. they form a functional system. (In short, for every animal b and every period τ of its existence, $m(b, \tau)$, if nonvoid, is a functional system.)

Consequently when a neurosurgeon cuts a living primate brain into two, he splits the plastic system into two, and thus also the corresponding mind of the animal. (See Corollary 4.7.)

Our last convention in this section is

DEFINITION 4.15. Let x be an object and b an animal endowed with a plastic neural system. Then

(i) x *is in the mind* of b iff x is a mental state or a mental process of b (i.e. if b has a plastic neural system engaged in a specific process containing x as a member or as a subset);

(ii) x *is in the mind*(or *is mental*) iff there is at least one animal y such that x is in the mind of y.

Remark 1 A mental state, or state of mind, such as one of persistent pain, or of perceiving a stationary thing, may be roughly constant. However, any such mental state is a state of activity of some neural system: no such activity, no mentation. In other words, mental states are more like states of motion than like static states such as the state of a gas in a rigid enclosure. *Remark 2* We have stated that mental states are neural states (or rather processes), not that every neural state has a mental "correlate". Although different mental states are different neural states (or rather processes), the converse is not true: a given mental state could be a process now in one itinerant neural system, now in another. Surely there are bound to be differences: even when doing routine sums, for all we know we may engage a different itinerant neural system each time – but we won't notice the difference introspectively (Bindra, 1976). (That is, calling N the set of neural states that happen to be mental, and M the set of mental states distinguishable introspectively, $M = N/\sim$, where $\sim$ is the relation of introspective equivalence.) *Remark 3* We have not characterized mental states independently of brain states, as dualists would like us to do. We have not done so for the following reasons: (*a*) mentalist predicates, such as 'sees red' and 'thinks hard', though indispensable, are coarse and vulgar rather than exact and scientific; (*b*) the whole point of the neurophysiological approach to mind is to get away from the ordinary knowledge approach and render mind accessible to science; (*c*) if mental events are characterized independently of brain events, then the identity theory turns out to be idle or false (Brandt and Kim, 1967). *Remark 4* The dualist claim that the mental must be nonphysical because it cannot be described with the help of physical concepts, such as those of pressure and conductivity, misses the point. Firstly, although such events as nightmares and feeling a sour taste are not describable in strictly physical terms, they are, at least in principle, describable in neurophysiological terms. Secondly, in ordinary life we may never care to describe mental states in neural terms, just as we do not care to describe such ordinary physical events as a car crash or the burning of a light bulb with the paraphernalia of physical theory. It would not be worth our while to do so; we are satisfied knowing that, if need be, such events, or some of their features, could be accounted for in scientific terms. *Remark 5 Pace* Armstrong (1968), mind and brain are not identical: there is no more brain-mind identity than there is lung-respiration identity. In our version of the so-called identity theory, the set of mental states is a subset of the states of the plastic neural system of the animal: see Figure 4.3.

2.3. *Psychosomatic Interaction*

The interactionist variety of dualism, from Descartes (1649) to Popper and Eccles (1977), holds that matter and mind, though heterogeneous and separate, act upon one another. This doctrine is untenable if only because

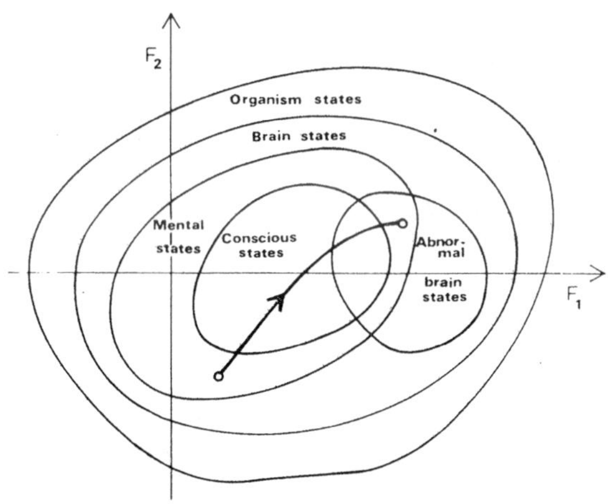

Fig. 4.3. Projection of the state space of a human being. Both axes represent neurophysiological properties; one of them is identical with a psychological property. The arc of curve represents a mental process, such as composing a melody, which is only partly conscious. One part of it is abnormal, e.g. consists of auditory hallucinations.

interactions are well defined only for concrete things (Vol. 3, Ch. 5, Def. 5.30). To speak of mind-body interaction is just as mistaken as to speak of shape-body, or property-thing, or composition-system, or behavior-animal interactions. However, it does make sense to speak of mental-bodily interactions provided this expression is taken to abbreviate "interactions among plastic (uncommitted) neural systems, on the one hand, and committed neural systems or bodily systems that are not part of the CNS, on the other". *Example* The interactions between cortical and subcortical regions of the CNS, between sensory and motor areas, between ideational neural systems and external receptors, between the visual cortex and the pituitary, etc.

Paradoxically enough, while the dualist has no right to claim that mental events influence nonmental bodily events (because he has no clear idea of such influence), the monist is entitled to this opinion. Indeed, since mental events are for him neural events of a kind, they can influence and even cause other events in any subsystem of the same body because of the integrative action of the CNS. In short, because mental events are neural events, and the causal relation is defined for pairs of events in concrete things (Vol. 3, Ch. 6, Def. 6.23), we have

COROLLARY 4.6. Mental events can cause nonmental events in the same body and conversely.

Consequently disturbances of nonmental (e.g. metabolic) biofunctions may influence mental states and, conversely, mental events, such as acts of will, may influence nonmental bodily states. This is what neurology, neurochemistry, psychiatry, education, and advertising are all about. *Example 1* Yogis learn to regulate their heartbeats, intestinal contractions and even rate of oxygen intake (cf. Miller, 1969). *Example 2* A blow in the head can "knock out the mind" erasing all memory of events immediately preceding the accident. *Example 3* By implanting electrodes in selected sites of the brain, and sending electric currents through them, the neurophysiologist can control some mental states and, through them, elicit or block behavior such as rage (cf. Delgado, 1969).

The most dramatic and revealing of all surgical interventions is of course the sectioning of the bridge between the two cerebral hemispheres, or corpus callosum (Sperry, 1964, 1966, 1977; Gazzaniga, 1967; Bogen, 1969). The split-brain patient behaves in many situations as if he were in two minds – as forecast by Fechner in 1860. For example, while his left hemisphere may wish to read, his right hemisphere may wish to go for a walk, and so a conflict arises within one and the same skull. This is not surprising in our framework: indeed, according to Postulate 4.3 and Corollary 4.4, there is a single mind as long as there is a single plastic supersystem. But if the latter is cut into two then two plastic systems emerge, each with its own mind or system of mental functions:

COROLLARY 4.7. Let b be an animal whose plastic nervous system is split into two detached parts, L and R. Then the mind of b during any time period τ posterior to the splitting divides into two disjoint functional systems:

$$m(b, \tau) = m_L(b, \tau) \cup m_R(b, \tau), \quad \text{with} \quad m_L(b, \tau) \cap m_R(b, \tau) = \varnothing.$$

Hence, *pace* Puccetti (1977), the results of commissurotomy do not support dualism. On the contrary, they show conclusively that no immaterial mind holds the hemispheres together. Dualism could be saved only by invoking divine intervention at the time of splitting the corpus callosum. God would have to replace the original single soul by two new, independent souls – each of which would have to remember something of the past history of the original soul. This looks like an easy solution but it is not for it poses the grave theological problems of which, if any, of the three souls deserves being saved, and what to do on Resurrection Day with three immortal souls and one body. Even assuming that this theological difficulty could be resolved, the dualist is in an awkward position. If he denies that the split brain patient has two minds, he defies experience. And if he acknowledges that a physical entity such as a surgeon's scalpel can slice a mind into two, he contradicts himself (Bunge and Llinás, 1978).

The chemical manipulation of the mind, though less spectacular, is no less clear cut an indictment of dualism. All drugs that act on mental states, from tea to cocaine, do so by modifying the CNS metabolism in some way or other, from changing properties of the neuron membrane to taking part in biochemical reactions. Thus by modifying membrane conduction this drug produces amnesia and that one facilitates recall; and by blocking the synthesis of a certain neurotransmitter this drug produces schizophrenia and that one cures it. (For a list of kinds of deliberate modifications of brain and behavior, see Omenn and Motulsky (1972).)

Finally, the purely behavioral or psychological manipulation of the mind is no less physical, because the sensory inflow and the behavioral output controlled by the teacher (or preacher, propagandist or psychiatrist) modulates neural connectivity, reinforcing certain connections and inhibiting others. In sum, the mental can be controlled, nay is being controlled, in various ways precisely because it is not immaterial: if it were, it would be immune to all attempts at manipulating it.

What holds for the physical control of the mind holds also, *mutatis mutandis*, for the mental control of the body, so often cited as proof of the superiority of Spirit over Matter. Consider three well known phenomena: hunger, anger, and yoga. The process involving the hunger feeling is normally this. Sensors in the digestive tract send signals to the hypothalamus, which in turn alerts the cortex. The motor cortex and the cerebellum control the muscales intervening in the search and reach for food, and in eating. The dualist thinks he can afford to skip all these links: he discards the process ensuing in the cortex's wanting the body to eat, and speaks of the direct action of the mind on the muscles. His account is far too simple – and unscientific. Saying that X ate because he *meant* to eat or had the *intention* of eating is just redescribing an ordinary situation in ordinary language terms, not giving an explanation of the process. And yet such accounts are as popular with ordinary language philosophers (e.g. Alston, 1974) as the account of behavior in terms of instincts was among ethologists half a century ago.

Consider anger next. The dualist is satisfied with "explanations" like this one: "She kicked him because she was angered by his words". The physiological psychologist, on the other hand, analyzes this ordinary knowledge account as a causal chain of bodily events: Hearing an utterance → Activation of psychons in the forebrain and of the adrenals → Action of forebrain and adrenals on motor center → Movement of limb. Likewise, when confronted with yoga exercises and biofeedback techniques for moving things without the intervention of any muscles, the physiological psychologist concludes that the autonomic nervous system is not as fully autonomous as we used to think, but can be subjected at least in part to control by cortical systems. There is no evidence for the action of an immaterial mind on the body and there can be no such evidence, for the ghostly is inaccessible to science. Epicurus knew this twenty-two centuries ago.

2.4. *Location of the Mind*

If the mind is immaterial then it makes no sense to ask where it is. The *prima facie* evidence for the nonspatiality of the mind is of this kind: when one thinks about something he does not feel his thoughts to be located anywhere. So much so, that in the past the heart and the liver were assumed to be the "seats" of the mind. However, as long as we remain on the ordinary knowledge level we cannot refrain from citing evidence for the thesis that the mind is coextensive with the body: e.g. when your foot hurts you feel pain there, not in your brain. So, ordinary experience is ambiguous: it supports now the thesis of the nonspatiality of the mental, now the thesis that the mind can wander throughout the body. This suggests that the nonspatiality thesis is confused, and that introspective evidence is insufficient or even irrelevant. We should examine both.

The thesis of the nonspatiality of the mental contains two different hypotheses. One is that mental events are *unextended*, the other that they occur *nowhere* – except of course "in" the mind, which is allegedly immaterial and therefore unextended. The latter hypothesis implies the former but not conversely. Indeed, mental events might be unextended yet occur somewhere – say in the brain. Likewise a physical event that takes no time occurs at some time or other. In our ontology there are no events in themselves but only events in some thing or other (Vol. 3, Ch. 5). Hence the question of the space "occupied" by an event is the problem of the extension of the changing thing. Thus not the *firing* of a neuron but the *firing neuron* is spatially extended.

Events occur wherever the "eventing" (changing) things may be (Vol. 3, Ch. 6, Sec. 4.2). In particular, mental events occur in some plastic neural system or other. So, in principle and often in practice, the neurophysiologist can locate a thought process, e.g. by eliciting it with electrical stimulation (Penfield and Rasmussen, 1950). Hence the ordinary language philosopher's claim that "it makes no sense at all" to talk about mental states and events as being located somewhere in the body has been experimentally refuted. Mental events are just as localized as the corresponding plastic neural systems. If the latter have a fixed location, so have the mental events they engage in. And if the neural system in question happens to be itinerant, the mental event itself will be itinerant – just as much as a light wave. In short, mental events occur in the brain. However, the mind, conceived of as the *set* of all mental events (Definition 4.14), is nowhere.

What about pain in limb, particularly a limb that has been amputated (phantom limb)? The short answer is that, just as we see stars in the sky, not in our brain, so we locate pains in several parts of the body, and sometimes either localization is wrong. A long answer is that the cortex of the adult human being has a map of the entire body (the somatotopic representation). It is here that the signals coming from various parts of the body are decoded and located. If they come from the stump of an amputated limb, the patient can feel them as originating in the missing part because he has formed since childhood a comparatively unalterable map. (There are alternative neurophysiological explanations of phantom experience.)

In sum, mental processes are located wherever the corresponding psychons are located; likewise they occur in time and take some time to happen. This generalization from contemporary neurophysiology – foreshadowed by the Hippocratic school – comes as no surprise for the materialist, for whom the *res cogitans* is a *res extensa*, namely the brain.

What about ideas: are they too in spacetime? The answer to this ambiguous question depends on the construal of 'idea'. If taken as ideation processes, ideas *are* in the brains that happen to think them up – and only there and at the time they are being thought. On the other hand the so called *product* of any such process, i.e. the idea in itself, is nowhere in spacetime because it does not exist by itself: we only feign it does. For example, although thinking of the number 3 is a brain process, hence one located in spacetime, the number 3 is nowhere because it is a fiction existing by convention or fiat, and this pretence does not include the property of spatiotemporality. What holds for the number 3 holds for every other idea – concept, proposition, or theory. In every case we abstract from the neurophysiological properties of the concrete ideation process and come up with a construct that, by convention, has only conceptual or ideal properties. (Cf. Vol. 1, Ch. 1.)

To sum up: Whereas the mental is in some head or other, ideas in themselves, having no being of their own, are nowhere.

2.5. *Mentalist Predicates*

At first blush the most formidable objection to psychophysical monism is this: "The materialist claims that all mental events are brain events, but he cannot offer a clear description, let alone a definition, of a mental event without the help of mentalist concepts such as those of self, privacy, and immediate accessibility. Consequently he cannot even state his identity thesis. In other words the materialist cannot consistently state the identity of phenomenal and physical predicates because he refuses to buy the former at their face value".

Actually the above is not an objection specifically addressed to the psychoneural identity thesis but one that could be raised against all science conceived of as a cognitive enterprise reaching beyond the ordinary language description of appearances. Our rejoinder is this:

(i) Science attempts to account for reality behind appearance, so it either does not employ phenomenal predicates or, if it does, regards them as derived not as basic. Surely phenomena, in particular mental phenomena, are experientially immediate, but they are neither ontologically nor scientifically primary: they are something to be explained.

(ii) Appearances (phenomena) can be explained by psychology, at least in principle, in strictly nonphenomenal terms. For example the various visual illusions can be explained either as results of incomplete or inadequate information, or as results of the fatigue of certain neurons, or the faulty connection of others.

(iii) A rigorous formulation of the monistic thesis must employ no ordinary phenomenal or mentalist predicates – or fuzzy expressions such as 'mental-neural correlation'. (At any rate ours does not.) The identity thesis is that every mental process is a brain process, not that every mentalist sentence is identical with some neurophysiological sentence: the identity is ontological not linguistic. (This important point has been obscured by the nonmaterialist version of the identity theory defended by Schlick (1925) and Feigl (1958).)

In other words, we must not demand that science descend to the level of common sense but, instead, strive to raise the latter to the level of the former. After all, this is what we do when shopping for breakfast cereal and, instead of allowing ourselves to be goaded by the advertisement that "X peps you up", we look at the protein, vitamin, and calorie content of X – or when shopping for a car and, rather than remain satisfied with the assurance that "X is spirited", we inquire into the acceleration of X. We should behave similarly with regard to mentalist predicates. More precisely, we should adopt

RULE 4.1 Whenever possible dispense with mentalist predicates: either replace them by, or define or deduce them with the help of, neurophysiological predicates.

Example of elimination "The idea of an immaterial mind controlling the body is vitalism, no more, no less; it has no place in science" (Hebb, 1974). *Example of reduction* The degree of introversion of a person equals the activity of his or her frontal-medial-septal-area hippocampal system (Gray, 1972a).

Rule 4.1 invites either of the following strategies with regard to mentalist predicates: elimination, definition, or deduction with the help of neurophysiological concepts. Of these the second, i.e. definition (or identification), has been commended by the so-called identity theorists (e.g, Smart, 1959). That is, they have proposed that, for any mentalist predicate M, there is a neurophysiological predicate N such that $M = N$, where $=$ is the relation of ordinary (or strict) identity. Ordinary language philosophers have claimed that this cannot be, for "it makes no sense" (Malcolm, 1964). The obvious rejoinder is that no scientific hypothesis whatsoever makes sense to them, for they insist on remaining within the confines of ordinary language, hence ordinary knowledge.

Another, related objection to the identity thesis, and one commended as definitive by Popper (Popper and Eccles, 1977), is as follows (Kripke, 1971). If the identity is to be taken just as seriously as "Heat is the motion of molecules", then it must be a necessary identity, in the sense that it must hold in all possible worlds – whatever these may be. However, this cannot be, because "it seems clearly possible" (to Kripke) that M (e.g. pain) could exist without the corresponding brain state N, or that the latter could exist without being felt as pain – whence the identity is contingent and therefore flimsy. Rejoinder: (*a*) scientists and science-oriented philosophers do not waste their time speculating about (logically) possible worlds: they want to explore the real world (Bunge, 1977a); (*b*) the difference between necessary (or strict) identity and contingent identity does not occur in ordinary logic or in science. In sum, the sophistic objection to materialism holds no water.

Surely our Rule 4.1 is not a summary of the state of the art but rather a programmatic injunction, i.e. one that can guide research. As such it can be confirmed. And it should be discarded (or simply forgotten) if found barren. But, far from barren, it is behind many a success in contemporary psychology and neuroscience.

Note that Rule 4.1 is the *weak* not the strong reductionist thesis: we favor the *partial* reduction (definition or deduction) of the mental to the neurophysiological, not the elimination of its emergent properties (cf. Bunge, 1977b, 1977c). Thus we do not state that a frog enjoys listening to a Beethoven quartet, or that what a human feels when listening to it is no different from a frog's sighting a fly. We only say that enjoying Beethoven is a process in a human auditory system educated to face such experience.

The partial reducibility (definability or deducibility) of mentalist predicates does not entail that *every* ordinary language sentence concerning mental events is translatable without further ado into a (possibly very complex) scientific formula. Nor, *a fortiori*, does it entail that every mentalist sentence is identical with some neurophysiological sentence. If this were so then there would be a difference only in degree between the scientific and the mythical accounts of the mental. In short, genuine psychoneural monism (unlike the linguistic "identity theory") does *not* assert that, given any mentalist sentence m, there is a neurophysiological sentence n such that $m = n$.

The best one can do with mentalist sentences is either to eliminate them altogether or to try and cleanse, deepen, and refine them. For example, the phrase 'The soul survives', which makes sense in some theologies, makes none in our framework, where *soul* does not occur at all and *mind* is defined as a set of neural processes. (Sets are neither alive nor dead.) And those ordinary language mentalist sentences that are transtatable into neurophysiologese – e.g. 'Her mind is at work', or 'I am having a yellow afterimage' – may be so only roughly. As for the converse translation, of neurophysiological and psychophysiological sentences into ordinary language (mentalist) sentences, it is, more often than not, impossible, as should be obvious from looking at any of the mathematical models of mental processes published over the past few years.

In short, the various languages employed to describe mental events – in particular mentalese, behaviorese and neurophysiologese – are not inter-translatable on the whole. This is because, save exceptions, their sentences do not express the same propositions. For example, 'She is happy', 'She is grinning', and 'Her pleasure center is active' are, though related, quite different. In fact, although they have the same referent (she), and each describes one aspect of one and the same process, they have different senses and therefore are inequivalent.

The inequivalence of most mentalist and neurobiological propositions is no obstacle to the building of bridges between them. The entire science of psychophysics is an attempt to disclose some such bridges, e.g. the relation between sound intensity and felt loudness. (Psychophysical laws are of the form: "For all x, if x is an animal of species K, then: If physical stimulus p impinges on x, then x feels mental phenomenon m". Such generalizations may eventually be deepened by expressing m in neurophysiological terms.) Neurology, too, looks for bridges between the mental (and also behavioral) states reported by neurological patients in phenomenal language (e.g. 'I suffer from migraine'), on the one hand, and sentences describing the corresponding malfunction of the CNS on the other.

(Neurological laws are of the form: "For all x, if x is human, then: If x is in neurological condition n, then x feels mental phenomenon m".)

So much for the general and basic concepts and assumptions of our version of the psychobiological theory of the mental. We proceed now to examining a few particular and rather typical kinds of mental process.

3. Sensation to Valuation

3.1. *Sensing and Perceiving*

All things react to external stimuli, but some react more selectively than others: they are said to detect them. The general concept of specific reaction, or detection, is elucidated by

DEFINITION 4.16 A system *detects* things or events of a certain kind (or is a *detector* of them) if and only if it reacts to them only.

Multicellular organisms have a variety of detectors grouped into systems called *selective systems* (or also, misleadingly, *recognition systems*). For example, the immune system can detect a large variety of molecules because it is composed of detectors (antibodies) of many kinds. And the CNS can detect a huge variety of internal and external events, though by no means all of them.

For an animal it is not enough to possess detectors: the animal must be able to do something about the events it detects. This necessitates amplification and transmission of signals in a uniform fashion, so that each signal can be "read" unambiguously by the CNS. This is what neurosensors or neuroreceptors do. A mammal has myriads of neurosensors: for motion, pressure, heat, rate of secretion, acidity changes, muscle contraction, novelty, etc. This is the general concept:

DEFINITION 4.17 A detector is a *neurosensor* (or *neuroreceptor*) iff it is a neural system or is directly coupled to a neural system.

Example The visual cortices of cat and monkey (and probably also of man) contain neurons specializing in detecting vertical lines and others horizontal lines (Hubel and Wiesel, 1959, 1962). These are called *feature detectors*.

In higher animals neurosensors come in systems, so we need

DEFINITION 4.18 A *sensory system* of an animal is a subsystem of the nervous system of it, composed of neurosensors and of neural systems coupled to these.

The sensory systems of the higher vertebrates are extremely complex systems involving not only sense organs but also portions of the cerebral cortex. Moreover they are closely linked to other subsystems of the CNS, particularly the motor system. And they are not just detectors of environmental inputs, like the thermocouple or the photoelectric cell. Indeed the sensory systems are under the constant action of the CNS by way of the gamma afferent fibers, so that their state depends not only upon external stimulation but also on the state (and history) of the CNS. This explains why we seldom perceive the same stimulus in the same manner. And it shows that (*a*) stimulus-response psychology is basically false, and (*b*) any model of perception contrived in imitation of purely physical or chemical detectors, such as photographic cameras, is bound to fail.

Next comes

DEFINITION 4.19 A *sensation* (or *sensory process*, or *feeling*) is a specific state of activity (or function or process) of a sensory system.

Feeling cold or hot, hungry or sated, tired or energetic, pain or pleasure, and so on, are functions of so many sensory systems. All sensations (or feelings) are then in the brain even though they are usually elicited outside it. Hence sensations can be either blunted or made more acute by manipulating the brain, be it electrically, chemically, or behaviorally. If they were in an immaterial mind we would have no access to them from the outside.

Sensing is not all: in vertebrates sensory information is processed or "interpreted" by special sensory "areas" in the brain cortex. In the primate brain each such sensory cortical "area" can be subdivided into three parts: the primary, the secondary, and the tertiary ones. Only the primary sensory cortical area should be regarded as a component of the corresponding sensory system. Presumably it loses plasticity as the animal develops. On the other hand the secondary and tertiary sensory areas seem to be plastic throughout the animal's life. We shall call them the *plastic* neural systems *directly* attached to the sensory system. We shall assume with Hebb (1968) that, whereas sensation is the specific activity of a sensory system including the primary sensory cortical area, perception is this activity together with the activity elicited by the sensory system in the plastic neural system directly attached to it. See Figure 4.4.

This figure summarizes the following features of human perception:

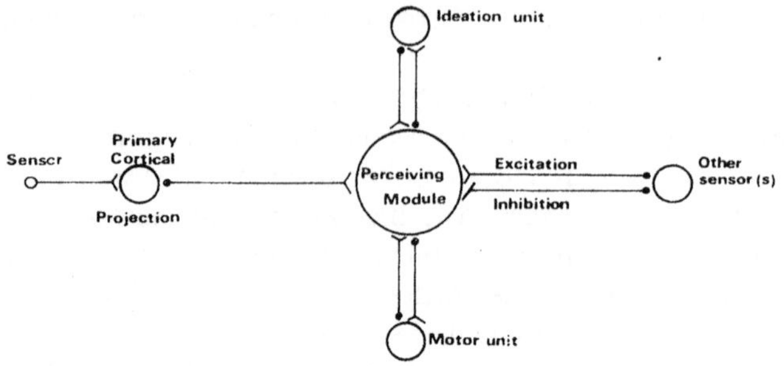

Fig. 4.4. Schema of a perceptual system: a modification of Bindra's (1976), in turn a modification of Hebb's (1949).

(i) the perceiving is done by a neural system located in a secondary sensory projection of the brain cortex;

(ii) the functioning of the central perceiving unit is strongly influenced by some motor unit or other as well as by ideational units and by sensory inflows of various modalities;

(iii) the central perceiving unit can be activated by any other units coupled to it (e.g. dreams, hallucinations, phantom limb experiences, eidetic images);

(iv) perception can guide motion as well as ideation.

We compress the preceding into

DEFINITION 4.20 (i) A *percept* (or *perceptual process*) is a specific function (activity, process) of a sensory system and of the plastic neural systems(s) directly coupled to it;

(ii) a *perceptual system* is a neural system that can undergo perceptual processes.

We assume – in line with Bindra (1976) – that, whereas a sensory system has a roughly constant neuronal composition, the corresponding perceptual system has a variable part. This explains why a given stimulus can activate different perceptions (as in the case of the duck-rabbit figure or of the Necker cube) and why different stimuli can give rise to the same perception (as in the case of size constancy). See Figure 4.5. In other words, two percepts are *equivalent* iff they are elicited by the same sensation and involve the same plastic neural system.

We shall assume that the perception of an external object is the distor-

Fig. 4.5. Sensation and perception. (*a*) Same sensation elicits different perceptions according to the plastic neural assembly it activates. (*b*) Different sensations give rise to same perception by activating the same plastic neural system.

tion it causes on the ongoing activity of a perceptual system. To be more precise, we assume

POSTULATE 4.6 Let ν be a perceptual system of an animal b, and call $\pi_s(\nu, \tau) = \{\mathbb{F}^\nu(t) \mid t \in \tau\}$ the specific process (or function) that ν engages in during the period τ when in the presence of a thing x external to ν, and call $\pi_s^\circ(\nu, \tau)$ the specific function of ν during the same length of time when x fails to act on ν. Then b perceives x as the symmetric difference between the two processes. I.e. the perception of x by b during τ is the process

$$p(x, \tau) = \pi_s(\nu, \tau) \Delta \pi_s^\circ(\nu, \tau).$$

This postulate accounts for the active or creative nature of perception and thus for the fact that one and the same perceptual system, when in different states, will perceive differently a given object. This hypothesis disagrees with the direct apprehension doctrine of empiricists and intuitionists (in particular Gestalt theorists). But it is consonant with the current neurophysiology of perception, particularly because it incorporates the spontaneity of the CNS denied by the causal theory of perception. External and internal stimuli do not set the CNS in motion but modulate or control its incessant activity. I.e. the environment enhances or dampens, and in general controls or modulates the activity of the CNS rather than causing it.

3.2. *Mapping Body and Environment*

What do we perceive and how? The naive realist's answer is: We perceive things as they are. Yet imagine Parmenides' homogeneous and eventless universe. An observer would be unable to perceive anything in it, because there would be nothing to discriminate from anything else, and because nothing would be stirring. He would be unable to see, because only things that bounce photons back can be seen; he would hear nothing, because only moving things can generate sound waves; and he would have no tactile sensations, because touching is disturbing – and being disturbed. The observer himself must be changeable, because perception is a process in an animal equipped with perceptual systems.

So, we perceive not things but events, and not all events but only some of those that affect us: the light bouncing off the surface of the lake and hitting our retina, the cry of the owl flying nearby and generating sound waves that set our ear drums in motion, the flame licking our hand, and so on. Whatever we perceive is an event or a sequence of events, and not just any event but one originating in a neurosensor or acting on the latter and, in any case, belonging to our own event space (or the set of events occurring in us). And our perceptions are in turn events in the plastic part of our own sensory cortex. They are not autonomous events but events that map or represent events occurring in other parts of the body or in our environment. Surely this mapping is anything but simple and faithful, but it is a mapping nevertheless. Thus we assume

POSTULATE 4.7 Let b be an animal equipped with a perceptual system c, and call $S(b)$ the state space of b, and $S(c)$ that of c. Moreover, let $E(b) \subset S(b) \times S(b)$ be the animal's event space, and $E(c) \subset S(c) \times S(c)$ its perceptual event space. Then there is a set of injections (one to one and into functions) from the set of bodily events in b to the set of perceptual events in c. Each map, called a *body schema*, depends on the kind of bodily events as well as on the state the animal is in. I.e. the general form of each body map is

$$m: S(b) \times 2^{E(b)} \longrightarrow 2^{E(c)},$$

where 2^X is the family of all subsets of X.

Remark 1 This assumption does not specify the body schemata and does not even state how many there are. Such specifications are tasks for psychobiology. *Remark 2* At first blush phantom limb experiences refute the postulate. They do not, because the body schemata are learned: young children do not have that experience. *Remark 3* The bodily events $x \in 2^{E(b)}$ can be said to be felt by animal b in state s iff b has a body schema m such that $m(s, x) \in 2^{E(c)}$ – i.e. if those events project on to the cortex c of b.

The external world maps are similar to the body schemata. Only, in this case the bodily events are caused by external events, and the maps are far more numerous than the body schemata. In fact there is an entire atlas of the external world, made up of dozens of maps – one for shapes, another for colors, another for motion, and so on. Our assumption is

POSTULATE 4.8 Let $E(e)$ be a set of events in the environment e of an animal b equipped with a perceptual system c, and call $S(b)$ the state space of b and $S(c)$ that of c. Moreover let $E(b) \subset S(b) \times S(b)$ be the animal's event space, and $E(c) \subset S(c) \times S(c)$ its perceptual event space. Then there is a set of partial maps k from sets of external events in $E(e)$ to ordered pairs ⟨state of b, set of bodily events in b⟩, and another set of partial maps p, from the latter set to sets of perceptual events. Furthermore the two sets of maps are equally numerous, and each map k composes with one map p to form an *external world map* of b in e, or ε. I.e.

$$\varepsilon: 2^{E(e)} \xrightarrow{k} S(b) \times 2^{E(b)} \xrightarrow{p} 2^{E(c)}.$$

DEFINITION 4.21 Let b be an animal with perceptual system c in environment e. Moreover call $S(b)$ the state space of b and $E(e)$ that of e. Then b, when in state $s \in S(b)$, *perceives* external events in $x \in 2^{E(e)}$ if and only if [these cause bodily events that are in turn projected on to the sensory cortex c, i.e. if] $k(x) = \langle s, y \rangle$ with $y \in 2^{E(b)}$ and in turn $p(s, y) \in 2^{E(c)}$. Otherwise the events in x are *imperceptible* to b when in state s [i.e. imperceptible events either do not cause any bodily events or cause them but do not get projected on to the perceptual system].

Remark 1 Just like the body schemata, the external world maps are not point to point but rather set to set correspondences. *Remark 2* There are many external world maps, several for each modality. The various visual maps integrate into one, which constitutes the visual space. Likewise the auditory maps integrate into the auditory space, and so on. *Remark 3* The various sets of external world maps, one for each modality, integrate in turn into the external world atlas. This integration can be explained as the simultaneous activation of the various perceptual systems. Such an activation can be achieved by sensory inflow to one or more receptors, and activation of other perceptual systems via associated neural systems. For example, a total or crossmodal perception of a familiar hand may be had just by its haptic perception, because the tactual receptors activate others. Such integration is learned. *Remark 4* Not only does the organism learn to integrate the activities of its various perceptual systems: it also learns to perceive in each modality. In particular, the young vertebrate learns to touch, to hear, to see, and to smell. Here again the psychobiological viewpoint can teach us something that dualism cannot. In fact from a neurobiological point of view learning to perceive is a process of self-organization of the synaptic junctions in the sensory cortex.

3.3 *Behavior: Definitions and Principles*

The set of motor outputs of an animal, whether global as in locomotion or partial as in grasping or grinning, moving the eyeballs or excreting, is called its behavior. In short, we make

DEFINITION 4.22 For any animal b,

(i) the *behavioral state* of b at time t is the state of motion of b at t;

(ii) the *behavior* of b during the time interval τ is the set of all behavioral states of b throughout τ.

When inherited, or else when learned and successful, behavior can become stable or recurrent. In this case it is often called a behavior pattern:

DEFINITION 4.23 A *behavior pattern* is recurrent behavior.

Animals belonging to different species can do different things, such as swimming of flying, burrowing or building nests, preying or fleeing from predators. And even when they do the same 'thing' (i.e. go through the same motions) they may do so in different fashions or styles. In short there are different kinds of behavior, and every animal species is characterized by some to the exclusion of others. We need then

DEFINITION 4.24 Let b be an animal of species K, and let A be the union of all animal species. Then

(i) the (*possible*) *behavior of type i of animal b*, or $B_i(b)$ for short, is the set of all (possible) behaviors of b associated with the ith biofunction (in particular neural biofunction) of b;

(ii) the *behavioral repertoire* of animal b, or $B(b)$ for short, is the union of all (possible) behavior types of b, i.e.

$$B(b) = \bigcup_{i=1}^{n} B_i(b);$$

(iii) the (*possible*) *behavior of type i of species K*, or $B_i(K)$ for short, is the union of all the (possible) behaviors of the members of K, i.e.

$$B_i(K) = \bigcup_{x \in K} B_i(x);$$

(iv) the *behavioral repertoire of species K*, or $B(K)$, is the union of all (possible) behavior types of K:

$$B(K) = \bigcup_{i=1}^{n} B_i(K);$$

(v) the *specific behavioral repertoire of species K* is the behavioral repertoire exclusive to members of K:

$$B_s(K) = B(K) - \bigcup_{X \subset A} B(X), \quad \text{with} \quad X \neq K;$$

(vi) *animal behavior* is the union of the behavioral repertoires of all animal species, i.e.

$$B = \bigcup_{X \subset A} B(X).$$

(All of the above definitions are understood to refer to a given stage in the development of the individual or the evolution of the species. As an animal develops its behavioral repertoire changes. Something similar holds for the evolution of behavior.)

Our first assumption is that behavior, far from being primary, is derivative:

POSTULATE 4.9 The behavior of every animal endowed with a nervous system is controlled ("mediated", "subserved") by the latter. I.e. for every behavior type B_i of animals endowed with a nervous system, the latter contains a neural subsystem that controls the motions in B_i.

An immediate consequence is

COROLLARY 4.8 Any change in (nonredundant) neural systems is followed by some behavioral changes.

Now, there are no two identical nervous systems – nor, for that matter, two physical systems identical in all respects. This generalization, together with Postulate 4.9, entails

THEOREM 4.1. No two animals behave in exactly the same way [even if they are conspecific].

The assumption that a nervous system may contain a plastic subsystem (Postulate 4.1), together with Postulate 4.9 and Definition 4.24, entails

THEOREM 4.2. The behavioral repertoire of an animal endowed with plastic neural systems splits into two parts: the one controlled by the committed (or prewired) part of the NS of the animal, and its complement, i.e. the behaviors controlled by the plastic components of the NS.

DEFINITION 4.25. The part of the behavioral repertoire of an animal, that is controlled by the committed part of its NS, is called its *inherited* (or *instinctive, stereotyped, modal* or *rigid*) repertoire, while the one controlled by the plastic part of its NS, its *learned* repertoire.

Remark 1 'Inherited' is not the same as 'inborn'. Only the adult animal is in possession of its full inherited behavioral repertoire. (That is, *Inborn* ⊂ *Inherited* ⊂ *Total.*) This distinction was not adopted by Chomsky (1968) when claiming that universal grammar is an "innate schematism". *Remark 2* Because natural selection eliminates most types of behavior resulting in backfiring, instinctive or programmed behavior looks marvelously suited to goals, i.e. purposeful. But, as Spinoza (1677, First Part, Appendix) saw, this is sheer anthropomorphism, and contemporary biologists and psychologists are finding less and less use for final causes in explaining behavior.

An immediate consequence of Theorem 4.2 and Definition 4.25 is

COROLLARY 4.9. The behavior of an animal deprived of plastic neural systems is totally stereotyped.

Most inherited behavior has been selected because it is adapted to a certain environment, i.e. it is biovaluable in it. (Not all: recall the moth's compulsion to circle about a flame.) But if the environment were to change radically, some of the behavior types, formerly adaptive, would be devalued. Joining both statements we have another axiom:

POSTULATE 4.10 Provided the environment does not change radically during the lifetime of an animal, most of its inherited behavioral repertoire has a positive biovalue for it.

The great biovalue of possessing a nervous system with plastic neural subsystems is, of course, its capacity to learn new behavior types and thus improve its chances of surviving environmental changes. So, we make

POSTULATE 4.11 Some of the inherited capabilities of an animal endowed with plastic neural systems are modifiable by learning.

So much for our behavioral principles. Now for drives. Explanations of the type "*X* eats (or sleeps, or mates) because *X* has an urge (or drive) to eat (or sleep, or mate)" are not illuminating as long as the drive concept remains an occult hypothetical construct. They become genuine explanations the moment drives are construed in physiological terms, e.g. as detection of imbalances (deficits or superavits) of certain physiological variables such as sugar or noradrenalin levels in the blood. In this case the hypothesis that behavior *X* is caused by drive *Y* can be put to the test, namely by manipulating the imbalance and watching behavior changes. The entire drive theory can be compressed into one definition and one assumption:

DEFINITION 4.26. A *drive* (or *motivation*) of kind *X* is detection of an imbalance in the *X* component(s) of the state function of the animal. More precisely: the intensity $D_X(b, t)$ of drive *X* in animal *b* at time *t* equals the absolute value of the difference between the detected and the normal values of *X* for *b* at *t*.

POSTULATE 4.12. For every drive in an animal there is a type of behavior of the animal that reduces that drive (i.e. that decreases the imbalance in the corresponding property).

The adoption of this hypothesis allows one to explain many behavior patterns without resorting to teleology. Thus a bird engaged in nest building exhibits inner-driven behavior (and, more precisely, a behavior driven by neuroendocrine imbalances), The animal does not allow random sensory stimuli to divert it from its job. (For other examples of the physiological approach to motivation, see Gross and Zeigler (Eds.) (1969).)

In Ch. 3 we stated that certain states and events are objectively valuable to the organism whereas others are not. We now go a step further and affirm that all animals are equipped with receptors enabling them to evaluate some incoming stimuli as nocent, favorable, or indifferent. This does not entail that all evaluations are correct, let alone that all animals are conscious of such evaluations and can make value judgments. Only a few higher mammals can form value judgments: in all others, and even in man most of the time, valuations are automatic. What distinguishes man from other animals, with regard to valuation, is that he can reason about values as well as evaluate reasons. Other animals do not have the proper CNS for performing such syntheses of reason and value: they behave without weighing behavior and, if they reason at all, they do not weigh reasonings.

In any case, saying that an animal is capable of evaluating items of a certain kind amounts to saying that it has a value system – a concept characterized by

DEFINITION 4.27. Let S be a set of items and b an animal. Further, let $\succsim_b$ be a partial order on S. Then the structure $\mathscr{V}_b = \langle S, \succsim_b \rangle$ is a *value system* for b at a given time iff

(i) b can detect any member of S and discriminate it from all other items in S;

(ii) for any two members x and y of S, b either prefers x to $y (x \succsim_b y)$ or conversely $(y \succsim_b x)$ or both $(x \sim_b y)$ at the given time.

Remark 1 This is a comparative concept of value. A quantitative concept is that elucidated by utility theory and occurring in decision theory. *Remark 2* Note the difference between the concept of psychovalue elucidated by Definition 4.27 and that of biovalue introduced in Ch. 3, Sec.2.2. The biovalue of a for b at t is the objective value that a *possesses* for b at t. On the other hand the psychovalue of a for b at t is the value that b *assigns* to a at t. These value assignments may be biologically mistaken, in which case psychovalues conflict with biovalues. The psychophysical dualist would regard this as a conflict between mind and body, while the monist would regard it as a conflict between the brain or a subsystem of it, on the one hand, and the rest of the body or a different brain subsystem, on the other. In any case, since such conflicts do occur, we had better not try to reduce one kind of value to the other even though both valuations are biological processes. *Remark 3* Value systems are known to change for one and the same organism, not only in the course of its development but also with circumstances. Therefore we incorporate this fact into our hypothesis:

POSTULATE 4.13. All animals are equipped with a value system, and those capable of learning can modify their value systems.

Finally we introduce the notion of choice. An animal presented with a set of alternatives may abstain from choosing among them or may pick one of them. If it tries out some of the options and finds out the hard way which lead to the most valuable outcomes, the animal will be said to have learned to choose. But in all cases there must be some freedom in order for there to be genuine choice. I.e. the animal must be able to choose any of the alternatives open to it, and some choices must be (at least initially) mistaken (i.e. disvaluable) for there to be choice learning. Hence we make

DEFINITION 4.28 Let $\mathscr{V}_b = \langle S, \succsim_b \rangle$ be a value system for an animal b at a given time, and call $A \subset S$ a set of alternatives open to b, i.e. belonging to the behavioral repertoire of b at the time. Then b *chooses* (or *selects*) option $x \in A$ iff

(i) it is possible for b to pick (i.e. to do) any alternative in A(i.e. b is free to choose);

(ii) b prefers x to any other options in A; and

(iii) b actually picks (i.e. does) x.

Note the difference between preference and choice: the former underlies and motivates the latter. Choice is valuation in action, or overt valuation – hence an indicator of valuation not a definition of the latter. And note also that not every choice implements a decision. Decisions are deliberate or reasoned (even if not always rational), and reasoning is the privilege of only a few animal species. Most choices, even in daily human life, are not preceded by any decision-making process.

4. Recall to Knowledge

4.1. *Memory and Learning*

Many a complex prebiotic system, such as a sword and a magnet, has memory of some past states. If only for this reason we must treat memory separately from, and even prior to, learning. A general definition is

DEFINITION 4.29 A system σ at time t has *memory of* (or *memorizes*) some of its past states iff the state of σ at t is a function(al) of those past states.

A system with memory can be called a *memory system*. A memory system records ("commits to memory"), stores, and recalls when suitably stimulated. (Actually the second property may be nonexisting: instead of keeping a record, trace or engram of a state, a system may acquire just the disposition or propensity of recalling the state.) All organisms are memory systems, if only because they have genes, and these record part of the history of the species. Besides genetic memory, all organisms have at least some sort of short span memory: otherwise they would not live long enough to reproduce. So, we can assume

POSTULATE 4.14 All animals have memory of some of their past states, and none of all of them.

There is probably a large variety of memory systems or mechanisms.

We may class them into fixed and itinerant. While invertebrates and lower vertebrates have probably only fixed memory systems (i.e. systems with constant neuronal composition), higher vertebrates have also itinerant memory systems, i.e. systems that are formed just for the occasion. The latter are particularly interesting in the case of man, where remembering seems to consist not in the reactivation of some fixed neural circuit but "in the fresh production or reconstruction of an item" (Bindra, 1976, p. 330). In any case, memorizing is not passive recording but a kind of activity of certain neural systems possibly involving the strengthening of synaptic connectivities (Anderson, 1972, 1973). Nor do we "store" past events in our memory, let alone in our mind: if we store anything at all it is in our brain.

Some memory systems have the ability to learn. According to Definition 4.9, learning is the acquisition of new neural functions, whence it involves plastic (not only wired-in) neural systems. This definition differs from the usual behavioral definition of learning as modification of behavior in response to stimulation. The behaviorist definition is concerned with the effects rather than the causes, and it could be applied to any thing, for any thing will behave differently if stimulated differently.

To emphasize the idea that learning is a modification of neural activity, we make

DEFINITION 4.30 Call E a kind of event or process in a neural system of an animal a involving a plastic subsystem, and S a kind of stimuli (external or internal) which a can detect. Then a has *learned* $e \in E$ in the presence of $s \in S$ during the time interval $[t_1, t_2]$ iff

(i) e did not occur in a in the presence of s before t_1;

(ii) after t_2, e occurs in a whenever a senses s [i.e. a has memorized e].

Since all behavior is controlled by some neural system (Postulate 4.9), the previous definition embraces the concept of behavioral learning, i.e. learning to perform a movement.

We shall also adopt

DEFINITION 4.31 The *experience* of an animal at a given time is the set of all it has learned up until that time.

That is, the experience of an animal is its accumulated learning, i.e. the collection of changes in its plastic neural systems. These include of course nonperceptual learning. On the other hand "learning" to avoid harmful stimuli is not learning proper according to our definition, for it may well consist in a process involving no plastic neural systems – as is indeed the case with invertebrates. The same holds for habituation, which consists in the weakening of existing (wired-in) synaptic connections with repeated stimulation.

Learning, even of the simplest task, is creative for it consists in the appearance of new neural connections and therefore new neural functions. An animal in the process of learning expands its neural state space. This expansion can be either quantitative (i.e. keeping the same axes) or qualitative (changing some axes). A baby's forming its first sentence is presumably an example of the latter type of expansion, whereas its adding one more word to its verbal repertoire is an instance of the former. In general, discovering and inventing are of the second kind, and only a few higher vertebrates seem capable of it.

4.2. *Expectation and Purpose*

So far we have dealt only with non-anticipatory systems, i.e. systems whose behavior does not depend on the expected outcome of their actions. We now introduce anticipation, a property only animals of some species have, namely those capable of expecting either reward or punishment when presented with certain stimuli or engaging in certain activities. We first adopt

DEFINITION 4.32 Animal b *expects* (or *foresees*) a future event of kind E when sensing an (external or internal) stimulus s while in state t, iff b has learned to pair s and t with an event of kind E.

Because foresight depends upon learning, the lower animals, which have no learning ability, have no expectations either. But if an animal can learn, then it does have expectations, and in turn expecting helps (motivates) learning. Animals capable of expecting can regulate the effort they put in doing something.

It seems that all vertebrates have organs serving an anticipatory or preparatory function. Examples: the salivary glands secrete saliva at the sight of food, and the glucoreceptor cells in the intestine signal to the pancreas the intake of glucose, which stimulates it to release insulin. In both cases one subsystem alerts another, and the anticipatory functions are autonomous or nearly so. No wonder that the mammalian brain too has anticipatory functions and more refined ones: they enable some animals to foresee and to prepare themselves for what may come. These animals are said to behave in a goal-seeking or purposive way. Because expectation presupposes learning, only learned behavior can be purposive and, being learned, it is also motivated: the outcome of the purposive action is expected to bring a reduction in drive. In sum, we make

DEFINITION 4.33 An action *X* of an animal *b* has the *purpose* or goal *Y* iff
(i) *b* may choose not to do *X*;
(ii) *b* has learned that doing *X* brings about, or increases the chances of attaining, *Y*;
(iii) *b* expects the possible occurrence of *Y* upon doing *X*;
(iv) *b* values *Y*.

Notice the conditions for purposiveness: freedom, learning, expectation, and valuation. Obviously machines, no matter how sophisticated, do not fulfill all four conditions, hence cannot be goal-seeking – except of course by proxy. Most animal behavior is non-purposive: it only looks goal-seeking because it is so often efficient, but this efficiency is the result of control mechanisms that work without the intervention of any plastic neural systems, hence without learning or expectation.

The claim that purposes defy scientific explanation and call for the hypothesis of an immaterial mind was downed by cyberneticians, who suggested a precise general mechanism of purposive action, namely the negative feedback loop (Rosenblueth *et al.* 1943, 1950). Since then neural modellers and physiological psychologists have been able to explain some specific purposive behavior patterns. According to such models purposes are not states or entities in the mind but certain patterns of neural activity. This view, rather than the mentalistic conception of purpose, agrees with evolutionary biology, which is concerned with explaining the emergence of goal-seeking behavior as a latecomer that has nothing to do with immaterial entities.

Having defined the notion of a goal we can now define that of a means:

DEFINITION 4.34 An action *X* of an animal *b* is a *suitable means* for attaining a goal *Y* of *b* iff in fact *b*'s performing *X* brings about, or increases the probability of occurrence of, *Y*.

4.3. *Thinking*

Forming concepts, propositions, problems, and directions are examples of thinking; so are comparing and transforming concepts, propositions, problems, and directions. We assume that thinking of any kind is an activity of some plastic neural system and, as such, one sensitive to minute chemical changes and changes in basic properties of neurons, such as membrane permeability. For example, humans unable to oxydize the amino acid phenylalanine cannot think at all, and normal subjects cannot think straight when in states of extreme stress – which are often states of hormonal imbalance – or when under the action of psychotropic drugs. Believers in the immateriality of reason must ignore the mounting evidence for the hypothesis that the content and level of performance of thinking depend upon such chemical and physiological variables – and, of course, to be consistent they must never resort to such stimulants as caffeine, nicotine and ethyl alcohol.

We shall deal here only with two basic thought processes, namely concept attainment and proposition formation. We shall conceive of the former as the process of forming kinds, such as the class of cats or that of triangles. And we shall conjecture that forming a concept of the "concrete" kind – i.e. a class of real things or events – consists in responding uniformly to any and only members of the given class. Thus we adopt

POSTULATE 4.15 Let C be a set of (simultaneous or successive) things or events. There are animals equipped with psychons whose activity is caused or triggered, directly or indirectly, by any member of C, and is independent of what particular member activates them.

DEFINITION 4.35 Let C be a class of things or events, and b an animal satisfying Postulate 4.15, i.e. possessing a psychon that can be activated uniformly by any and only a member of C. Then *b attains a concept* $\theta_b(C)$ of C (or *conceives* C, or *thinks up* C) iff the activity (process, function) stimulated by a C in that psychon of b equals $\theta_b(C)$.

Consider next the formation of propositions. We conjecture that the operation involved is that of psychon pairing. For example, "Children are helpless" pairs "Children" off to "are helpless". Two psychons, possibly cortical columns, are activated sequentially, one for each concept. The sequential activity of the two component system constitutes thinking the corresponding proposition. The idea is encapsulated in

POSTULATE 4.16 Thinking up a proposition is (identical with) the sequential activation of the psychons whose activities are the concepts occurring in the proposition in the given order.

POSTULATE 4.17 A sequence of thoughts about propositions is (identical with) the sequential activation of the psychons whose activities are the propositions in the sequence,

This holds for complex propositions (such as conjunctions) as well as for "trains of thought", in particular inferences. *A fortiori,* it holds for rational thinking, or thought controlled by certain master thoughts, or rules, such as "Abhor contradiction", "Check what follows", and "Supply evidence or counterexample".

In line with our semantics (Vols. 1 and 2) we have distinguished constructs (such as concepts and propositions) from thoughts about them as well as from linguistic expressions. Unlike constructs, thoughts are brain processes. Therefore no two thoughts can be exactly identical. Nobody thinks twice of the number 5, or of the Moon, in exactly the same manner, if only because we never go through exactly the same states. However, we can assume that all the processes of thinking of the number 5 (or of any other construct) fit the same neural pattern, i.e. are equivalent in an essential respect. This notion is elucidated by

DEFINTION 4.36 Two thoughts are *equivalent* iff they consist in thinking of the same constructs. I.e.

$$\theta_a(C) \sim \theta_b(C') \quad \text{iff} \quad C = C' \text{ for any animals } a \text{ and } b.$$

It would have been desirable to define a construct as a set of equivalent neural events instead of presupposing the concept of a construct, for it would root constructs on neural events while preserving their abstract quality. Unfortunately we have been unable to find the proper equivalence relation that would allow one to construct such a definition.

4.4. *Cognition and Decision*

All cognition is of some object, concrete or conceptual, and it consists in some information about its objeet – complete or partial, true or false. Cognition can be behavioral (e.g. knowing how to ski), perceptual (e.g. knowing the song of the lark), or conceptual (e.g knowing that the lark song is related to its reproductive cycle or its territorial defense). If behavioral, cognition is a disposition of the motor control system to steer motions of a certain kind; if perceptual it is a potential net of percepts, and if conceptual a potential system of concepts. In either case cognition is a brain process. We summarize the preceding in

DEFINTION 4.37 Let a be an animal. Then

(i) if b is a learned behavior type (or pattern), *a knows how to do* (or perform) b iff b is in the actual behavioral repertoire of a;

(ii) if c is a construct, then *a knows c* iff a thinks up (or conceives) c;

(iii) if e is an event, then *a has knowledge of e* iff a feels or perceives e, or thinks of e.

We can use the concept of knowledge to elucidate that of decision:

DEFINITION 4.38 Let x be an arbitrary member of a set A of alternatives accessible to an animal b with the value system $\mathscr{V}_b = \langle S, \succsim_b \rangle$. Then b *decides to choose x* iff

(i) b has knowledge of every member of A;

(ii) $A \subseteq S$ (i.e. b prefers some members of A to others);

(iii) b in fact chooses x.

A possible decision mechanism is this. If the animal perceives or thinks of a number of objects, and is motivated to choose among them, then he will first compare them. Presumably each object will be apprehended by one psychon at a time, and the excitation levels of the various psychons will differ. If one of them is greater than the others, that one will prevail and will send the strongest (or the only) signal to a volition psychon.

The ability to make decisions is restricted to animals capable of knowing. But not all knowledge is of the same grade, nor all valuation correct. When they are, they constitute the basis of rational decision:

DEFINITION 4.39 A decision made by an animal is *rational* iff it is preceded by

(i) adequate knowledge and correct valuations, and

(ii) foresight of the possible outcomes of the corresponding action.

DEFINITION 4.40 A *rational animal* is one capable of making some rational decisions.

It is quite possible that rational decisions are not the exclusive property of man: some subhuman primates seem capable of making rational decisions. And it is sure that no animal is rational all the time.

4.5. *Creativity*

All higher vertebrates are creative, and man is so superlatively. This may be conceded while at the same time it is doubted whether the notion of creativity is clear. The notion is clear if construed as *neural novelty*, i.e. emergence of new neural systems or new functions of existing neural systems – due e.g. to new connections. Such neural novelties may or may not have behavioral manifestations. If they do one speaks of *adaptive behavior*, in particular imitation and invention, according as the novelty is in the particular animal or in the entire species; otherwise one speaks of mental creativity. So, we make

DEFINITION 4.41 Let a be an animal of species K with behavioral repertoire $B(K)$ at time t. Then

(i) a *invents* behavior type (or pattern) b at time $t' > t$ iff a does b for the first time, and b did not belong to $B(K)$ until t';

(ii) a *invents* construct c at time $t' > t$ iff a knows c for the first time at time t', and no other animal of the same species knew c before t';

(iii) a *discovers* event e at time $t' > t$ iff a has knowledge of e for the first time at time t', and no other animal of the same species had such knowledge before t';

(iv) a is *creative* iff a invents a behavior type or a construct, or discovers an event before any other member of its species;

(v) a is *absolutely creative* iff a creates something before any other animal of any species.

Now the hypothesis:

POSTULATE 4.18 Every creative act is the activity, or an effect of the activity, of newly formed neural systems.

Although creativity is maximal in humans, it is likely to be found among all mammals and birds – if only the theological and empiricist prejudice against the concept of creativity is overcome and the phenomenon is investigated scientifically, particularly among young animals placed in unusual circumstances. We make bold and assume that all higher vertebrates are creative:

POSTULATE 4.19 All animals endowed with plastic neural systems are creative.

However, so far as we know only humans are absolutely creative, i.e. capable of inventing or discovering certain "things" before any other animals. We shall use this assumption in defining humanness in Sec. 5.

What is the ontological status of the so-called creations of the human mind? To obtain a sensible answer we must first distinguish the creative process – whether in science, technology, art, or any other field – from its public materializations, such as books, artworks, and machines. Every creative process is a process in some brain or other (Postulate 4.18). In a few cases such mental processes are externalized as physical things or processes which, when perceived by competent observers, elicit in their own brains processes similar to those undergone by the creators. Thus a musical piece can be whistled, performed on a musical instrument, listened to, written down or read. Novels and theories, artifacts and constitutions are similar. What becomes of a musical piece between performances or listening sessions, or after everybody has forgotten it? It does not exist. If the memories of the piece have faded in all brains, and if all scores and discs have been destroyed, the music itself is no more and has no chance of resurrection – except for the most unlikely chance of being reinvented. But if some traces of it do remain – either in brains or on paper, discs or tapes – then the musical piece has the possibility of "coming alive" again. Something that cannot be played on an instrument, an audio system or something similar, and cannot even be hummed, is not a musical piece.

I submit that the same holds, *mutatis mutandis*, for all cultural objects. Thus a sculpture that nobody looks at is just a chunk of matter – and so is a philosophical treatise that nobody reads. There is no immortality in cultural creations just because they can be externalized ("embodied") and catalogued. Only that endures which is being recreated – reperceived, refelt, rethought, re-enacted. Deserted libraries, museums, art galleries, and laboratories, are just cultural cemeteries. Nothing is more dependent and vulnerable – less autonomous – than the "world" of culture. (Let the current contempt for basic science prevail, and soon there will be no scientific culture left.) Culture lives neither in cultural artifacts, such as books, nor in an immaterial and autonomous world of ghosts: it lives in the brains of those who care about it – who cultivate it.

Ideas, then, do not exist by themselves any more than pleasures and pains, memories and flashes of insight. All these are brain processes. However, nothing prevents us from *feigning* that there are ideas, that they are "there" up for grabs – which is what we do when saying that someone "discovered" such and such an idea. We pretend that there are infinitely many integers even though we can think of only finitely many of them – and this because we assign the set of all integers definite properties, such as that of being included in the set of rational numbers. Likewise we make believe that every deductive system contains infinitely many theorems – and this because, if pressed, we can prove any of them. All such fictions are mental creations and, far from being idle or merely entertaining, are an essential part of modern culture. But it would be sheer animism to endow such fictions with real autonomous existence and causal efficacy – the way idealist philosophers from Plato on have been doing, or the way Popper (1968, 1972, 1974, 1977) does. A differential equation does not exist by itself and therefore is harmless; only thinking of it can have some effects. In short, ideas in themselves are fictions and as such have no physical existence.

5. SELF TO SOCIETY

5.1. *Awareness, Consciousness, Will*

While many mentalists equate consciousness with mind, and behaviorists have no use for either concept, physiological psychologists, neurologists and even ethologists are feeling the need for both. For example, the mere description of yoga experiences, such as the voluntary control of heartbeats, uses the concept of initial conscious control. Unfortunately the concept of consciousness is still rather vague – particularly among those who advocate its use but refuse squarely to define it (e.g. Popper and Eccles, 1977).

We start by distinguishing consciousness from awareness. We say that an animal is (or fails to be) *aware* of what goes on in its environment or in itself (in particular, of what it does or is done to it). Only if the animal is aware of some of its own brain processes (not necessarily mental ones) shall we say that it is *conscious*. More precisely, we make

DEFINITION 4.42 If b is an animal,
(i) b is *aware* of (or *notices*) stimulus x (internal or external) iff b feels or perceives x – otherwise b is *unaware* of x;
(ii) b is *conscious* of brain process x in b iff b thinks of x – otherwise b is unconscious of x.

Whereas awareness requires only a sensory system, consciousness requires also intelligence, i.e. the ability to think, and moreover to think of brain events. The lower animals can be aware but not conscious, and a conscious subject may be unaware of some external stimuli. So, we make

POSTULATE 4.20 All animals are aware of some stimuli and some are also conscious of some of their own brain processes.

DEFINITION 4.43 The *consciousness* of an animal b is the set of all the states of the CNS of b in which b is conscious of some CNS process or other in b.

In other words, consciousness of brain event x is direct knowledge of x. (This is what introspection is all about.) Moreover we may hypothesize that a conscious event is a brain activity consisting in monitoring (recording, analyzing, controlling, or keeping track of) some other brain activity, much as a voltmeter measures the electromotive force between two points of an electrical network. I.e. we adopt

POSTULATE 4.21 Let P be a subsystem of the CNS of an animal b engaged in a mental process p. Then the CNS of b contains a neural system Q, other than P and connected with P, whose activity q equals b's being conscious (thinking) of p.

This hypothesis is far from outlandish: after all, every biosystem is a control biosystem, i.e. it monitors (watches and checks) its own activity. By sensing and correcting its own functioning, the biosystem assures normality or return to it. So, it should come as no surprise that the brain, the most complex of biosystems and the supreme control of the animal, is capable of controlling its own activity.

Since by hypothesis two distinct and interconnected neural systems are involved in all conscious states, it is perfectly possible for them to interact. Thus if a subject is conscious of a routine mental task, he or she may bungle it: the monitoring interferes with the execution of the well learned task. If on the other hand the subject is beginning to learn a task, he had better be both aware and conscious of it: the monitoring system will guide the establishing of the psychon. To account for such phenomena there is no need to invoke an immaterial consciousness (as Sperry, 1969, Popper and Eccles, 1977, and so many others have done). If one wishes to speculate on the causal efficacy of conscious events, then he ought to do it within the scientific framework of changing things instead of the mythical framework of disembodies entities.

Since awareness and consciousness are activities of a biosystem, they can be dimmed or heightened by appropriately inhibiting or stimulating the neural components involved. Such degrees are nothing but the intensities of the activities of the corresponding neural control systems. If consciousness were a separate entity we would not know how to explain such variations and, in particular, the daily obliteration and resurgence of consciousness.

It is a well known law that, as learned tasks are repeated, they tend to become automated, i.e. unconscious:

POSTULATE 4.22 In the course of the life of an animal capable of learning, learned behavior, if initially conscious, becomes gradually unconscious.

When automatisms prove insufficient to solve a problem or execute a task, it proves convenient to become conscious or to heighten the level of conscious mental activity. A conscious animal is better equipped to cope with novelty, for it can evaluate its own thoughts and actions, criticizing and correcting them if necessary. (For the adaptive advantages of consciousness see Griffin, 1976.) Given the biovalue of consciousness as a device for instant adaptation, it is no wonder that man has been so successful, for his CNS is the only one capable of generating conscious states almost at will.

Consciousness, then, is not an entity but a set of states of a highly evolved CNS. Therefore to speak of 'states of consciousness' is sheer reification: there are only conscious (and unconscious) states of the brain. (Recall Fig. 4.3 in Sec. 2.2.) Likewise it is mistaken to speak of the Unconscious (or the Subconscious) as an entity, in particular one capable of influencing Consciousness (another entity). There are only brain events, some conscious and the others unconscious, and because they are concrete events they can influence others. What holds for consciousness and unconsciousness holds also for Freud's Id and Superego. There can be no mental entities within mental entities, because mental entities are nonentities. Surely there are controls, e.g. feedback systems, in the brain. Some such controls may "censor" (inhibit) certain brain activities, e.g. block the flow of information to the system doing the monitoring, or the activity of the psychon(s) doing the thinking. For example one may have learned an item A and, in the course of mental work, think an item B incompatible with A. If A is deeply ingrained, it may "repress" or "suppress" B, i.e. inhibit the psychon that does B. But if B is as vivid as A, further psychons may be activated until the "contradiction is resolved" in favor of one of the original options or of a third one. Physiologically, this is nothing but an interplay among competing neural feedback circuits.

Our next topic is volition, or conscious purposeful behavior:

DEFINITION 4.44 An animal act is *voluntary* (or *intentional*) iff it is a conscious purposeful act – otherwise it is *involuntary*.

The will is not an entity but a neural activity: x wills y iff x forms consciously the purpose of doing *y*. Nor is the will a mysterious faculty of an immaterial mind, but a capacity of highly evolved CNSs, namely "a control of behavior by the thought process" (Hebb, 1968, p. 75). When you pick a fruit from a tree or a book from a shelf, and are aware of what you are doing and moreover have a definite purpose in doing so, what happens is that certain psychons in your associative cortex – most likely psychons located in the frontal lobes – activate certain motor centers (in particular the gyrus), which in turn control the movements of the limbs. Thus *intentions can cause behavior.* Curiously enough this is also a thesis of the dualist philosophers who defend the "intentionalist" (as opposed to the scientific) explanation of human behavior. However, they have no right to do so because causal relations obtain only between events, and the dualists refuse to regard intentions as (neural) events.

Voluntary acts can be free or compelled. The general who decides to launch an attack may act freely, but those of his soldiers who go unwillingly to battle act voluntarily though under compulsion. Free will is volition with free choice of goal, with or without foresight of possible outcome. We make then

DEFINITION 4.45 An animal acts of its own *free will* iff
(i) its action is voluntary and
(ii) it has free choice of its goal(s) – i.e. is under no programmed or external compulsion to attain the chosen goal.

Free will has of course been either affirmed or denied for theological or philosophical reasons. In our theory it follows from Postulate 4.20 and Definitions 4.44 and 4.45:

THEOREM 4.3 All animals capable of being in conscious states are able to perform free voluntary acts.

If consciousness is not exclusively human, neither is free will. And both are legitimate subjects of scientific research.

5.2. *Person and Self*

There are a number of concepts of a person. We are interested in this one:

DEFINITION 4.46 If *b* is an animal endowed with a plastic neural system capable of mentation (i.e. with a nonvoid mind), then

(i) the *personality* of b is the functional system composed of all the motor and mental functions of b;

(ii) a *person* is an animal endowed with a personality.

Remark 1 It follows that not only humans but all animals with mental abilities are entitled to be called 'persons'. *Remark 2* Personality is a whole-body property: there is no special organ of personality, since it is defined as the union of behavior and mentation. *Remark 3* For the same reason the destruction or ablation of considerable portions of the CNS, as well as the amputation or the paralysis of limbs, ensue in personality changes. Irreversible and massive destruction of the brain destroys personality altogether, in which case the animal ceases to be a person both psychologically and legally. *Remark 4* Learning shapes and often enriches personality; it can even produce radical personality changes. *Remark 5* People who undergo shattering experiences (war, concentration camp, prison, or ideological conversion) may acquire whole new personalities. *Remark 6* In deep sleep or in coma we lose much if not all of our personality: we become nonpersons or nearly so. (But of course when reawakening we recover – or rather reconstruct – our personality.) *Remark 7* Since the behavior and the mentation of animals endowed with plastic neural systems depend partly on their environment, one and the same animal is bound to exhibit different personalities in different environments – e.g. a tyrant at home and meek at work. (For the various determinants of personality see Gray, (1972b).) *Remark 8* There is no personal identity or enduring personality any more than there is digestive or cardiovascular identity: the behavior and the mentation of a higher vertebrate are more changeable and vulnerable than any other bodily function. (For an early criticism of the idea of personal identity see Hume, 1739, Part IV, Sec. VI.) *Remark 9* Since mind is not incorporeal stuff but a collection of brain functions, and since there are no two strictly identical brains, "identical" twins reared in the same family have different minds, hence are different persons instead of constituting (as Shaffer, 1977, has claimed) a single mind with two bodies. *Remark 10* Since split brain patients have two minds (Corollary 4.7), they also have a double personality, i.e. they are two different persons united anatomically. *Remark 11* If an individual were to get a new brain by transplant, the personalities of both donor and recipient would disappear, and a third personality would emerge. Indeed, the brain in its new encasement would receive new stimuli and fail to receive some of the customary stimuli. And, because it would be controlling a different body, it would effect different movements, so it would function differently. The original selves would be lost. The same holds, *a fortiori*, for the transplant of each hemisphere into a different skull: two different persons would be manufactured – possibly neither of them viable. In short, tampering with a minding body involves tampering with its mind.

Finally we come to the notion of self, which in our theory is elucidated by

DEFINITION 4.47 An animal

(i) has (or is in a state of) *self-awareness* iff it is aware of itself (i.e. of events occurring in itself) as different from all other entities;

(ii) has (or is in a state of) *self-consciousness* iff it is conscious of some of its own past conscious states;

(iii) has a *self* at a given time iff it is self-aware or self-conscious at that time.

Remark 1 This concept is more restrictive than that of mind: all animals possessing a self (i.e. all selves) mentate but the converse is false. *Remark 2* The self is not an entity but a state of an entity, namely an advanced brain. Therefore in the context of our theory to say that the self has or uses its brain (Toulmin, 1972; Popper and Eccles, 1977) amounts to saying that certain brain states have a brain. *Remark 3* Possibly animals other than humans have self-awareness but, so far as we know, only humans have *self*-consciousness, at least when normal and past their infancy. The young child is self-aware but not self-conscious. Self-consciousness is generally believed to appear at about seven years of age (Piaget) and is sometimes conjectured to originate as internalized speech, hence as a spinoff of social life (Vygotskii, 1962; Luria, 1969). It is therefore as old as language, and both have evolved (and continue to evolve) together with society. But the subject of sociality deserves a new subsection

5.3. *Social Behavior*

A bird's territorial call and a criminal's "antisocial" action are no less social than a bee's dance or a baboon's scouting on behalf of its troop, even though the former actions tend to reinforce solitude while the latter strengthen gregariousness. (Cf. Hinde, 1974.) The notion in question is elucidated by

DEFINITION 4.48 An animal engages in *social behavior* iff it acts on, or is acted upon by, other individuals of the same genus.

The condition of congenericity excludes preying and flight from predators from the social behavior repertoire but on the other hand includes reproductive behavior. Needless to say, social behavior may be automatic, in particular genetically programmed, as is the case with insects.

Although not all animals are gregarious or social (i.e. live in communities), all are social in the above sense, if not actually at least potentially. We assume therefore

POSTULATE 4.23 The behavioral repertoire of every animal includes types (patterns) of social behavior.

Like all behavior, social behavior is controlled both internally – in particular by the NES – and externally. For example, aggressiveness depends upon endogenous variables such as certain hormones (in particular testosterones) and the excitation of certain brain systems (e.g. the anterior hypothalamus), as well as upon exogenous variables, such as scarcity of food, cold, and crowding. Consequently biological determinism, which holds that all social behavior is determined exclusively by endogenous variables, holds only for animals whose social behavior repertoire is inherited, hence stereotyped. The social behavior of bees and ants, the ritualized courtship of birds, and the patterns of aggression and submission of cichlid fish and even of wild dogs, seem to be of that kind. On the other hand primates seem to inherit only those patterns of social behavior that concern reproduction and the care of the young: all others are learned. What primates do inherit is (*a*) the *need* for social intercourse and (*b*) the *capacity* to form and change rules of social behavior, and to form and undo social groups.

Animals get together to form families or communities, temporary or permanent, loosely or tightly knit, when such behavior improves their lot. The simplest such system is the family, and the simplest family is the mother–infant couple. The root of such a system is of course the care of the young. Note that families, though formed naturally – i.e. on the strength of natural bonds – are not biosystems. Only the components of a family are alive, i.e. metabolize, are capable of reproducing, etc. A family is a social system. We shall make a detailed study of social systems in the next chapter, but will offer now a definition that will be useful for our sociobiological considerations:

DEFINITION 4.49 A system σ is a *sociosystem* (or *social system*, or *social group*) iff

(i) the *composition* of σ is a set of animals of the same order (not necessarily the same species);

(ii) the *environment* of σ is the set of things, other than the components of σ, that act on or are acted on by the latter;

(iii) the *structure* of σ is the social behavior repertoire of the members of σ.

Every society is a sociosystem but the converse is not true: industrial plants, schools, and clubs are sociosystems but not societies. A society is a self-reliant sociosystem:

DEFINTION 4.50 A sociosystem is a *society* iff it is self-sufficient [i.e. does not depend entirely upon other sociosystems].

This definition will be refined in the next chapter.

5.4. *The Root of Social Cohesion*

What elicits the formation of a social system and keeps it together despite the somewhat divergent interests of its members? Our answer is this: rearing of the young in the case of the family, sharing (or participating) in the case of other sociosystems, and social pressure (peaceful or forceful) in all.

Cooperation need not be conscious: it can be automatic, as among social insects. Animals can cooperate either in their own individual interest (as is the case with sexual partners and with partners in a foraging expedition) or in the interest of their group – as in the case of an ant colony or a wolf pack. The general concept is elucidated by

DEFINITION 4.51 If a and b are animals, then a and b *cooperate* with one another iff the social behavior of each is valuable to the other or to a third animal.

When cooperation concerns things (e.g. goods) of some kind, it is called *sharing*, and *participation* when it concerns activities. More explicitly:

DEFINITION 4.52 Let σ be a social system with composition $\mathscr{C}(\sigma)$, environment $\mathscr{E}(\sigma)$, and structure $\mathscr{S}(\sigma)$. Further, let $T \subset \mathscr{C}(\sigma) \cup \mathscr{E}(\sigma)$ be a set of members of σ or of items in the environment of σ, and $A \subset \mathscr{S}(\sigma)$ a kind of activity in σ. Then for any component x of σ,

(i) x *shares* in T iff x cooperates with other member(s) of σ in acting on items in T;

(ii) x *participates* in A iff x cooperates with other member(s) of σ in doing A.

Now our hypothesis that association is based on cooperation:

POSTULATE 4.24 A set of conspecific animals forms a social system if and only if each of them cooperates with some other members of the same set [i.e. shares in the resources of the society or participates in some of its activities].

The early evolutionists stressed competition at the expense of cooperation, and both Freud and Lorenz regarded aggression as instinctive and pervasive. We have learned since that the struggle for life involves both. For example, flocking and the formation of colonies are effective defenses against predation (Hamilton, 1971). Group grazing, hunting in packs, and active defense involve cooperation – not to speak of higher forms of sociality such as division of labor and games. No cooperation, no sociosystems.

Only some forms of cooperation are purposeful, even fewer conscious. Purposeful cooperation deserves a name of its own:

DEFINITION 4.52 For any animals a and b,

(i) a behaves *altruistically* (or with *solidarity*) towards b iff a engages in purposeful social behavior that may be of value to b though not directly or immediately to a;

(ii) a and b are *reciprocally altruistic* (or *solidary*) iff a is altruistic towards b and conversely.

A rough measure of solidarity is this. Let a and b be animals, and $N(b)$ the set of needs of b (e.g. food, shelter, or care), and $G(a, b)$ the set of items a is prepared to give b. (Both $N(b)$ and $G(a, b)$ are taken during the same period of time.) Then the solidarity of a towards b is $G(a, b) \cap N(b)$. A quantitation of this concept is the following. *The degree of solidarity* of a towards b (over the same period) is

$$s(a, b) = |G(a, b) \cap N(b)|/|N(b)|.$$

(The more you give needed items the more solidary your behavior. Items that are not needed do not count.) The reciprocal solidarity between a and b is of course $s(a, b) + s(b, a)$. This measures the total number of items exchanged between a and b without regard to their value.

A more precise concept is easily constructed by assigning a value to each item needed and got. Such value need not be biological or psychological: in the case of higher animals it can be social, i.e. the items being traded off may benefit the social group as a whole. If only for this reason one can hardly subscribe to the cost-benefit analysis of altruism championed by Hamilton (1964) and Wilson (1975). This rests on the assumption that genes are so smart and species-conscious that they know it is good business for an individual to lay down its life for three or more sibilings, for in this case the chances are that its full genetic equipment (or rather a replica of it) will be saved for the benefit of posterity. Altruism requires a highly developed CNS capable of becoming aware that a conspecific is in need. Skipping levels, the way some sociobiologists do, won't do. (See Ruse (1979) for a balanced assessment of sociobiology.)

In addition to the inter individual actions that keep sociosystems going, there are the group influences on the individual that prevent it from rending the social fabric. Social pressure is not a mysterious action of the whole on the part but rather the action (direct or indirect) of the various components of a social group on those that deviate from the norm or mode. One way of elucidating the concept is by means of

DEFINITION 4.54 Let F be a function representing a property of members of a social system σ, and assume that the distribution of F in $\mathscr{C}(\sigma)$ is bell-shaped with average $\bar{F}$ and scatter d. Then, for any member x of σ,

(i) x *conforms* with respect to F iff $|F(x) - \bar{F}| < d$;

(ii) x *deviates* with respect to F iff x does not conform with respect to F.

We shall assume the obvious, namely that every social group contains some deviants, and that the group pressure on an individual increases with its deviance:

POSTULATE 4.25 In every social system (*a*) there are deviants in some respects, and (*b*) some of the fellow members of the deviant components subject them to a pressure that is a monotonically increasing function of the deviance.

In most vertebrate societies social structure has a rather weak reaction on individual behavior. Indeed flocks and troops of various kinds split and even dissolve altogether under altered environmental conditions. In hierarchical primate societies, on the other hand, the over-all social structure seems to be overpowering as long as the social system lasts. This makes for social stability – as well as for the occasional rebellion. The deviant, underprivileged, or just young, are often outcast. In a few cases they may find a more favorable habitat offering new opportunities and new challenges. In other cases the deviants may join forces to attempt a restructuring of their group. In any case, just as genic variety is a motor of biological evolution, so social deviance is a motor of social evolution.

5.5. *Communication*

All social animals, and many which are not, can exchange information of some kind, and some of them using a great variety of means. We shall adopt some of the standard concepts used in describing animal communication (Morris, 1938; Smith, 1965). They are all included in

DEFINITION 4.55 (i) An *animal signal* is a physical process executed or controlled by an animal, perceptible to other animals and capable of altering their behavior;

(ii) the *message* carried by an animal signal is an encoded representation of events in the CNS of the signalling individual;

(iii) the *significance* of a message to a recipient is the change in the latter's CNS caused by the signal carrying the message;

(iv) an animal *understands* a message if the events triggered in its CNS by the carrier signal are similar to those in the animal that emitted the signal;

(v) two animals *communicate* iff they understand the messages of the signals they exchange.

Remark 1 An animal signal may be blunt, as in pushing, or subtle, as in singing. *Remark 2* Communication is always among animals, though it is not always direct. For example, we do not communicate with computers but through them. *Remark 3* Although communication can be valuable (or disvaluable), it is not necessarily purposeful. For example, the signals exchanged by social insects, being genetically programmed, are not purposeful even though they are functional (biovaluable). *Remark 4* The psychoneural concept of significance (Definition 4.55 (iii)) differs from both the behavioral notion (Morris, 1938; Paivio, 1971) and the semantical one (Vol. 2, Ch. 7).

Not every set of signals is a language. For example, the set of human gestures is not a language. A language is a system of learned signals and moreover one that allows an animal to compose an unlimited number of distinct messages. A single signal suffices to build a language if allowed to concatenate with itself; on the other hand bees communicate without having a language: they do not learn their signals from scratch and they cannot generate an unlimited number of signals out of a few simple ones. The following convention summarizes all this:

DEFINITION 4.56 Let S be a finite nonempty set, and $\circ$ a binary operation in S. Further, call $\mathscr{L} = \langle S^*, \circ \rangle$ the free semigroup on S [i.e. the set of concatenates of members of S]. Then $\mathscr{L}$ is a *language* iff

(i) S^* is a set of learned animal signals;

(ii) there is at least one animal capable of understanding some of the messages carried by members of S^*.

Remark 1 This definition allows for private languages such as those deaf children can invent (Goldin-Meadow *et al.*, 1977). Incidentally, Wittgenstein and his followers have denied that private languages could exist. *Remark 2* Chimpanzees can learn certain sign languages (Gardner and Gardner, 1969). This finding has weakened the hypothesis that language is exclusively human – weakened rather than refuted, for the chimpanzees in question were taught a man-made language, namely the American Sign Language or ASL, whereas humans *create* languages. Moreover, although ASL has as much expressive power as any natural language, chimpanzees cannot make full use of it because they cannot think all human thoughts. *Remark 3* By using borrowed languages, chimpanzees can express thoughts of their own and thus refute the dualist's contention that it is "really senseless to conjecture that animals *may* have thoughts" (Malcolm, 1973). We are now sure, on the basis of experiments with teaching artificial languages to chimpanzees, that animals can think. *Remark 4* Because chimpanzees can be trained to communicate not only by means of ASL but also with the help of plastic chips (Premack, 1971) and via computers (Rumbaugh *et al.*, 1973), man can no longer be defined as *the talking animal.* On the other hand we can evidently postulate that humans (and chimpanzees and possibly other animals as well) have an innate linguistic ability or potentiality or disposition. This is not to say that man is born with a knowledge of certain basic structural features common to all natural languages. What we are born with is the vocal tract, the Wernicke and Broca areas, and subsidiary neural systems – as well as a social environment, carrier of a tradition, that stimulates the acquisition and development of verbal languages.

5.6. *Protoeconomy, Protoculture, Protopolity*

In the next chapter we shall argue that every human society, no matter how primitive or evolved, is composed of four main subsystems: kinship, economy, culture, and polity. We submit that all four are already found *in nuce* in some prehuman societies.

That prehuman societies have a kinship system, i.e. a social network based on reproduction, seems obvious. Nor is there any doubt that some have also an economic system, i.e. a social network based on work. As for politics in the large sense of management of social activities, it is well known than in certain animal societies law and order, as well as defense, are well organized. (Cf. Gray Eaton (1976) for the case of the Japanese macaque society.) Less well known, but true nonetheless is that some prehuman societies have a culture of sorts, i.e. a system whose members engage in social activities that are mental rather than biological, productive, or managerial. The patterns of such activities can moreover be transmitted throughout the community as well as handed over to the next generation – i.e. traditions are formed, spread, and are kept. Such a transmission is effected by imitation and occasionally also by teaching, hence with the help of signals. This has been found to be the case among Japanese macaques (Kawai, 1965) and chimpanzees (Menzel *et al.*, 1972). In both cases the initiation of new traditions called for some bold individuals – creative deviants – that sought newness rather than fearing it.

Because economic, cultural and political systems are already found in prehuman societies, although in a primitive form, we should characterize the corresponding concepts:

DEFINITION 4.57 If σ is an animal society, then

(i) the *economy* of σ is the subsystem of σ whose members engage in the active and organized transformation of the environment of σ;

(ii) the *culture* of σ is the subsystem of σ whose members engage in mental activities that control (or are controlled by) some of the activities of other members of σ;

(iii) the *polity* (or *political system*) of σ is the subsystem of σ whose members control (or are controlled by) the social behavior of other members of σ.

A particular case is that of tripartite societies composed of absolutely creative animals, i.e. animals capable of discovering and inventing certain things before any other animals (Definition 4.41). We submit that this is the case of all humans, from *Homo erectus* through *habilis* and *sapiens* to *sapiens sapiens*, i.e. us. That is, we assume

POSTULATE 4.26 An animal society is *human* iff
(i) some of its members are absolutely creative, and
(ii) it is composed of an economy, a culture, and a polity.

Remark 1 We are defining humanness in psychological and sociological rather than purely biological terms. In this we follow the tradition of anthropology and prehistory, and reject Neo-Darwinism. *Remark 2* Clause (i) suggests that the economy, the culture and the polity, far from being rigid, are plastic – as plastic as their members. Man is not only an economic, cultural and political animal, but also one that can alter very quickly almost any feature of this triad without having to wait for genic mutations or environmental cataclysms: man is the supreme creator and destroyer.

We now have all the preliminary material necessary to build an ontology of human society – a task for the next chapter.

6. Concluding remarks

Wind is the motion of air: there is no wind apart from moving air. Likewise there is no metabolism over and above metabolizing systems, or social change over and above social systems. In every factual science, states are states of concrete entities, and events are changes of state of such objects. It is only in traditional psychology and philosophy of mind that talk of behavior independent of behaving animals, and of mentation alongside or above the brain, is tolerated, nay encouraged by the dualist myth. This separation between the mind and that which does the minding – between function and organ – has kept psychology and the philosophy of mind far apart from neurobiology and has prevented a full utilization of the scientific approach to the mind-body problem. The philosophy of psychophysical dualism is responsible for the backward state of research into that problem. Anyone interested in the progress of the science of mind must therefore reject the prehistoric myth of the autonomous mind.

The rejection of psychophysical dualism does not force one to adopt eliminative materialism in either of its versions – i.e. that mind and brain are identical, that there is no mind, or that all things have mental capacities. (This last version is of course indistinguishable from panpsychism or animism.) Psychobiology suggests not just psychoneural monism – the strict identity of mental events with brain events – but also emergentism, i.e. the thesis that mentality is an emergent property possessed only by animals endowed with a very complex plastic nervous system. This ability confers its owners such decisive advantages, and is related to so many other properties and laws (physiological, psychological, and social), that one is justified in asserting that the organisms endowed with it constitute a level of their own – that of *psychosystems*. However, this is not saying that *minds* constitute a level of their own, and this simply because there are no disembodied (or even embodied) minds, but only *minding bodies*. (The dual notions of embodiment and disembodiment are rather vague and typically dualistic, hence – *pace* Margolis (1978) – have no place in a science-oriented ontology.)

To repeat the same idea in different words: One can hold that the mental is emergent relative to the merely physical without reifying the former. That is, one can hold that the mind is not an entity composed of lower level things – let alone a thing composed of no entities whatever – but a collection of functions (activities, processes) of certain neural systems, that individual neurons do not possess. And so emergentist (or systemist) materialism, unlike eliminative materialism, is seen to be compatible with overall pluralism, or the ontology that proclaims the qualitative variety and mutability of reality. (Cf. Vol. 3, Ch. 5.)

Our espousing emergentist (or systemist) materialism does not entail claiming that it has already solved the mind-body problem. It has and it won't, for emergentist materialism is a philosophy providing only a scaffolding or general framework for the detailed scientific investigation of the many problems one lumps under the rubric 'the mind-body problem'. It behoves neuroscientists and psychologists to attack these problems – as scientists not as amateur philosophers let alone theologians. (Likewise emergentist materialism provides only a general framework for the detailed investigation of the problems of inanimate matter and of life, both of which are just as bottomless as the problem of mind.)

However, philosophy is far from behaving as a nonparticipant spectator of the scientific investigation of the mind-body problem. The dualist philosophy of mind has for centuries actively blocked the scientific approach to the problem. On the other hand emergentist (or systemist) materialism contributes to the scientific investigation of it by (*a*) dispelling confusions, (*b*) exposing myths, (*c*) suggesting that all problems about mental states or processes be construed as problems concerning brain functions, and therefore (*d*) encouraging the building of neural models of mental functions. As a matter of fact emergentist (systemist) materialism is the philosophy of mind behind physiological psychology, psychochemistry, and neurology. It is the only one that has been fruitful, that does not support quixotic reductionism, and that is compatible with a systemist and dynamicist ontology as well as with all the sciences.

CHAPTER 5

SOCIETY

This chapter is devoted to human society. Whereas to biologism human society is just one more animal society, according to spiritualism it is nothing of the sort because it is guided by ideas and values. We take the alternative view that human society is an animal society with many and remarkable novel properties, only a few of which it shares with animal protoeconomies, protocultures, and protopolities. We assume that man is neither an animal at the mercy of its genetic makeup and its environment, nor a free spiritual being akin to divinity: man is, instead, the primate that works and strives to know, that builds, maintains, and transforms social organizations far beyond the call of the gene or the environment, and that creates artistic, technological, and intellectual cultures, and also plays. Man is *faber* and *sapiens, oeconomicus* and *politicus, artifex* and *ludens*.

No aspect of mankind can be properly understood unless one keeps in mind the fivefold character of human nature – biological, psychological, economic, cultural, and political. The very existence of human society – of its economy, culture, and polity – has biological roots that cannot be cut, and psychological constraints that cannot easily be bent. There is no philosophizing on an empty stomach and, from a certain point on, no subsistence without new thinking, cooperating, and organizing. The biological, psychological, economic, cultural and political aspects hang together.

Human society is composed of animals but is not rigidly determined by biological factors, as shown by the variety of types of society. Moreover social behavior reacts on the animal basis by selecting socially or culturally valuable traits. This process may even oppose that of natural selection, as when the handicapped are protected. Thus social selection modifies genetic composition. For example whereas urbanization and crowding select resistance to stress and foul air, and perhaps also pushiness and resistance to bureaucacy, it abates other features. Such a social selection of biological and psychological traits modifies the genetic composition of the population, a change that has in turn social consequences. In sum there is a tight relationship between the biological and the social.

Consequently mankind does not evolve exactly as other species do: the theory of evolution does not apply to it nowadays the way it did a couple of million years ago. In fact we should distinguish three modes of evolution: "blind", animal, and social. "Blind" evolution is evolution by mutation, recombination and natural selection; it is the way molecules and plants evolve. Animal evolution combines genic change with behavior: the environment selects not only phenotypes but also behavior patterns, and animals adapt not only phylogenetically but also ontogenetically by learning adaptive behavior. Finally social evolution proceeds not only by genic change and behavioral change, but also by social creation and selection.

In all three cases the individual proposes and its environment decides – metaphorically speaking. But in the cases of animal and social evolution the individual learns and the environment, rather than being given to it, is partly chosen by and and partly even remade by it. And in the case of social evolution there are not only biological and psychological factors but also economic, cultural, and political ones. Moreover, thanks to language and social life, some of the learned skills and some of the products of human activity are transmitted to the new generations: here we can speak of the (social) inheritance of acquired characters. But even this inheritance is plastic, for we are free either to disregard or to enrich our heritage.

In the present chapter we shall examine the social aspect of man, centering our attention on social relations and the resulting social structure. Our aim will be to clarify and interrelate some of the key concepts of the social sciences – those which are so basic that no particular social science deigns to clarify them. We shall do it from a point of view that supersedes not only biologism and spiritualism, but also individualism and holism. This alternative is of course the systemist one. By regarding human society as a concrete system of a particular kind, and by analyzing the various subsystems and supersystems of society, we hope to attain some clarity about a traditionally obscure subject. Needless to say, our discourse will be extremely general: it will aim at some of the so-called cultural universals. Therefore we shall not be able to explain or predict any specific social events. We shall at the most supply a scaffolding for building specific theories in the social sciences.

1. Human Society

1.1. *Definitions*

Our Postulate 4.26 characterizes a human society as an animal society containing some absolutely creative members, and composed of an economy, a culture, and a polity. That postulate served the purpose of defining (implicitly) humanness rather than human society. We shall now define the latter in some detail, assuming that biology, psychology and psysical anthropology can take care of the biological characteristics of *Homo*.

We submit that the chief characteristic of human society, in contradistinction to other animal societies, is that it possesses jointly all the following properties:

(i) some members of every human society do *labor*, i.e. engage habitually in the deliberate transformation of part of their environment (*homo faber*);

(ii) the workers use *tools* made in standardized patterns and work with them according to rules or *techniques* that they have invented, improved, or learned;

(iii) some members of every human society *manage* (direct or control or contribute to either directing or controlling) the activities of others: they organize labor or games, learning or fighting;

(iv) some members of every human community engage habitually (though perhaps not exclusively) in *cultural activities* – cave painting, story telling, tool designing, healing, reckoning, teaching skills, advancing knowledge, etc. (*homo culturifex*);

(v) all members of every human society devote some time to playing – sex, dancing, play-acting, racing, playing hide and seek, games of chance, etc. (*homo ludens*);

(vi) all members of every human society *communicate* with some other members of the same society or of other societies by means of standardized signs, in particular a language;

(vii) every member of every human community *shares* information, services or goods with some other members of the same community;

(viii) all the members of every human society *learn* attitudes, skills and information not only from their parents but also from other members of their community (by imitation or formal teaching);

(ix) every human society is divided into *social groups*, such as families and occupational groups;

(x) every human society lasts as long as every member of it *participates* to some extent in several social activities and benefits from such participation.

Surely other animals too work – e.g. bees and beavers. But they make no tools. Still others use natural tools, such as branches, stones or thorns, but only occasionally and in any case not fashioned by them or at least not to standardized patterns. Again, some birds and mammals play, and still others have evolved coded sign systems (though not languages proper). However, only humans organize (manage) some of their own activities, *and* teach behavior patterns and feeling and thinking patterns, *and* create novel objects (artifacts, works of art, organizations, words, etc.). In sum only humans possess, socially if not individually, *all ten* properties listed above.

The preceding characterization of human society, though sketchy, suffices for the moment. We proceed to express it in terms of concepts elucidated in the previous chapters:

DEFINITION 5.1 A concrete system σ is a *human society* (or *community*) iff σ is an animal society such that

(i) the *composition* $\mathscr{C}(\sigma)$ of σ is included in the set of human beings (i.e. the genus *Homo*, not necessarily our own species);

(ii) the *environment* $\mathscr{E}(\sigma)$ of σ contains some of the items necessary for the survival of some of the components of σ;

(iii) the *structure* $\mathscr{S}(\sigma)$ of σ equals the disjoint union of two sets of relations, S and T – called *social relations* and *transformation relations* respectively – such that

(a) S includes the relation of biological descent and all other (kinship) relations derived from it;

(b) every member of the set S of social relations is accompanied by communication (information flow);

(c) S includes the relations of sharing (goods and services) and participating (in social activities);

(d) S includes a nonvoid set $M \subset S$, called *management*, such that every member of M is a relation in some Cartesian power of the membership $\mathscr{C}(\sigma)$, and representing some habitual actions of members of σ upon fellow members, consisting in controlling certain relations in S (e.g. courtship and learning) or in T (e.g. hunting and ploughing);

(e) the set T of transformation relations includes a nonempty set $W \subset T$, called *work*, such that every element of W is a relation from a subset of $(\mathscr{C}(\sigma))^p \times (\mathscr{E}(\sigma))^q$, with $p, q \geq 1$, into a nonvoid proper subset A of $\mathscr{E}(\sigma)$, representing the habitual transformation, by some members of σ, of certain things in $\mathscr{E}(\sigma)$ (e.g. iron ore) into things in A (e.g. iron tools) called *artifacts*;

(iv) the social relations S and the transformation relations T generate equivalence relations inducing partitions of the membership of σ into social groups (social differentiation);

(v) σ is self-supporting as long as it lasts, i.e. is capable of satisfying its needs by work.

According to this definition not every bunch of humans, not even every sociosystem (Definition 4.49), constitutes a human society: to qualify as such a collection of humans must share an environment, transform the latter deliberately (by work, not just by wearing it out), hold social relations and communicate among themselves, be divided into social groups, and constitute a self-reliant unit. Thus a factory, a school and an army, though subsystems of a society and therefore sociosystems, are not societies. Further, the social relations among the members of the society must include man-transforming relations M such as educating, organizing and coercing, and they must be set up and maintained with the help of communication.

So much for a quick characterization of the concrete systems we call 'human societies', that make up the membership of the set of all human societies – past, present, and future. Let us emphasize that they are systems, not just either things or aggregates of such – let alone supraindividual entities of an immaterial nature. Note also that, although societies are composed of organisms, they have no organismic properties: they do not synthesize proteins, reproduce by cell division or conjugation, etc. In fact societies have bulk or systemic properties, some of which are peculiar to them:

DEFINITION 5.2 Let P be a property. Then

(i) P is a *societal property* iff there exists (now or at any other time) a human society possessing P;

(ii) if P is a societal property of some society σ, then P is a *resultant* property of σ iff P is also a property of some components of σ;

(iii) if P is a societal property of some society σ, then P is an *emergent* property of σ iff P is not a resultant property.

Example 1 The total food consumption per year of a given society is a resultant property of it for being the mere additive aggregation of the individual consumptions. Not so the total energy consumption, let alone the total production, for they are social. Likewise the family structure, the political organization, and the cohesiveness of a community are emergent societal properties. *Example 2* All members of a society possess properties that are neither individual nor societal, such as shyness and conformity (or its dual, deviance). Take the latter, analyzed by Definition 4.54. In it 'F' represents a property of an individual, the average $\bar{F}$ a resultant property of a sociosystem, and the standard deviation d of the distribution of F values in the society represents an emergent property of the latter. (So much so that if d is small we can say that the sociosystem is conformist.)

Our first hypothesis is that the concept of an emergent societal property is not idle:

POSTULATE 5.1 Every society has emergent properties.

We hasten to note, though, that 'emergent' does not mean "unexplainable" or "unpredictable". Firstly, because "emergence" is an ontological not an epistemological category. Secondly because it is the business of science not only to acknowledge emergence but also to incorporate it into theories and thus render it understandable and sometimes also predictable.

1.2. *Subsystem, Supersystem, Specific Function*

Every society, no matter how primitive, is composed of various subsystems. And almost every contemporary human society is a subsystem of a more inclusive social system, such as a province, a nation, a regional system, or the world system. So, we need

DEFINITION 5.3 Let σ be a human society. Then

(i) any social system σ' such that σ' is a subsystem of σ (i.e. $\sigma' \lesssim \sigma$) is a *social subsystem* of σ;

(ii) any society σ' such that σ is a subsystem of σ' (i.e. $\sigma \lesssim \sigma'$) is a *social supersystem* of σ.

Example The health system of a modern community is a subsystem of it characterized by the following coordinates:

Composition: physicians, nurses, administrators, clerks, janitors, in-patients, out-patients, etc.;

Immediate environment: hospital buildings, furniture, instruments, ambulances, families of the members of the system, etc.;

Structure: relations of diagnosing and prognosticating, medicating and operating, advising or otherwise acting on patients, keeping their families informed, organizing work, teaching, and supplies, keeping the wards clean, etc.

Not every component of a social system is a social subsystem of it. Thus the persons composing a social system are biosystems not sociosystems. Nor is every subset of the membership (composition) of a social system a social system itself. For example the employees of an industrial plant and the students of a school need not be social systems. It is only when organized, and acting as a whole in certain respects (e.g. going on strike), that a bunch of persons becomes a system. *A fortiori* the working class and the student population of a society may not constitute systems.

Consider now a homogeneous set of subsystems of a given society, such as the collection of all farms, all refuse collection and disposal systems, or all schools. Such sets are not systems but *kinds* of system. An arbitrary such set will be called an F-system, where F is short for the function(s) characterizing the member subsystems in contradistinction to others. More precisely, we put forth

DEFINITION 5.4 Let σ be a human society and call

$$S(\sigma) = \{\sigma' \in \Sigma \mid \sigma' \lesssim \sigma\}$$

the collection of all social subsystems of σ. Further, let F be a certain set of social relations or transformation relations, i.e. $F \subset S \cup T$, and let

$$F(\sigma) = \{\sigma' \in S(\sigma) \mid F \subset \mathscr{S}(\sigma')\} \subset S(\sigma)$$

be the collection of subsystems of σ where the F relations hold.
Then

(i) $F(\sigma)$ is called the *F-sector* of σ;

(ii) $G(\sigma) = \bigcap_{\sigma' \in S(\sigma)} \mathscr{S}(\sigma')$ is (are) called the *generic function*(s) of the subsystems of σ;

(iii) $F_s(\sigma) = \bigcup_{\sigma' \in F(\sigma)} \mathscr{S}(\sigma') - G(\sigma)$ is (are) called the *specific function(s)* of the F-sector of σ.

To put this definition to work think, e.g., of the retail, the school, the manufacturing, and the postal sectors of a community: see Figure 5.1.

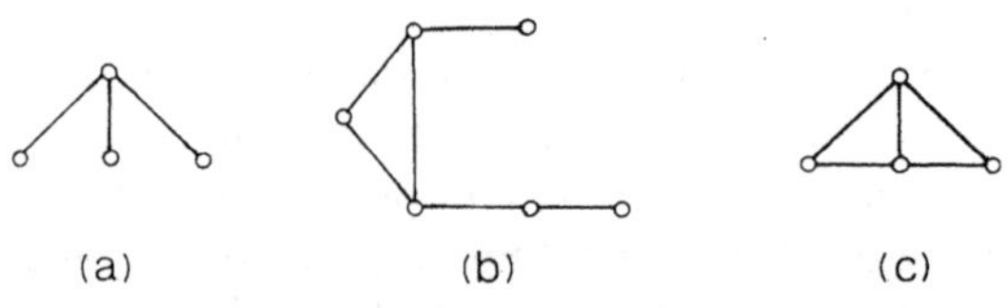

Fig. 5.1. Three sectors: (*a*) school system (connected indirectly via school board) or retail marketing system (connected indirectly via wholesale distributor); (*b*) manufacturing system (where the inputs to some subsystems are outputs to others); (*c*) postal network (controlled by central post office).

In all these cases the subsystems concerned perform certain peculiar functions that set them apart from the subsystems in other sectors; but they also share a number of functions proper to all social subsystems, such as consuming or transforming energy, producing waste products, and communicating with other subsystems of the community.

Primitive societies have relatively few sectors – e.g. the domestic work sector, the hunting sector, the manufacturing sector, the cultural sector (which is actually a system), and a few others. As society evolves some sectors become extinct while new ones emerge. Industrial societies are characterized by the large number and size of their sectors. They are also characterized by an increasing systematization of their sectors, i.e. an increasing coupling among the members of their sectors. Thus in a monopolistic economy every economic unit is a member of a tightly knit system. The general concept is elucidated by

DEFINITION 5.5 The F-sector $F(\sigma)$ of a society σ is the *F-system* of σ iff the membership of $F(\sigma)$ constitutes the composition of a subsystem of σ.

Finally we assume

POSTULATE 5.2 Every human society has a number of sectors. Moreover (*a*) every member of a human society belongs to at least two sectors of it and no individual belongs to all sectors at the same time, and (*b*) some sectors of every human society are systems.

Remark 1 In other words, there is some division of labor in every society, however primitive; yet the division of labor in human societies is never as thorough as in insect societies. *Remark 2* The above axiom, together with the epistemological postulate of knowability, entails the methodological theorem that every society can be ananalyzed into a number of sectors and, in particular, subsystems. What is nowadays called *large systems analysis* is mainly such an analysis, or recognition and characterization of the sectors and subsystems of a given society – such as, e.g., the transportation system of a city, and the marketing system of a region.

1.3. *Institution*

The word 'institution' is ambiguous: sometimes it denotes an individual sociosystem of a certain kind, such as a given school, and at other times it denotes a set of social actions and norms, such as marriage and religious cult. We are interested here in yet another acceptation of the word 'institution', namely the one used in speaking of the Family, the Mail, the Bureaucracy, or the Law.

Social philosophers are divided over the nature of institutions, hence over the proper way of studying them. The two classical doctrines in this field are holism and individualism. The holist regards an institution such as the Law as a disembodied corpus of rules or laws that individuals enforce, obey, or violate. According to this view the institution hovers above the persons and moreover is superior to or more valuable than them. No wonder holism, in some from or other, is inherent in totalitarian ideologies. The individualist rejects such an abstraction and regards the Law as the set of judges, lawyers and policemen – with total neglect of the subjects of the Law, namely the common citizens. When confronted with a sentence such as 'The Law compelled X to do Y', the individualist translates it into 'Judge (or policeman) U forced X to do Y'. But of course this is a poor substitute for the original sentence, as the person U in charge of enforcing the law behaves as he does because of what he is, namely a member of a certain legal system with a code of legal procedure. Different individuals might have behaved in roughly the same way. On the other hand the same individuals, in a different legal system, might behave differently. In conclusion, the legal system in question is neither the set including the individuals U and X nor a set of supraindividual rules. Hence although both holism and individualism have their points, neither is competent to account for the behavior of the individual in society. Holism fails because it postulates totalities transcending the components, and individualism fails because it ignores the totalities.

Our systemist view is as follows. A legal system is a subsystem of some society, i.e. it is a part of a human community set off by certain peculiar social relations of the M type, such as those of judging and passing sentence. And what is called 'the Law', with an awesome capital, is the collection of all legal systems. In general, we shall construe an institution as the set of all the F-sectors of all societies, regardless of their spatial and temporal propinquity. Thus the collection of all school sectors is called Education, the set of all trade unions Organized Labor, the set of all postal systems the Mail, and so on. (Recall Definition 5.4 of an F-sector and Definition 5.5 of an F-system.) We make then

DEFINITION 5.6 Let $\Sigma = \{\sigma_1, \sigma_2, \ldots, \sigma_m\}$ be a set of human societies, and call F_{ik} the kth sector of society σ_i, i.e.

$$F_{ik} =_{df} = F_k(\sigma_i) = \{\sigma' \in S(\sigma_i) \mid F_k \subset \mathscr{S}(\sigma')\},$$

where F_k designates a certain set of social or of transformation relations. Then the family of F-sectors

$$\mathscr{F}_k = \{F_{ik} \mid F_{ik} \text{ is the } k\text{th sector of } \sigma_i \;\&\; \sigma_i \in \Sigma \;\&\; 1 \leq i \leq m\}$$

is called the *F-institution* of Σ.

In our construal, then, an institution is neither an individual sociosystem nor a Platonic set of rules, norms or goals governing people from above. But, since an institution is a *family of sets* of concrete things, it is tolerably abstract. (A sheer abstraction is one that, unlike an $\mathscr{F}_k$, does not boil down to concrete particulars. Sheer abstractions belong in logic and pure mathematics but are intolerable everywhere else.) To endow such a set of sets with properties of concrete individuals, such as life (or worse: an eternal life), a mind of its own (or worse: a deranged mind), aggressiveness, or the possibility of conflicting with similar types (e.g. other institutions), is to indulge in reification.

Note that not everything that passes for an institution in ordinary language qualifies as such according to our definition. For instance Work, War, Marriage, and Money do not because they are not families of sets of social subsystems.

It might be objected that our definition of an institution misses an essential ingredient, namely the goals and norms of the institution. Thus when applied to the case of the Law regarded as an institution, our convention seems to make no room for the codes of law and legal procedure that regulate the relationships at work in any legal system. The part such rules play is so important that some social philosophers and scientists have gone to the point of identifying an institution with the set of its norms or rules. This, though exaggerated and definitely Platonic, does contain a grain of truth: it is obvious that, unless certain rules are observed, the institution (or rather each of the subsystems forming it) decays or even becomes extinct. However, it does not follow that a set of institutional rules (e.g. the bylaws of an association) is an Idea dangling above the corresponding social subsystems, let alone identical with them.

The institutional rules are just the way the corresponding subsystems are thought (rightly or wrongly) to function optimally in the interest of society or of some social group. If preferred, those rules are prescriptions to operate the subsystems in an efficient manner – i.e. to attain the goals of its members, or rather of those whom the system is supposed to serve.

Undoubtedly such institutional rules can be rendered explicit for the most part, just as the rules for doing properly some job or playing some game can be formulated explicitly up to a point. But the writing down of a rule does not confer autonomous existence upon it: this may only suggest, wrongly, that it has one – perhaps because, when inscribed on a clay tablet, a stone slab, or a papyrus, it can outlast its author. The upright and merciless judge enforces rules just as the foreman has an operating manual observed.

The actions abiding by the rules of an institutional game may be called 'rule directed' as long as we do not suggest thereby the idealist thesis that rules exist independently of the rulers, the rule enforcers, and the ruled – anymore than the laws of nature are separate from the things satisfying them. The rules of social behavior are inherent in the very relationships holding among the members of the system in question. (Much the same holds for any deviations from that behavior: the delinquent may be an outlaw in the sense that he does not observe the code of law, but he is as much a member of the legal system as the judge – so much so that there is no need for a legal system where there is no occasion for delinquency.) In short the institutional rules, and also the patterns of breaking such rules and the rules for punishing these violations, are incorporated into the very network of social relationships that hold the system together. And it is not "society" that sanctions them or punishes any deviations from them, as the holist would have it, but rather some members of society, namely those in charge of regulating the functioning of the social system. By conceiving of institutional rules as patterns of the optimal or at least desired functioning of social systems, it becomes easier to understand how those rules originated, where they fail, and in what respects they could be improved upon if not repealed altogether.

So much for a preliminary conceptual framework. Let us now proceed to analyze more closely some of the typical concepts of the social sciences, starting with that of work.

2. Social subsystems and supersystems

2.1. *Work*

Work, one of the peculiarities of human behavior, is a kind of activity – not just an action but a regular one. An activity, whether performed by animal or by machine, may be defined as a patterned sequence of actions aiming at some preset goal. Thus, far from being erratic, an activity proceeds in such a way that, normally, the difference between attainment and target is gradually decreased (Wiener, 1948, Ch. IV). See Figure 5.2. We compress these remarks into

DEFINITION 5.7 A system performs an *activity* iff it executes a sequence of goal-directed actions.

For a system to perform an activity it is necessary and sufficient that it be a control system and that it, or whatever controls it, have a definite goal. In the case of work, the worker himself, together with his tools, constitutes a control system.

Unlike machines, humans are versatile or multipurpose systems: they can perform activities of very many kinds. This multiplicity can be represented as follows. Call ${}^{\sigma}A_{mn}$ the action performed by system σ on the mth aspect (property) of the nth thing in the environment of σ. Then the total activity of σ is displayed by its activity matrix ${}^{\sigma}A = \|{}^{\sigma}A_{mn}\|$. This is an $M \times N$ matrix, where M is the number of properties and N the number of things influenced by σ. An empty entry corresponds to a nonexisting feature of the corresponding thing. If an entire community, or a sector of it, with membership S, is engaged in a set of activities, we may define the aggregate activity of the community or sector as the matrix

$$A = \bigcup_{x \in S} {}^{x}A,$$

every entry of which is the set theoretic union of the partial entries with the same indices.

Work is a special kind of activity, namely one thought to be useful. We define it as follows:

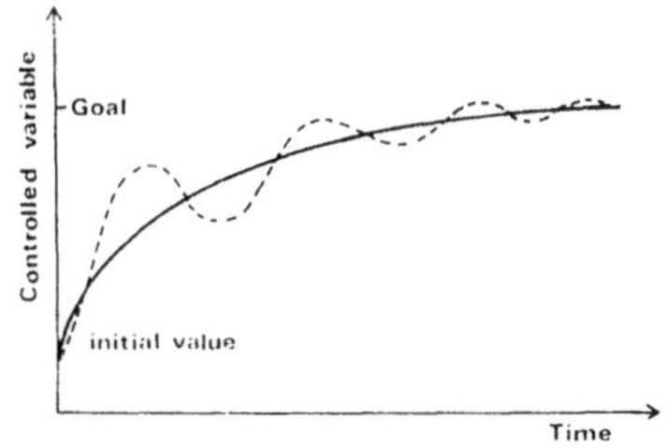

Fig. 5.2. A system attaining asymptotically a goal state in either of two typical ways: (*a*) monotonically (full line) or (*b*) by trial and error (dotted line). Note in the latter case the phenomenon of overshooting (momentarily overreaching the goal).

DEFINITION 5.8 If x is a system, then x *works* iff there exists a thing y and an animal z (equal to or different from x) such that

(i) x exerts an activity on y (in a way other than by eating y) and by so doing produces an appreciable change in the state of y, and

(ii) the thing y in the new state brought about by the activity of x is of value to z.

We distinguish three kinds of work, namely labor (L), cultural work (K), and managing (M). Labor aims at transforming nonhuman things into more valuable things; cultural work aims ultimately at modifying brain functions; and managing consists in controlling both labor and cultural work. See Figure 5.3. Let us first examine labor.

2.2. *Labor*

We start off by making

DEFINITION 5.9 Let σ be a society with composition $\mathscr{C}(\sigma)$, environment

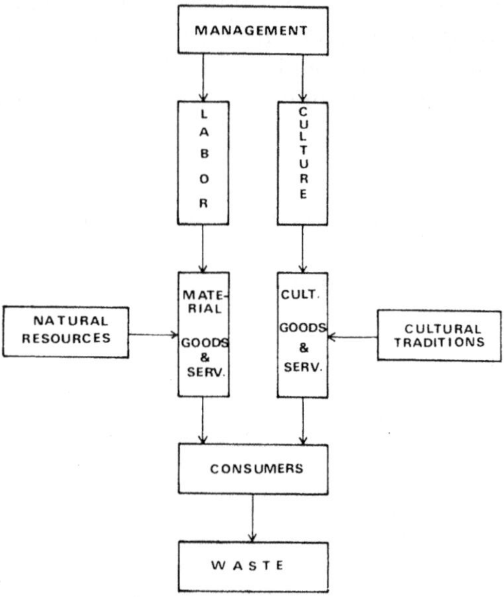

Fig. 5.3. The three main kinds of work: labor, culture, and management of both. The outputs of labor: material goods and services. The outputs of cultural work: cultural goods (e.g. books) and services (e.g. education).

$\mathscr{E}(\sigma)$, and structure $\mathscr{S}(\sigma) = S \cup T$, and call L a subset of the transformation relations in σ. Then

(i) the element $L_i \in L$, with $1 \leq i \leq r$, is called the ith kind of *labor*, or *primary work*, done by members of σ, iff L_i is an activity performed on nonhuman things, the outcome of which is a set of things or a change of state in things, that is valuable to members of σ;

(ii) the domain of L_i is $(\mathscr{C}(\sigma))^p \times (\mathscr{E}(\sigma))^q$, where $p, q \geq 1$, and the range $A_i \subset \mathscr{E}(\sigma)$ of L_i is called the *product* of L_i, and each element of A_i an *artifact*;

(iii) the totality of products [goods and services] of the labor performed in σ, i.e. $\mathscr{A}(\sigma) = \bigcup_i A_i$, is called the *material production* of σ, whereas the complement $\mathscr{N}(\sigma) = \mathscr{E}(\sigma) - \mathscr{A}(\sigma)$ of $\mathscr{A}(\sigma)$ to the environment of σ is called the *natural environment* of σ;

(iv) the subset ${}^1P_i \subset \mathscr{C}(\sigma)$ occurring in the domain of the labor-relation L_i is called the set of *primary producers* of A_i, or the ith *labor force* of σ;

(v) the union ${}^1P = \bigcup_i {}^1P_i$ of primary producers of σ is the *labor force* or *primary manpower* of σ.

Remark 1 Our definition includes not only the production of material goods but also the performance of services. *Remark 2* It is not implied that labor be manual: it may well consist in controlling automated machinery. *Remark 3* Our concept of labor or primary work does not coincide with the standard concept of primary industry: the primary producers or laborers referred to above may be hunters or fishermen, farmers or industrial workers, miners or carpenters or repairmen. *Remark 4* Nor does our concept of an artifact coincide with the popular one: it is a technical one. Thus domestic bacteria, such as those used to make butter and cheese, as well as cultivated plants and domesticated animals, being products of artificial selection, are artifacts in our sense of the word. Machines are only a very special kind of artifact.

The set $L = \{L_i \mid 1 \leq i \leq r\}$ of kinds of labor cannot be empty for any society, under penalty of extinction. It may happen that a community subsists for a time on spoils of some kind, but even the most bountiful spoils of war become exhausted after a while. Moreover the set L must contain more than one member, for each kind of labor 'leads to' other kinds. Thus felling trees is done to employ timber for one or more purposes (activities), and it cannot be done efficiently, except by beavers, without axes or saws – which have to be manufactured. We compress these remarks into

POSTULATE 5.3 In any human society

(i) there is more than one kind of primary work, i.e. $|L| > 1$;

(ii) there are workers engaged (permanently or temporarily, exclusively or not) in every kind of work done in the society, i.e. $P_i \neq \varnothing$ for every $1 \leq i \leq r$.

Finally we make

DEFINITION 5.10 Let $\mathscr{E}(\sigma) = \mathscr{N}(\sigma) \cup \mathscr{A}(\sigma)$ be the immediate environment of a society with labor force ${}^1P$ and labor (primary work) relations in L. Then the *material force of production* of σ is the ordered triple $\langle {}^1P, \mathscr{N} \cup \mathscr{A}, L \rangle$.

Should any coordinate of this triple be empty, there would be no material force of production. In particular, an inactive (unemployed) labor force is as ineffectual as a nonexisting stock of natural resources. However, a scarcity in the latter can be compensated for by material production thanks to exchange. But someone, in some society or other, has got to produce food: there is no such thing as a postagricultural world society. Likewise there is no postindustrial society but only one in which most of the labor force is engaged in performing services rather than in agricultural or industrial work. No labor, no human society.

Let us move on to the other components of the manpower of any human society.

2.3. *Cultural and Managerial Work*

Let us now characterize the kind of work done by artists, bards, composers, performers, actors, story tellers, writers, teachers, scholars, scientists, engineers, and physicians of all descriptions, whether competent or not:

DEFINITION 5.11 Let σ be a society with composition $\mathscr{C}(\sigma)$ and social relations in S, and call $K \subset \mathscr{S}(\sigma)$ a proper subset of the social structure of σ. Then

(i) the element K_j of K, for $1 \leq j \leq s$, is called the jth kind of *cultural* (or *secondary*) work done by members of σ iff K_j is an activity capable of evoking feelings or thoughts, or supplying ideas intervening in the primary production of σ;

(ii) the subset ${}^2P_j \subset \mathscr{C}(\sigma)$ occurring in the domain of the cultural work relation K_j, is called the jth set of *cultural* (or *secondary*) producers of σ;

(iii) the union ${}^2P = \bigcup_j {}^2P_j$ is called the *cultural* (or *secondary*) *manpower* of σ.

Remark 1 Although cultural work aims at changing the way people feel or think, K is not disjoint from T. Indeed, all cultural work involves transforming some thing or other, if only the state of the surrounding air. Moreover there are useful works of art, such as artistically designed pieces of furniture, as well as useful works of science, such as scientifically controlled instruments. *Remark 2* According to clause (i), writing a book that may be published is doing a spot of cultural work; even writing or performing a piece of rock "music" is perpetrating some cultural work. Not so engaging in praying, dreaming, or transcendental meditation – which is sleeping anyhow. Likewise drawing the blueprint for a house or a machine or the pattern for a dress is engaging in cultural work, whereas drawing exclusively for one's pleasure, or humming a melody, are not. To be cultural an activity must be social or socializable as well as artistic or intellectual.

We shall return to the matter of cultural work in Sec.3.2. Let us now characterize the relations aimed at influencing social behavior in an immediate way, rather than through the subtle medium of culture:

DEFINITION 5.12 Let $M \subset S$ be a subset of the social relations in a human society σ. Then

(i) the element $M_k \in M$ is called the kth kind of *managerial* (or *organizing* or *tertiary*) *work* done by members of σ iff M_k is an activity contributing to controlling some primary or secondary work done in σ;

(ii) the subset ${}^3P_k \subset \mathscr{C}(\sigma)$, for $k = 1, 2, \ldots, t$, occurring in the domain of the managerial relation M_k, is called the kth set of *tertiary* (*managerial, organizing*) *workers* of σ;

(iii) the union ${}^3P = \bigcup_k {}^3P_k$ is called the *managerial force* (or *tertiary manpower*) of σ;

(iv) the subset ${}^3P_L$ of P^3 devoted to controlling the primary production of σ is called the *production management* of σ, and its complement ${}^3P_k = {}^3P - {}^3P_L$ the *culture management* of σ.

In other words, we call *managerial* all the kinds of work aimed at controlling production, whether agricultural or artistic, industrial or intellectual. Planners, plant managers, executives, members of socialist farm or factory councils, commune leaders, judges, bureaucrats, censors, and in general decision makers and transmitters of all kinds belong to management as construed above. Moreover one and the same individual may belong to two or more manpowers, as is often the case in primitive societies.

We assume

POSTULATE 5.4 Every human society has both a cultural manpower (i.e. ${}^2P \neq \varnothing$) and a managerial one (i.e. ${}^3P \neq \varnothing$).

From this platitude it follows that, under risk of destroying society, it is impossible to do entirely without either the cultural force or the managerial force. What one should aim for is maximizing the harmonious cooperation among all three work forces, i.e. easing the tensions and therefore eliminating the need for keeping repression forces. In particular, rather than try quixotically to eliminate management one should strive to make it work in the interest of material and cultural production rather than in its own interest. Such a blending of management with productive forces obtains in a well run modern industrial plant, marketing system, communications network, hospital, or school – in short wherever some people work according to the canons of advanced (scientific) technology. Such sociosystems deserve a name of their own:

DEFINITION 5.13 A system τ is a *technosystem* iff

(i) the composition of τ includes rational beings and artifacts;

(ii) the environment of τ includes components of a society;

(iii) the structure of τ includes production, maintenance or use of artifacts.

A modern American household full of gadgets does not qualify as a technosystem unless it is the seat of a family concern where intellectual work plays a significant role. On the other hand a modern defense system is a technosystem, for military experts are supposed to do some intellectual work as part of their duty.

The composition or population of any human society is made up of an active part – those who do work of some kind – and an inactive one. All those persons – housewives, shepherds, fishermen, hunters, craftsmen, industrial workers, white collar workers, physicians, engineers, poets, musicians, philosophers, etc. – who participate in the transformation of raw materials into artifacts, do artistic or intellectual work, or take part in organizing material or cultural production of goods or services, make up the active population or human resources. In short, we have

DEFINITION 5.14 Let σ be a human society. Then

(i) the *active population*, or *human resources*, of σ is the union of its manpowers: $A = {}^1P \cup {}^2P \cup {}^3P$;

(ii) the *inactive population* of σ is the complement of A to the membership of σ, i.e. $I = \mathscr{C}(\sigma) - A$.

2.4. *The Main Subsystems of a Human Society*

Neither the active population nor the inactive one constitute systems; nor are the subsets ${}^1P$, ${}^2P$, and ${}^3P$ systems. Thus the managerial force of a society is hardly a system: managers are components of either the economic, the cultural, or the political systems of a society. Each of these systems draws its components from all three manpowers: see Figure 5.4. Thus industry, trade and government employ artists, intellectuals and professionals of various descriptions; and both manual and clerical workers take part as auxiliaries in cultural production as well as in political management. We have then the division

$${}^iP = {}^iP_L \cup {}^iP_K \cup {}^iP_G \quad \text{for} \quad i = 1, 2, 3$$

of each manpower into one part engaged in primary production, another engaged in cultural production, and another engaged in political control. Whereas the upper index i identifies the specific output of the sector, the lower indices name its various inputs.

Correspondingly we distinguish certain subsets of the collection S of social relations: we call S_L, S_K, and S_G respectively the social relations associated with primary production, cultural production, and political administration. Finally, we group similarly the work relations: we call

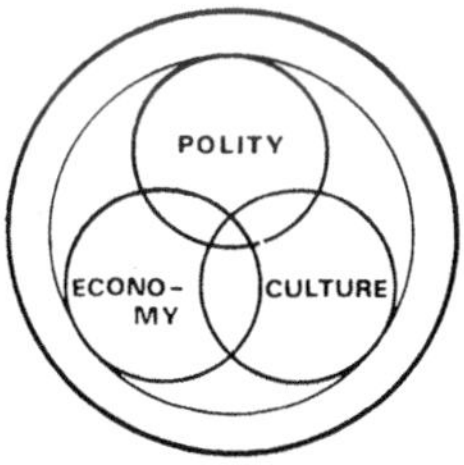

Fig. 5.4. The main subsystems of any society. Outer circle: whole society. Inner concentric circle: kinship system. Eccentric circles: *economy* (gathering, hunting, fishing, manufacturing, trading, financing); *culture* (artistic, technological, scientific, humanistic creation and their diffusion); and *polity* (government of all kinds, nongovernmental political activity, and the military). The overlaps suggest that some persons participate in two or even three systems. E.g. a scientist belongs to the cultural system by virtue of his research and teaching; to the economic system if he writes textbooks; and to the political system if he acts as consultant to the government.

L_K and L_G repectively the kinds of primary work done in cultural production and in government by members of the labor force; K_L and K_G the kinds of cultural work involved in primary production and in politics respectively; and M_L, M_K and M_G the kinds of management involved in primary production, cultural production, and politics respectively. We are now ready for

DEFINITION 5.15 Let σ be a society with labor force ${}^1P$, cultural force ${}^2P$, and managerial force ${}^3P$. Further, let the subscripts L, K and G identify whatever is associated with primary production, cultural production, and political administration respectively. Then

(i) the subsystem of σ represented by

$$\varepsilon_\sigma = \langle {}^1P_L \cup {}^2P_L \cup {}^3P_L, \mathscr{E}(\sigma), S_L \cup (L \cup K_L \cup M_L)\rangle$$

is the *economy* (or *economic system*) of σ;

(ii) the subsystem of σ represented by

$$\kappa_\sigma = \langle {}^1P_K \cup {}^2P_K \cup {}^3P_K, \mathscr{E}(\sigma), S_K \cup (L_K \cup K \cup M_K)\rangle$$

is the *culture* (or *cultural system*) of σ;

(iii) the subsystem of σ represented by

$$\pi_\sigma = \langle {}^1P_G \cup {}^2P_G \cup {}^3P_G, \mathscr{E}(\sigma), S_G \cup (L_G \cup K_G \cup M_G)\rangle$$

is the *polity* (or *political system*) of σ;

(iv) ε_σ, κ_σ, and π_σ are the *artificial* main subsystems of σ.

The previous definition allows us to restate Postulate 4.26 in a more complete way:

POSTULATE 5.5 (i) Every human society is composed of four subsystems: the kinship, the economic, the cultural, and the political systems; (ii) every sociosystem in a human society is part of at least one of its four main subsystems; (iii) every individual in a society belongs to the composition of at least one of the four main subsystems of the society.

Remark 1 By conceiving of the economy, the culture and the polity as systems, one avoids the sterile philosophies of holism and of individualism. For the same reason one obtains the possibiity of modeling the entire economy (or culture or polity) of a society as a system with definite composition and structure – provided one includes in the latter the links of the economy (or the culture or the polity) with the other subsystems. *Remark 2* All three artificial subsystems share the same natural and artificial environment; in particular no culture and no polity can function in a physical vacuum. Hence extreme internalism (e.g. cultural idealism) is just as inadequate as extreme externalism (e.g. ecological or geographic determinism). *Remark 3* Each manpower is distributed among the three artificial subsystems. In particular, not even the most primitive economic system fails to employ some cultural workers and some organizers – even if they are at the same time primary producers – and not even the most anti-intellectual of political systems dispenses altogether with intellectuals, e.g. as planners and censors. *Remark 4* It is possible for one and the same individual to belong to more than one of the main subsystems. It is impossible for an individual to fail to belong to at least one of the main subsystems – i.e. complete marginality is impossible within a society. *Remark 5* A significant change in any of the three coordinates of each subsystem (membership, environment, structure) will affect the system as a whole. Examples: an increase, or decrease, in the percentage of skilled workers or of intellectuals; an increase, or decrease, in the consumption of energy *per capita*. When the change is both drastic and fast, it is called a *crisis* – of growth or of decline. *Remark 6* Because each of the three artificial subsystems is a subsystem of the same society, it is linked to the other two. Such couplings would be impossible if the systems in question (the economy, the culture, and the polity) were sets of individuals (individualism) or Platonic ideas (idealistic holism). *Remark 7* Disregarding or even minimizing the importance of any of the four main subsystems of a society leads to lopsided views, hence inadequate doctrines of political action. The same holds for neglect of the interactions among the subsystems.

2.5. *The Supersystems: Nation and World*

We regard a *nation* as a social system, or rather supersystem, composed of all the kinship, economic, cultural, and political systems coexisting and interacting within a territory. Unlike the city-states of Antiquity, modern nations are composed of several societies, each of which occupies a given geographical region within the national territory (which need not be in one piece). Each component society contributes its own kinship, economic, cultural, and political subsystem to the whole. The artificial subsystems are coupled to one another forming a national economy, a national cultural system, and a political system. None of these national systems need be homogeneous: on the contrary, variety makes for viability. (The ideology according to which homogeneity is a supreme desideratum is called *nationalism*, and is particularly vociferous in demanding racial purity, cultural homogeneity – i.e. poverty – and political uniformity – i.e. dictatorship.) If the integration among the various societies composing a nation is weak, then the nation is so only in name, as was the case with the Holy Roman Empire most of the time. If the integration is too strong, then the nation is cut off from other nations. The best coupling strength is middling.

The relations among or within the three main artificial subsystems of a nation are of the following kinds: (*a*) economic – e.g. society *X*'s supplying energy to society *Y*; (*b*) economicocultural – e.g. research scientists's developing a new variety of wheat; (*c*) economicopolitical (e.g. an economic group's lobbying for a bill favoring its interests); (*d*) cultural (e.g. intercity library loans); (*e*) politicocultural (e.g. a governmental agency's encouraging or obstructing technological innovation); (*f*) political (e.g. the armed forces seizing control of certain governmental agencies).

All of these relations are intersystemic. Yet in the final analysis they rest on person to person relationships, where the persons concerned act out their social roles. Thus when analyzing a social relation *S* between two subsystems of a nation we may ultimately explain *S* in a way such as this: System *b* bears relation *S* to system *c* if and only if there is an individual compoment *x* of *b* and an individual component *y* of *c*, such that *x* bears *S* to *y*, where *S* is an inter personal bond such as, e.g. writing a note asking for some item of property on loan. (Clearly the individuals concerned may be represented by robots, but the latter are built to act not in their own interest but in that of their master.) In other words, the relations in a supersystem such as a nation can be analyzed, at least in principle, in terms of interpersonal relations, as long as the roles of the persons are identified – roles which are not further reducible to properties of isolated persons. If such roles and the associated responsibilities are forgotten, i.e. if societal relations are mistaken for private relations, not only a faulty analysis results but a law may be broken. In short, although every social system is ultimately composed of individuals, these happen to engage in relations that are incomprehensible without the hypothesis that the system has supraindividual properties. (Recall Definition 5.2.)

Economists, sociologists, politologists, and social historians are becoming increasingly interested in whole regional blocs and international systems, thus recognizing that the nation is often either a subsystem or a supersystem, and in either case an inadequate unit of study. (The model for social history is Braudel's *La Méditerranée* (1949). A recent addition to the systemist approach is Wallerstein's *The Modern World-System* (1974).) Indeed there are no isolated nations left it there ever were. Every modern nation bears economic, cultural or political relations to other nations, and this on four levels: person to person, subsystem to subsystem, nation to nation, and international organization to international organization. (It goes without saying that all system to system relations are designed and implemented by individuals. However, these individuals can often be replaced by others enacting the same roles or functions. Not that their personalities are unimportat; only, these serve or disserve their roles or functions.) Let us take a quick look at international relations. (Cf. Deutsch, 1968; Fritsch, 1974; Bunge *et al.*, Eds., 1977.)

In addition to some 150 nations there are today about 10,000 international organizations. The latter include (*a*) international governmental organizations such as the World Health Organization, (*b*) international nongovernmental organizations such as the International Union for the History and Philosophy of Science, and (*c*) multinational corporations such as the General Motors Co. Since every other social system is either a subsystem of some international organizations or closely related to some of them, it would be utopian to plan anything on a national scale without taking the international context into account. This context is the *international system*, composed by all nations and all international organizations; its environment is no less than the biosphere; and its structure the set of relations (economic, cultural, political, economicopolitical, politicocultural, etc.) among its various components. These relations are rather entangled. For example, nation A trades with nation B and has close political ties with nation C, a rival of B; consequently the trade relations are affected by the political ones and reciprocally. Politologists tend to say that this is a case of relations among relations or cross relations. Since this notion is fuzzy, we prefer to analyze the same facts using our systemic framework, namely as follows.

Every international actor (nation, intergovernmental organization, multinational corporation, etc.) has three chief subsystems, every one of which can interact with any of the subsystems of the other two. See Figure 5.5. By completing this diagram we realize that two members of the international system can engage in 9 *sets* of binary relations, to wit: $E_1 - E_2$, $E_1 - C_2$, $E_1 - P_2$, $C_1 - E_2$, $C_1 - C_2$, $C_1 - P_2$, $P_1 - E_2$, $P_1 - C_2$, $P_1 - P_2$. For n members of the supersystem, the total number of sets of binary international relations is $N = (9/2)n \cdot (n - 1)$. For the entire interna-

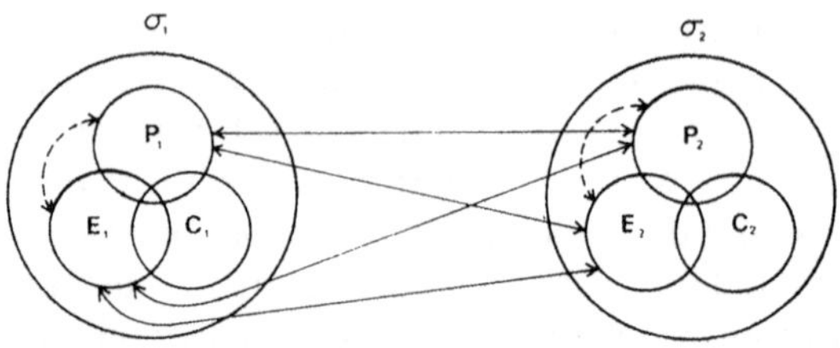

Fig. 5.5. Example of international economic and political relations between two nations σ_1 and σ_2. Note the two intranational relations besides the four international ones: the former are indicated by dotted lines.

tional system today, $n \cong 10{,}000$, whence $N \cong 450$ million sets of international relations. (The number computed on the assumption of the ghostly relations among relations is far more staggering, namely of the order of 10^{15}, or 7 orders of magnitude greater than ours.) This is a forbidding figure yet still modest by comparison with the number of pairwise relations among the atoms in a wedding ring, namely 10^{38}. The politologist can get around this hurdle just the way the physicist does, namely by taking into account only the relations among nearest neighbors. In this way whole blocs emerge, and then the relations among blocs can be studied – just as with any other supersystem analyzed into lower level subsystems. Such a systems analysis paves the way to the construction of theoretical (mathematical) models including precise functional relations among the variables. (Cf. Bunge, 1977h.)

So much for a preliminary study of the main subsystems and supersystems of human society. Let us now take a closer look at each of the three artificial subsystems of every society.

3. Economy, Culture, and Polity

3.1. *The Economic System*

According to Definition 5.15 the economic system of a society is composed of its primary workers (${}^1P_L$), cultural workers (${}^2P_L$), and managerial workers (${}^3P_L$). The primary workers are engaged in primary production (L), the cultural workers are in charge of the cultural input to primary production (K_L), and the managerial workers organize both (M_L). Thus the machinists, electricians, truck drivers and maintenance workers of an industrial plant belong to ${}^1P_L$; the engineers, quality controllers and draughtsmen, to ${}^2P_L$; the executives, accountants and typists, to ${}^3P_L$; the foremen often belong to all three sets but their specific function is as members of ${}^3P_L$. All these workers share the same environment and bear (production) relations S_L among themselves in addition to whatever other relations they may hold. The input–output schema of an economic system is shown in Figure 5.6.

Of all the aspects of the economic system, those of greatest philosophical interest are perhaps technology and the artifacts it produces and employs. Both raise numerous philosophical problems of all kinds, hardly investigated (cf. Bunge, 1977g). Let us say a word or two about artifacts, i.e. things fashioned by rational beings with the help of some technique, whether primitive or advanced. An artifact is not just one more thing but a thing belonging to a kind that did not exist before the emergence of man or some other rational being. Artifacts may be regarded as constituting a whole new level of reality, namely the *artiphysis*. The set of all concrete things may now be classed as follows:

- Things
  - Natural
  - Social
  - Artificial
    - Nonliving
      - Economic(tools, machines)
      - Cultural(books, records, etc.)
    - Living
      - Domestic plants, fungi, and bacteria
      - Domestic animals

Far from being contemptible entities whereof a philosopher does not deign to speak, artifacts are at the heart of human society and have properties absent from natural things. To understand this point it may help to

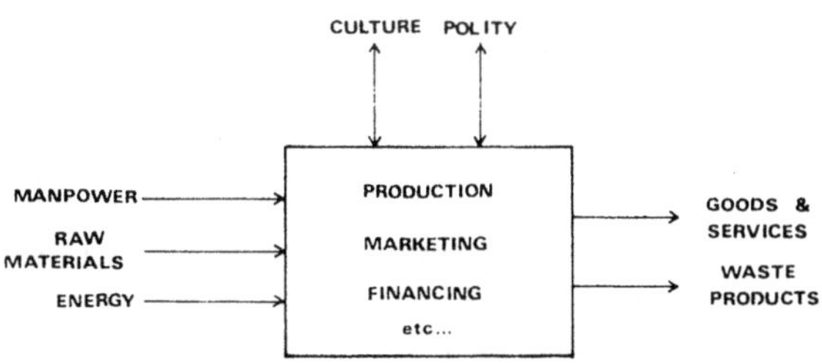

Fig. 5.6. Schema of the economic system. Inputs: natural resources (raw materials and energy) and human resources. Outputs: goods, services, and waste products. Some of the outputs are fed back into the system (e.g. as tools). Controls: culture (mainly techniques and advanced technology), and politics.

recall the various items entering the production, servicing, use, and circulation of artifacts. They are chiefly the following: (*a*) the idea or design of the artifacts; (*b*) the idea of the possible use(s) of the artifact; (*c*) a plan or design for the manufacture of the artifact; (*d*) the raw material and the energy required to produce the artifact, (*e*) the actual production process; (*f*) the marketing of the finished product; (*g*) the control of the entire process. Of all these seven ingredients only one, the fourth, is supplied by nature. Because all the others are ultimately in the hands or brains of people, the artiphysis is totally at their mercy. Men, or similar social animals, create it and can destroy it. Artifacts have no will of their own and, indeed, no existence of their own: even the most sophisticated among them are but tools. And technology has no autonomy either: it is cultivated only in societies that care for it. A band of technoclasts could kill every technologist and destroy every artifact on Earth in a few days. So, to claim that technology and the set of artifacts have a life and a momentum of their own, that they cannot be stopped, is bizarre to say the least.

We summarize the foregoing in

POSTULATE 5.6 (i) Every human society has members able to design, manufacture, repair, or use artifacts; (ii) every nonliving artifact [tool or machine] serves only the purpose(s) of some animal or other.

It follows that no machine, not even a highly evolved robot, has a purpose of its own: whatever the goal seeking mechanisms built into it, they have been designed and ultimately manufactured by rational beings in their own interests. Hence machines stand no chance of taking over unless of course programmed to do so. It follows also that, when a nonliving artifact imitates rational animals in certain regards, it does so by proxy not spontaneously. A further consequence is that artifacts can neither evaluate nor think on their own. That they do not evaluate is obvious since they have no goals other than those built into them by their designers. (This vicarious evaluation capacity may reside in the hardware, as when a machine is designed to seek or to avoid certain stimuli or else to rank them. Or in the software, as when a machine is programmed to weigh apples or alternative routes for mail delivery. In either case delegated evaluation may trigger a vicarious decision – e.g. that of discarding an underweight apple or an inconvenient route. When speaking of 'the vicarious abilities of machines' the technologist underlines 'abilities' whereas the philosopher emphasizes 'vicarious' – unless of course he takes the former's word.)

As for our claim that machines cannot think, it is quite obvious in the light of the psychobiological view on thought as an activity of certain neural systems (Definition 4.35). Here again machines, unlike rational beings, have no purposes – in particular cognitive goals – of their own, so they can properly be said to act by delegation. Surely machines can help us solve certain problems, but they cannot discover any because formulating a problem is not a rule directed activity and, as long as we do not master a rule, we cannot design or program a computer to follow such a rule. (This is also why there can be no self-programming machines: if a machine were to have problems of its own we would quickly scrap it.) Also, machines can be used to process the assumptions of certain theories, but they do not help create them – again for the same reason, namely because theory construction is not a rule directed activity but a thoroughly creative one. In short tools, however marvelous, are nothing but tools.

3.2. *The Cultural System*

In Definition 5.15 we characterized the cultural system of a society as being composed not only of cultural workers (${}^2P_K$) but also of primary workers (${}^1P_K$) and managerial workers (${}^3P_K$). The former are of course in charge of cultural production (K), whereas the other two groups are attached to them either in an ancillary capacity, as in the case of the maintenance workers at a university, or as organizers, as in the case of university administrators. Another example: the technicians and the typists working for an experimental biologist belong in ${}^1P_K$, while the director of the scientist's laboratory is in ${}^3P_K$, and occasionally also in ${}^2P_K$.

All the members of a cultural system share the same environment and are linked by certain relations S_K. But not all of the ingredients of a cultural system are on the same footing. Its pivots are: (*a*) the workers engaged in direct cultural pursuits, i.e. ${}^2P_K$; (*b*) the part of the natural or social environment that is the object of contemplation, study or rational action; (*c*) the part of the stock of artifacts composed of cultural artifacts such as pictures, books, microscopes, and the like, and (*d*) cultural work proper – e.g. research, writing, and instructing.

Ours is a restricted concept of a cultural system. A wider concept includes the entire cultural force of the society, i.e. not only the persons working for the sake of culture but also those engaged in an auxiliary capacity by the economic and political systems, such as the technologists in a quality control unit or the sociologists in a census bureau. However, this cultural supersystem is much less tightly knit or integrated. Thus the chemical engineer working in an oil refinery is more strongly attached to the economic system than to the cultural system *stricto sensu* – unless of course he is a research chemist, in which case he will belong to both the economic and the cultural systems, and may accordingly be torn by a conflict of loyalties. In either case the input–output schema of the cultural system is sketched in Figure 5.7.

This schema serves as a reminder that, whether in the narrow or in the comprehensive construal, a cultural system is a *concrete* system – not just "a body of meanings, values and rules". More precisely, a culture is a concrete system composed of living rational beings engaged in activities of various kinds, some of which go beyond the call of biological need, and all of which are social because they draw from tradition and influence social behavior. I.e. the creation and the diffusion of culture are social processes, shaped by society and in turn having an influence on the economy and the polity.

Our concept of a culture as a concrete system is at variance with the idealistic notion of a disembodied system of ideas, values, and rules, which is the concept favored by most anthropologists. Actually this other notion is that of a *type* of culture rather than of a particular culture. Thus where a cultural system (in our sense) includes certain actions, such as writing and inventing, a culture in the anthropological sense includes only the patterns of such actions – e.g. the rules that are supposed to regulate those actions. Nor does our concept of culture coincide with that of an ideal superstructure – made up of ideas, values, and norms – mounted on the material structure (economy and polity). Unlike these dualistic concepts of culture, ours is monistic.

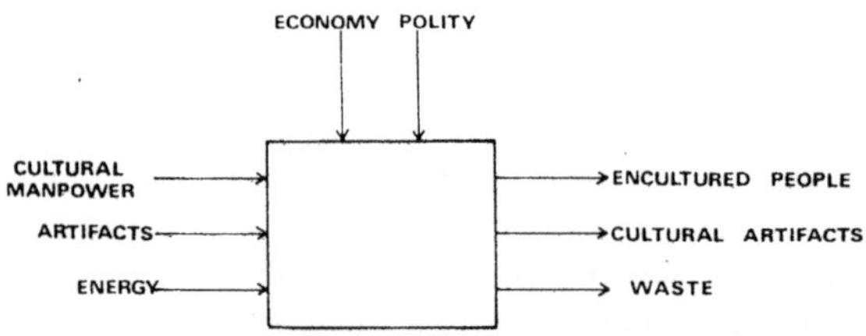

Fig. 5.7. Input-output model of the cultural system. Inputs: manpower (pivotal and auxiliary), artifacts (paint brushes, paper, pencils, books, archives, etc.), and energy (light, heat, etc.). Outputs: cultural artifacts (musical scores, sculptures, journals, etc.), people transformed by the latter, and waste products. Some of the outputs are fed back into the system. Controls: economy and polity.

The cultural systems of primitive societies are at least as monolithic as their economic and political systems, in the sense that they are not composed of subsystems. The emergence of civilization was accompanied by an explosion in the division of work, in particular cultural work. The single shaman was succeeded by a cohort of healers, priests, bards, teachers, painters, and later on skilled craftsmen and even scribes. Accordingly the cultural system split into a number of subsystems or subcultures – the religious system (centered around temples), the educational system (centered around schools), and so on. These various subcultures complement one another in certain respects but are at odds in others, if only because they all compete for a finite number of people as well as for a finite amount of material resources. In any case they interact. The mutual actions among the main subsystem of contemporary culture are shown in Figure 5.8.

Both the internal mechanisms of each cultural subsystem and its interactions with other members of the cultural system drive its changes, however slight and slow. The successive states of a cultural system constitute the *cultural evolution* (or *history*) of the society concerned. Cultural evolution modulates biological evolution and both proceed by selection. But, unlike biological evolution, cultural evolution is not purposeless. Cultural innovation and diffusion, be it technical or artistic, scientific or ideological, is partly random, partly deliberate.

A flourishing culture is one that teems with newness – valuable newness

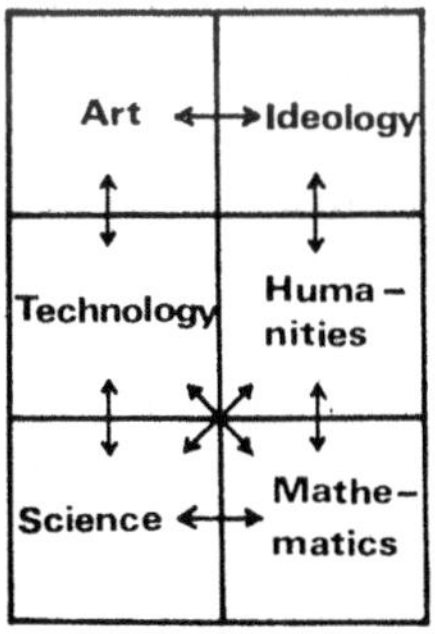

Fig. 5.8. The contemporary cultural system and its subsystems. Each box represents both creation and diffusion. The arrows represent both information flows and controls. Art and ideology, which have the most numerous memberships in traditional societies, have the fewest information links with the remaining subcultures. (From Bunge, 1977g.)

not just oddity – and is free to develop it without too much interference from the economic and political systems. A declining culture is one that ceases to value discovery and invention, and takes refuge in lamaistic repetition or in withdrawal from reality. The management of a cultural system – the politics of culture – can encourage creativity, channel it, or nip every creative bud. A dead culture is still a culture and one that can remain in that state for a long time. But because of the functional relationships among the three main artificial subsystems of every society, any major change in the economy or in the polity is bound to have cultural repercussions.

The cultural system, then, coevolves with the economic and political systems. However, the modes of evolution of these subsystems may be quite different, and neither is merely an image of the other. Thus cultural growth is compatible with zero economic growth and political stagnation, provided the political system does not interfere destructively with cultural evolution. Moreover, while economic development is limited by natural resources, there are no such limits to cultural growth: the more we know the more fresh problems we can pose and solve.

Yet there are of course economic and political constraints on cultural evolution. For one thing any given society can support only so many professional composers, mathematicians, philosophers, and so on. But at least here, unlike the case of the economy, a solution is in sight, namely automation and the amateur pursuit of cultural interests. In fact the former is freeing many hands and brains from routine work, and the good society should turn that added freedom into leisure and creation instead of unemployment. A revaluation of culture should encourage people to engage in cultural work for the fun of it rather than primarily as a source of income. Everyone should be able to master whatever cultural skill he or she likes, and to practice it during the increasingly long leisure hours rendered possible by automation. This need not encourage substandard cultural production. On the contrary, it is the increasing substitution of extrinsic motivation (pay, promotion, fame) for intrinsic motivation that is lowering the quality of cultural work. (For the negative effects of extrinsic rewards on children see Lepper *et al.*, 1973.)

To conclude. Our view of culture is at variance with idealism ("The spirit is the driving force") as well as with economism ("Culture is but a byproduct of the economy"). If every culture is a subsystem of a society then it has its own dynamics, hence some degree of autonomy, and it also interacts with the other subsystems, namely the economy and the polity.

Therefore it is neither totally free and omnipotent nor totally enslaved and powerless. Just as some members of the economic system exercise economic power, and some members of the political system wield political power, so some members of the cultural system exercise cultural power, particularly if they are entrenched in certain cultural subsystems, whether private or governmental. For example, the school system of a community exerts some cultural influence on all the inhabitants of the community – often as strong an influence as that exerted in the past by organized religion. This influence is not restricted to purely cultural matters. A creative cultural organization may study, discuss, and even propose and circulate blueprints for economic and political action. Such proposals do not amount to action but may arouse and guide it. After all people are being moved by thinking individuals all the time. Why let obsolete ideologies do all the pushing when it is possible to draw rational and feasible blueprints – i.e. to ground social action on the social sciences?

3.3. *The Political System*

The political system of a society is the subsystem of the latter that controls (to some extent) economic and cultural work as well as other types of social behavior. This control is ultimately exercised by individuals (political leaders and bureaucrats) in the interests of a few or of the majority, arbitrarily or according to rule. There is no such thing as "a government of laws, not of men" – the modern substitute for the fiction of government in the name of a deity.

Since even infants are affected by certain political actions, one might regard the entire population of a society as composing its political body. However, for practical reasons it is convenient to restrict the composition of the political system to its actors or active members, namely the citizens with political rights or responsibilities: all others are politically marginal or passive. The structure of a political system is of course the set of social relations among the political actors, such that effect a political control. Every political *regime* (e.g. parliamentary democracy) is characterized by certain types of political relations.

The political system of a modern society, unlike that of a primitive society, has a distinguished subsystem, namely the *state* or *government*. In turn, the state of an advanced society may be analyzed into a number of subsystems, such as the defense system, the health system, the educational system, and so on. Some such subsystems constitute a relatively hard core ("technical", "nonpolitical", nonideological) that remains nearly invariant under changes in political leadership. For better or worse many such a governmental subsystem grows and strengthens to the point where it escapes almost all control. If it possesses physical force, as is the case with the army and the police, it may end up by controlling the entire political system and, through it, the whole society.

The differences among political regimes are differences in the distribution of political power, i.e. in the participation in political action. Consider a community formed by three social groups to be called A, B, and C. Assume further that A and B are mutually disjoint and that C is the political control group. Moreover all the C's have effective control: they cannot be outvoted on every important issue. Under these conditions there are exactly three possibilities: see Figure 5.9. The left circle illustrates the case of aristocratic (or theocratic) government, where the control class does not mesh with the other groups. The middle circle illustrates the various kinds of democracy (representative, participatory, etc.). The right circle illustrates anarchy: no control proper. The degrees or measures of the participation of A's and B's in C are respectively

$$\alpha = |A \cap C| / |C|, \qquad \beta = |B \cap C| / |C|.$$

In addition to such objective measures of political control, the politologist is interested in the manner in which the political system is "perceived" by people. That is, politics is not only a matter of objective political structure and power but also one of social psychology. This was seen by the founder of political science, in particular when he remarked that rebellions are caused either by the desire for equality or the desire for superiority (Aristotle, *Politica*, Bk. V, Ch. 2). Likewise Marx emphasized the

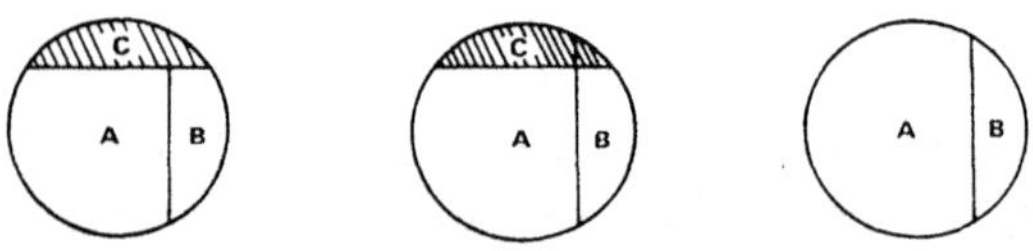

Nil participation:
No A's or B's participate in C:
$A \cap C = B \cap C = \varnothing$
$\alpha = \beta = 0.$

Partial participation
Some A's and B's participate in C:
$A \cap C \subset A,\ B \cap C \subset B$
$0 < \alpha, \beta < 1.$

Full participation
All A's and B's participate in C:
$C = A \cup B$
$\alpha + \beta = 1.$

Fig. 5.9. Three modes of political participation (or regime).

difference between the objective and the subjective conditions for a revolution. Let us take a quick look at two key concepts in political psychology: those of adhesiveness, and feeling of unfairness.

Whether or not a society is highly cohesive is of less consequence to its "survival" than the way that cohesiveness is "perceived". A social fabric with a low actual cohesiveness may be "seen" as highly cohesive (because e.g. of the repression of dissent); and conversely a highly cohesive society may be "seen" not to be so (because e.g. of a vocal nonconformist minority). The key concept here is that of attachment of an individual to a group or system. We may say that an individual feels attached to a social group or a social system if he believes that the benefits (material, cultural, emotional, etc.) he derives from sticking to it are greater than the price he has to pay for doing so. In other words, the *adherence* or *attachment* of individual x to a given group or system G_i is $A_i(x) = B_i(x) - C_i(x)$, where B and C are the benefit and cost functions respectively. There is positive attachment to G_i if the benefits outweigh the costs (i.e. $A_i(x) > 0$). Otherwise there is either rejection ($A_i(x) < 0$) by x of G_i or indifference ($A_i(x) = 0$) of x to G_i. The *total adherence* or *attachment* of individual x to a society σ formed by n social groups G_i is

$$A(x) = \sum_{i=1}^{n} A_i(x),$$

and the total *adherence* or *attachment of the membership* of σ to σ is

$$A(\sigma) = \sum_{x \in \mathscr{C}(\sigma)} A(x).$$

By choosing a suitable normalization factor, $A(\sigma)$ may be made to range, say, between -1 and 1. For example, the 1978 elections showed that the total adherence of the French people to their society is 0 (or $\frac{1}{2}$ on a [0, 1] scale).

Because adherence is psychosocial it can change swiftly in the light of information, misinformation, and criticism. When such means fail to enhance adherence, some structural social change is impending. Those in power can meet a decrease in popular attachment in either of two ways: with concessions or with coercion. The extent of either depends upon the imbalance between the actual adhesion value and the minimum value required for the social structure to remain unaltered.

Our second concept of interest is that of equity "perception". The political behavior of any political agent is partly determined by his "perception" of the equity or inequity of his station in society and his evaluation of the equity (or inequity) ideal. The imbalance between the two will contribute to his attachment or disaffection to the prevailing social structure. Therefore the social scientist is interested not only in actual equity but also in perceived equity, desired equity, and desirable or feasible equity. Let us take a quick look at these concepts.

Anything that can be distributed or allocated, from goods and services to leisure hours to political participation, can be assigned with various degrees of fairness or equity. Now, what is fair in one society may be unfair in another. Thus whereas in certain societies an average of 10 schooling years would be fair for all, in others this would constitute too heavy a burden, hence be unfair to the majority, and in still others it would be insufficient, hence just as unfair to the majority. This notion of an objectively desirable value of a property (such as the number of schooling years) must be distinguished from that of a subjectively desired value (desideratum). Whereas the latter may be unrealistic and incompatible with fairness, the former should in principle be determinable by studying the state of development and resources of the society. Such an objectively determinable value (or distribution, as the case may be) we shall call an *equity goal*, in contrast with the multiple *equity desiderata.*

We may assume that for each feature or variable there is one measure of equity or, dually, inequity. Whatever the feature may be, the objective inequity is some function f of the difference between the actual value X of the variable and its goal value G, i.e. $i(X) = f(X - G)$. *Example 1* The variable has a single value at a given time, as is the case with the mean number of schooling years. In this case a suitable inequity indicator is just the difference between G and X divided by the greater of the two. *Example 2* The feature in question is a distribution, such as that of arable land among the families of a society. If the goal is a straight line (e.g. the area of land is proportional to the number of families), then the Gini index is a suitable measure of inequity: see Figure 5.10. *Example 3* The goal distribution is symmetric – as is the case of the distribution of working hours *vs.* age. In this case inequity equals skewness, of which there are various measures.

The acuteness of the perception of inequity, as well as the desire for equity, depend on a number of factors in addition to actual inequity. One of them is social mobility. In a primitive society equity is taken for granted, whereas in a highly stratified society inequity may go just as un-

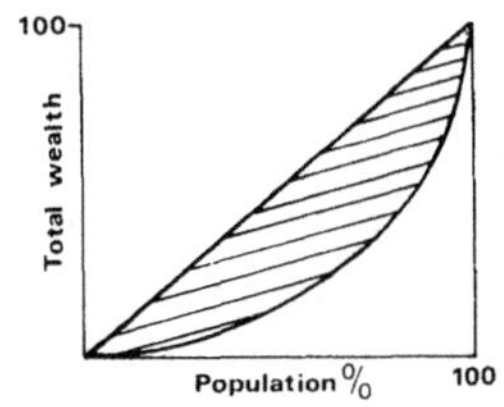

Fig. 5.10. Inequality in the distribution of wealth. The Gini index is the difference between the area under the straight line (equality) and the area under the actual curve.

questioned for being sanctified by ideology. On the other hand where social mobility is appreciable many persons feel status differences as iniquitous and also avoidable either by individual or group action. If the former proves ineffectual, the time is ripe for dissent and even sedition. We may then adopt the following version of the *Aristotle-de Tocqueville-Marx law*: Extreme inequity ends up by being perceived and negatively evaluated, that diminishes the attachment to the social structure, and this in turn leads to either indifference or active rebellion – in either case to a deterioration of the political system.

Needless to say, inequity may be economic, cultural, or political. Inequity felt in just one respect is neither necessary nor sufficient for a group to rebel against the social fabric. Any group that fits fairly well in the prevailing social (economic, cultural, and political) order – i.e. that is not marginal – is likely to be conservative even though that group may not be the principal beneficiary of that order. What do seem to be necessary and sufficient for members of a social group to rebel against the prevailing social order are (*a*) to feel being discriminated against in all three respects (economically, culturally, and politically), (*b*) to feel free from the traditions attached to the given social order (i.e. to be cultural not just economic or political outcasts), and (*c*) to be able to organize themselves into a system.

Finally we come to the notion of freedom. There are as many kinds of freedom as there are kinds of social behavior. Since human social activities occur in either of the three main artificial subsystems of society, we may group the species of freedom into three genera: economic, cultural, and political. Freedom to work (or to loaf) is economic freedom; freedom to learn is cultural freedom – and so is the freedom to remain ignorant; and freedom to participate in making decisions affecting the entire community is political freedom – and so is the freedom to remain politically inactive. Given the variety of freedoms and the fact that each has its opposite number, we should be wary of characterizing a society as free without further ado. We should specify in what respects and to what extent the members of a society are free or unfree. But such a specification presupposes an exact concept of freedom. Let us try and supply one, albeit one that is of interest to politology, namely that of *potestas agendi*, or freedom *to*, rather than that of freedom *from* (hunger, cold, fear, etc.). We may characterize the power to act by

DEFINITION 5.16 Let A be a set of alternatives, each consisting in engaging in an activity or in abstaining from it. Further, let x be an arbitrary member of a society σ. Then

(i) person x is *free* to do $y \in A$ in σ iff y is a possible choice for x & x can effectively do y & x values y & the benefit x expects to derive from doing y is greater than the cost to x of doing y;

(ii) society σ is *free in respect A* iff every member of σ is free to do any $y \in A$;

(iii) society σ is *free* iff σ is free in every respect, i.e. for any set A of alternatives (or choices).

Our definition of power or freedom to act rests on the notions of alternative or choice, of capacity or competence to perform an act, of value, and of cost and benefit – all of which are presumably elucidated in some discipline or other. It exhibits the dependence of freedom of action upon laws and circumstances: it construes freedom as depending on determinacy (whether causal, stochastic, or other) rather than on indeterminacy (hence lawlessness). In a lawless universe, in fact, there would be no freedom of any kind, since in order to be free do to something an agent must be able to count on certain regularities, at least those involving his possibilities, valuations, and actions.

We shall assume the rather platitudinous

POSTULATE 5.7 Every person and every society are free in some respects, none in all.

A completely free society – i.e. one in which every person were totally free – would be impossible, because the exercise of one person's freedoms is limited by that of his neighbors'. And a thoroughly enslaved society would be just as impossible, because every person needs to exercise some freedoms if only to survive and to contribute his or her share to society. Consequently neither extreme libertarianism nor extreme authoritarianism is a feasible blueprint for an enduring society – let alone an endearing one.

The matter of personal freedom constitutes the center of ideological debate in our time, yet it cannot be argued with profit on the ideological plane. On the other hand the systemic view of the political system can shed some light on that debate, namely as follows. A normal person is a component of some society or other. To survive, he or she must keep some measure of individuality or autonomy as well as engage in some synergic (harmonious) interactions with other components of the system, for otherwise the latter would break down. Indeed, exaggerating the autonomy of the individual would lead either to his or her breaking away from the system, or to subjecting the latter to the tyrannical control of the individual at the expense of the freedom of the other components. And overstressing the synergy of the whole may lead to depriving their components of creativity and thus to impoverishing society. So, if we care for both the system (not necessarily the "establishment") and the individual, we had better devise a political regime combining the good points of individualism with those of collectivism. But the matter of political regimes will have to be deferred to Vol. 7 of this *Treatise*.

4. Social structure

4.1. *Root Social Relations*

We have drawn a distinction between a relation in general and a connection (or link, coupling or bond) in particular (Ch. 1, Sec. 1.2). In the case of social relations we distinguish social bonds, such as exchange relations, from nonbonding social relations such as that of belonging to the same occupational group. If a bond holds between two individuals, then at least one of them will behave differently from the way he or she would behave if not so bonded. On the other hand a nonbonding relation makes no difference to the relata. (Only the realization of belonging to the same social group may induce a feeling of either solidarity or hostility and thus influence social behavior. In the jargon of sociologists, the "perception" of a relation is bound to affect behavior.)

Among all the social bonds we now single out the root or basic ones, i.e.

those on which all the others depend – such as the kinship relations and the relations of economic, cultural, and political power. We may characterize them as follows:

DEFINITION 5.17 A *root* (or *basic*) social relation is a social relation that is
(i) bonding,
(ii) person to person (rather than system to system), and
(iii) an interaction (rather than a one-sided action).

We assume that all social relations are either basic or generated by relations of this kind:

POSTULATE 5.8 If σ is a society, then
(i) for any nonbonding person-to-person relation X in the structure of σ, there is a bonding person-to-person relation Y, in the same structure, such that, for any components a and b of σ, if Xab then there are components c and d of σ such that Ycd, where c and d are possibly equal to a and b respectively;
(ii) for any (bonding or nonbonding) system-to-system relation U in the family of subsystems of σ, there is a bonding person-to-person relation V in the structure of σ such that, for any subsystems x and y of σ, if Uxy, then there is at least one component a of x, and another b of y, such that Vab.

Example 1 Slave owners a and b, living in different places and times (hence not bonded), are socially equivalent ($\sim$) by virtue of being masters (M) of slaves (S): $a \sim b =_{df} (\exists x)(\exists y)(Sx \ \& \ Sy \ \& \ Max \ \& \ Mby)$. *Example 2* Factory a supplies concern b with merchandise (or goods) c, or $Sabc$ for short. This boils down to the following: members of a produce c, and members of a (the same or different) deliver c to members of b. In obvious symbols: $Sabc =_{df} (\exists x)(\exists y)(\exists z)(x \in \mathscr{C}(a) \ \& \ y \in \mathscr{C}(a) \ \& \ z \in \mathscr{C}(b) \ \& \ Pxc \ \& \ Dycz)$.

Social bonds may in turn be classed into voluntary and involuntary. Voluntary relations are those the individual can enter by choice – e.g. friendship, participation in community activities, and even – in industrial societies in times of full employment – job relations. Examples of involuntary social ties are the filial relation, the relation of being trained at a nursery school, the prisoner–captor relation, and the bondages of slavery and serfdom. We assume that, just as bonds determine nonbonding relations, so involuntary ties determine some of the voluntary ones. For example, being a prisoner, a serf, or a slave, determines most of the other relations a person can bear to other individuals. The mere existence and the basic character of involuntary relations suffices to brush aside the philosophy of voluntarism, according to which the individual will is the source of all behavior. The truth lies between the extremes of voluntarism and total determinism: man can act of his own free will (Theorem 4.3) but not always (Postulate 5.7).

Of all the involuntary social bonds family relations are the most basic because they are determined by reproduction, which is a biofunction. The family is thus situated at the biosocial interface. The root of the family is then the division of the sexes. In other words, the same-sex relation $\sim_s$ splits the total membership $\mathscr{C}(\sigma)$ of a society σ into two disjoint sets: males (M) and females (F). In short, $\mathscr{C}(\sigma)/\sim_s = \{M, F\}$. A human family of the nuclear or restricted kind may be defined as an aggregate of human coresidents among whom the relations of mate (or spouse) or of parenthood obtain. The relation μ of mate (or rather its graph) is included in the cartesian product $M \times F$. The relation π of being parent of somebody is a binary relation in $\mathscr{C}(\sigma)$. And the relation ρ of coresidence too may be construed as a binary relation in the same set. This is all we need to form

DEFINITION 5.18 Let σ be a society with sexual division $\mathscr{C}(\sigma)/\sim_s$, mate relationship μ, parenthood bond π, and coresidence tie ρ. The *nuclear family* $F(b)$ of the component b of σ is the sociosystem constituted by the mate(s) and the children of b who live with b:

$$\mathscr{C}(F(b)) = \{x \in \mathscr{C}(\sigma) \mid (\mu xb \vee \pi bx) \mathbin{\&} \rho xb\},$$
$$\mathscr{E}(F(b)) \supset \mathscr{E}(\sigma) \mathbin{\&} \mathscr{E}(F(b)) \cap \mathscr{C}(\sigma) \neq \varnothing,$$
$$\mathscr{S}(F(b)) = S \cup T, \quad \text{with } S = \{\mu, \pi, \rho\}.$$

A member of a nuclear family may have more than one spouse (as in the cases of bands, polygamy, and polyandry) provided all his or her mates are coresident. But each person belongs to exactly one nuclear family. (A childless widow constitutes a one person nuclear family but may belong to an extended family.) The set of nuclear families of a society is the composition of the nuclear kinship system of the society. That set is split into disjoint sets by the relation of belonging to the same family. The importance of this relation and of the partition it induces justifies making

DEFINITION 5.19 Let σ be a society. Then

(i) two components of σ are *family-wise equivalent* ($\sim_f$) iff they are components of the same family:

For any $x, y \in \mathscr{C}(\sigma)$, $x \sim_f y =_{df} (\exists z)\ (z \in \mathscr{C}(\sigma)\ \&\ x \in \mathscr{C}(F(z))\ \&\ y \in \mathscr{C}(F(z))\}$;

(ii) the *nuclear kinship system* of σ is the subsystem whose members are in some family or other, so that its membership is

$$\mathscr{C}_n(\sigma) = \{x \in \mathscr{C}(\sigma) \mid (\exists y)\ (y \in \mathscr{C}(\sigma)\ \&\ x \in \mathscr{C}(F(y))\};$$

(iii) the *family structure* of σ is the set of all the nuclear families of σ, i.e. the partition of $\mathscr{C}_n(\sigma)$ by the relation $\sim_f$ of family equivalence:

$$\mathscr{F}(\sigma) = \mathscr{C}_n(\sigma)/\sim_f.$$

Because of the biological root of the family, the family structure of a society constitutes its fundamental structure. So it is the basis of other social partitions. For example, a social class is a certain set of families, hence a subset of the family structure rather than a collection of individuals. So much for the so-called blood relations.

Other important social relations are those said to be "based on" work, i.e. those people enter by virtue of their work. Such are the "vertical" relations between a blue collar worker and a member of the management of his enterprise, and the "horizontal" relation between two workers on the same echelon in an organization. See Figure 5.11.

DEFINITION 5.20 Two persons hold a *bonding relation of production* iff either

(i) both individuals take part in the production of a set of products (goods or services) either as primary (L) or as secondary (K) workers; or

(ii) one of the individuals produces material or cultural goods while the other manages (controls, oversees) him or her; or

(iii) the two individuals comanage at least one producer.

In other words, the relations of production can be of coproduction, management, or comanagement. They are all, whether directly or indirectly, "based" on primary work insofar as, without the latter, there would be no

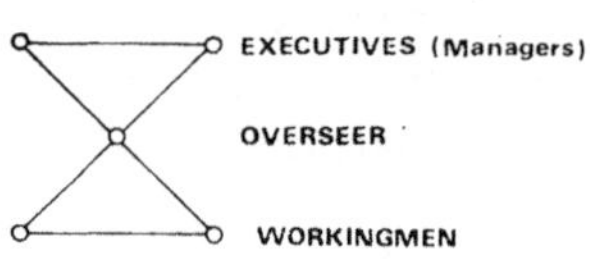

Fig. 5.11. Fragment of a particular web of social relations of production in an economic system.

production at all. And without the relations of production the social structure would be far poorer. Note, however, that not every relation of production is bonding. For example, the relation of belonging to the same occupational group, i.e. of doing work of the same type, is not bonding. It is only when individuals with the same occupation gather to form a guild, trade union, or some other organization "based on" work, that they enter into bonding relations. But these relations are no longer economic: they are political in our sense of the word, for their main impact is on social behavior rather than or production.

Whether economic, cultural, or political, social relations influence behavior. If such influence is decisive, we call it 'power' or 'control'. The social influences of all kinds fall under

DEFINITION 5.21 Let σ be a society where an activity of type $A \subset \mathscr{S}(\sigma)$ is performed. Then for any components x and y of σ, x *influences* y *in respect* A iff the activity of kind A performed by y while it communicates with x differs from that performed by y when it does not.

This definition makes room for indirect influences, i.e. those mediated by third persons. Thus even if a worker never sees a stockholder of his company, the two influence each other via the company management. (However, the influence relation we have just defined is not transitive – nor is it intransitive.)

Concerted influence is group pressure: this is not a mysterious influence exerted by a group as a totality, but the simultaneous and convergent influence of several individuals of group on a single individual. More precisely, we make

DEFINITION 5.22 Let $s \subset \mathscr{C}(\sigma)$ be a subset of the membership of a society σ. Then s exerts (group) *pressure* on a component x of σ in respect A iff each member of s influences x in respect A.

Power or *control* may be said to exist when influence is far stronger in one direction than in the opposite direction. It is possible to quantitate influence and power, hence to define a relation of power equivalence inducing the partition of the membership of a society into groups wielding the same power in some respect (Bunge, 1974d). However, we must rush on to other matters. The point of bringing in the above considerations is that, in order to disclose the social groups in a society, we must spot bonding social relations with the help of which (nonbonding) social equivalence relations can be defined, that induce the splitting of the membership of a society into homogeneous social groups. This is the method that we shall use to display the structure of a society.

4.2. *Social Structure*

Each social bond $B_k \in \mathbb{B} \subset S$ in a society generates a social equivalence relation $\sim_k$, which in turn induces a partition $\mathscr{P}_k = \mathscr{C}(\sigma)/\sim_k$ of the membership of the society σ into a family of homogeneous sets (social groups). There are as many such partitions as social equivalence relations. The collection of all such partitions we call the social structure of the society. More precisely, we propose

DEFINTION 5.23 Let σ be a society with composition $\mathscr{C}(\sigma)$ and call

$$\sim \; = \{\sim_k \mid \sim_k \subset \mathscr{C}(\sigma) \times \mathscr{C}(\sigma) \;\&\; \sim_k \text{ is an equivalence relation } \&\; 1 \leq k \leq n\}$$

the set of social equivalence relations definable on $\mathscr{C}(\sigma)$ on the basis of social bonds. Then

(i) the kth *primary social structure* of σ is the kth partition of $\mathscr{C}(\sigma)$, i.e. $\mathscr{P}_k(\sigma) = \mathscr{C}(\sigma)/\sim_k$, with $\sim_k \in \sim$;

(ii) the ith *social* group of the kth primary structure of (or *ik cell* of σ for short) is the ith member S_{ik} of

$$\mathscr{P}_k(\sigma) = \{S_{1k}, S_{2k}, \ldots, S_{ik}, \ldots, S_{mk}\}, \quad \text{with} \quad \bigcup_i S_{ik} = \mathscr{C}(\sigma) \quad \text{and} \quad \bigcap_i S_{ik} = \varnothing;$$

(iii) the *overall primary social structure of* σ is the family of all the partial social structures of σ, i.e.

$$\mathscr{P}(\sigma) = \{\mathscr{P}_k \mid \mathscr{P}_k = \mathscr{C}(\sigma)/\sim_k \;\&\; \sim_k \in \sim \;\&\; 1 \leq k \leq n\}.$$

The overall social structure of a society is displayed by the matrix $\|S_{ik}\|$, every entry of which represents a social group or cell. (For the mathematics of set matrices see Bunge, (1974d).) The corresponding quantitative concept is got by just counting the population of each cell at a given time, i.e. by finding out how the total population $|\mathscr{C}(\sigma)| = N$ of the society is distributed among its various groups. Calling $N_{ik} = |S_{ik}|$ the population of cell S_{ik}, we can form the *population density matrix of* σ:

$$\|D_{ik}\| = (1/N)\,\|N_{ik}\|, \quad \text{with} \quad \Sigma_i N_{ik} = N.$$

A primary social structure is a partition of the membership of a society into groups or cells of individuals. But in some cases the groups themselves may band together to form larger groups or even systems. For example, the collection of families may be split into social classes, educational groups, income groups, political factions, and so on. Again, the collection of enterprises may be split into agricultural, industrial, and service enterprises – or into state owned, cooperative, and privately owned, and so on. Any such further partition of a primary partition will be called a secondary structure:

DEFINITION 5.24 Let $\mathscr{P}_k(\sigma) = \mathscr{C}(\sigma) / \sim_k$, with $\sim_k \in \sim$, be a primary social structure of a society σ. Then

(i) the kmth *secondary structure* of σ is

$$\mathscr{P}_{km}(\sigma) = (\mathscr{C}(\sigma) / \sim_k) / \sim_m, \quad \text{with} \quad \sim_m \in \sim \quad \text{and} \quad m \neq k;$$

(ii) the *overall secondary social structure* of σ is the totality of partial secondary structures of σ:

$$\mathscr{P}\mathscr{P}(\sigma) = \{\mathscr{P}_{km}(\sigma) \mid \mathscr{P}_{km}(\sigma) = (\mathscr{C}(\sigma) / \sim_k) / \sim_m \,\&\, \sim_m \in \sim \,\&\, m \neq k \,\&\, 1 \leq k, m \leq n\}.$$

Obviously not all such partitions are sociologically significant. For example, it makes no sense to group enterprises according to cultural background, or to partition the set of families into privately managed, cooperatively managed, and state managed. The sets resulting from such unnatural or insignificant partitions are either disregarded or set equal to the empty set. In particular, $\mathscr{P}_{kk}(\sigma) = \varnothing$. And, as with the case of the primary social structure, we can count items and build the density matrix of the overall secondary structure of a society.

Because both the primary and the secondary structures of a society are sets of sets, it might be wondered whether they stand for properties of the society. They do, for they may be regarded as values of property-representing functions, namely

$$\mathscr{P}_k : \Sigma \longrightarrow \{\mathscr{C}(\sigma) / \sim_k \mid \sigma \in \Sigma\} \quad \text{and}$$
$$\mathscr{P}_{km} : \Sigma \longrightarrow \{(\mathscr{C}(\sigma) / \sim_k) / \sim_m \mid \sigma \in \Sigma\},$$

where $\mathscr{C}(\sigma)$ is the membership of σ and Σ the totality of social systems. Furthermore, far from being arbitrary these functions occur in societal laws.

The concepts of overall primary and secondary social structures allow us to exactify the vague notion of social fabric: The *social fabric* of a society σ is the set of all primary and secondary social structures of σ, i.e.

$$\mathscr{P}(\sigma) \cup \mathscr{P}\mathscr{P}(\sigma).$$

Finally, one word about social class and class structure. Not all social groups are social classes, and therefore not all societies partitioned into groups need be stratified. A social class or caste is in a control relation with some other groups in the society. The control can be economic, political, military, administrative, or cultural. Such dominant groups originate in a variety of manners, from military conquest to gradual concentration of power. A typical case is that of medieval aristocracies. Small bands of soldiers, initially hired to defend a territory from marauders or from war lords, ended up by using military force to gain full economic and political control. What had been a horizontal division of society became a vertical one, i.e. a stratification or hierarchical order of social classes. This mechanism is quite general: the concentration of power in a social group turns it into a class. And, because power can be economic, cultural, or political, we need to distinguish three kinds of social class. That is, one and the same society may be divided into economic, cultural, and political classes. Occasionally these three partitions coincide, but in general they do not. For example, in a theocracy there are essentially two classes: the priestly caste, which wields total economic, political and economic power, and the laity. On the other hand in a contemporary Western democracy there are essentially two economic classes (the owners and the non-owners of the means of production), two cultural classes (the producers and the consumers of cultural goods and services), and – at least in theory – a single political class (the citizenry). We compress all this into

DEFINITION 5.25 Let σ be a society, and s and t two groups of σ. Further, let $\sim_A$ be a relation of social equivalence in respect A, where A stands for 'economy', 'culture', or 'polity'. Then

(i) group s *controls* group t in respect A iff every member of t is controlled in the respect A by some member(s) of s;

(ii) an *A-social class* is a social group that A-controls, or is A-controlled by, other social group(s) in the same society;

(iii) the *A-class structure* of σ is the partition of the composition of σ by $\sim_A$, i.e.

$$C_A(\sigma) = \mathscr{P}_A(\sigma) = \mathscr{C}(\sigma)/\sim_A;$$

(iv) the *class structure* of σ is the family of all partial class structures of σ, i.e.

$$C(\sigma) = \{\mathscr{P}_A(\sigma) \mid A = \text{economy, culture, or polity}\}.$$

If the quotient of $\mathscr{C}(\sigma)$ by $\sim_A$ fails to partition the composition of σ into disjoint subsets, then there is no A-control in the society, and thus the society is A-classless. In other words, σ is *classless* in the respect A iff $\mathscr{C}(\sigma)/\sim_A = \{\mathscr{C}(\sigma)\}$. It follows that an A-classless society is possible only if each of its components participates extensively in activities of kind A. The problem of participation will be addressed in the next subsection.

4.3. *Differentiation and Cohesion*

No society is homogeneous in all respects: every society is differentiated, i.e. split into social groups – if only sex and age groups. A natural though admittedly coarse measure of the difference between two social groups A and B is the numerosity of their symmetric difference $A \triangle B = (A \cap \bar{B}) \cup (\bar{A} \cap B)$. Hence we obtain a measure of the social differentiation (or variety) of a society by adding up the cardinalities of the pairwise differences among all its social cells, i.e. the $|S_{pq} \triangle S_{rs}|$. And, since it is convenient to have the resulting number comprised between zero and one, we divide the total by a suitable normalization factor. The end result (Bunge and García-Sucre, 1976) is

DEFINITION 5.26 Let σ be a society with social structure matrix $\|S_{ik}\|$, where $S_{ik} \in \mathscr{C}(\sigma)/\sim_k$ is the ith cell ensuing from the partition of $\mathscr{C}(\sigma)$ by the social equivalence relation $\sim_k \in \sim$. Further, let N be the total population of σ at a given time, and n the number of social equivalence relations (i.e. $N = |\mathscr{C}(\sigma)|$, $n = |\sim|$). Then the *degree of social differentiation* of σ with respect to $\sim$ at the given time equals

$$\delta(\sigma, \sim) = [2N(N-1)n^2]^{-1} \sum_{\sim_q \in \sim} \sum_{S_{pq} \in \mathscr{C}(\sigma)/\sim_q} \sum_{\sim_s \in \sim} \sum_{S_{rs} \in \mathscr{C}(\sigma)/\sim_s} | S_{pq} \triangle S_{rs}|.$$

This formula simplifies to

$$\delta(\sigma, \sim) = [n(N-1)]^{-1} \sum_k (n_k - 1),$$

where n_k is the number of cells produced by the kth partition, i.e. $n_k = |\mathscr{C}(\sigma)/\sim_k|$.

Minimal differentiation occurs when each partition produces a single class, namely $\mathscr{C}(\sigma)$ itself – i.e. when everyone is everything. In this case $\delta = 0$. At the other end of the spectrum there is just one individual in each social group, i.e. $|S_{ik}| = 1$ and $n_k = N$ for each pair (i, k). The corresponding value is $\delta = 1$. We take it that neither of these extreme values is realistic. The intermediate case is that in which all cells are uniformly populated, i.e. $|S_{ik}| = m$ and $n_k = N/m$, with $1 < m < N$ for every k. The corresponding degree of differentiation is

$$\delta(\sigma, \sim) = (N - m)/m(N - 1) \longrightarrow 1/m \quad \text{for } N \longrightarrow \infty.$$

A person who belongs to a social group is said to *participate* in it. Participation calls for differentiation: there is no occasion to participate if there are no different social groups. (However, participation is also nil if differentiation is maximal, since in the latter case no individual belongs to more than a single social group.) The individual participations aggregate to form a systemic property that may be called the *overall participation* in a society. This is made up of the overlaps of all the social groups, i.e. of $S_{pq} \cap S_{rs}$. More precisely, we make

DEFINITION 5.27 Let σ be a society whose composition is partitioned by the n equivalence relations contained in $\sim$. Then

(i) the *degree of participation of group* S_{pq} in σ is

$$\pi(S_{pq}, \sigma, \sim) = \sum_{\sim_s \in \sim} \sum_{S_{rs} \in \mathscr{C}(\sigma)/\sim_s} |S_{pq} \cap S_{rs}| \,/\, |S_{rs}|;$$

(ii) the *degree of marginality* (or *alienation*) of group S_{pq} in σ is the complement to unity of its degree of participation, i.e.

$$\mu(S_{pq}, \sigma, \sim) = 1 - \pi(S_{pq}, \sigma, \sim);$$

(iii) the degree of *overall participation* in σ is

$$\pi(\sigma, \sim) = F \sum_{\sim_q \in \sim} \sum_{S_{pq} \in \mathscr{C}(\sigma)/\sim_q} \sum_{\sim_s \in \sim} \sum_{S_{rs} \in \mathscr{C}(\sigma)/\sim_s} |S_{pq} \cap S_{rs}|,$$

where F is the normalization factor ensuring that the maximal value of π is 1;

(iv) the degree of *overall margination* (or *alienation*) in σ is

$$\mu(\sigma, \sim) = 1 - \pi(\sigma, \sim).$$

From the findings of social scientists we infer that all societies are differentiated, though none maximally, and that there is some measure of participation in all. That is, we propose

POSTULATE 5.9 For every society σ, $|\sim| > 1$, and $0 < \delta(\sigma, \sim) < 1$, $0 < \pi(\sigma, \sim) < 1$.

Every society is more or less cohesive: were it not, it would not be a system. Social cohesion is of course one of the emergent societal properties: it is a systemic property not possessed by any of the individuals composing a society. Cohesion can be spontaneous, forced, or a combination of the two. Spontaneous cohesion, the one that does not result from coercion, is undoubtedly the more important of the three. When strong it ensures the stability of the social fabric; when weak it calls for either structural change or the violent repression of any attempt to effect structural changes.

In Ch. 4 we assumed that cohesion derives from sharing (of things and services) and participation (in activities): recall Postulate 4.24. We shall presently discuss a measure of spontaneous cohesion based on that assumption (Bunge and García-Sucre, 1976). No participation at all, i.e. extreme segregation, results in instability, which ends up in fragmentation. Yet maximal participation has a similar dissolving effect, for a society of equals in every respect is a collection of competitors in no need of mutual assistance, i.e. in no need of belonging to the society. In other words, if every individual is either nothing or everything in a society, i.e. utterly marginal or ubiquitous, then cohesion is minimal. Maximal cohesion must then occur somewhere in between, when each individual shares several, though by no means all, cells with fellow members. In sum, cohesion must result from moderate participation. We proceed to formalize this intuitive idea.

The more members two social groups share the stronger they are bonded. However, these bonds operate only if the groups are different, in particular functionally complementary to one another. More precisely, the effective bonding between two social groups A and B is the numerosity of their intersection times that of their symmetric difference, i.e. $|A \cap B| \cdot |A \triangle B|$. A low degree of differentiation (i.e. a small $|A \triangle B|$ value) may be compensated for by a high degree of participation (i.e. a large overlap value $|A \cap B|$) and conversely. Certainly differentiation by itself does not contribute to cohesion, but the various partitions (structures) may intertwine in such a manner that cohesion results: i.e. cells situated in different columns of the social structure matrix may overlap partially. By adding up all the pairwise bonds among the social groups of a society we obtain its cohesion:

DEFINITION 5.28 Let σ be a society whose composition is partitioned by the n social equivalence relations contained in $\sim$. Then the degree of *spontaneous social cohesion* of σ with respect to $\sim$ equals

$$\kappa_s(\sigma, \sim) = [2N(N-1)n^2]^{-1} \sum_{\sim_q \in \sim} \sum_{S_{pq} \in \mathscr{C}(\sigma)/\sim_q} \sum_{\sim_s \in \sim} \sum_{S_{rs} \in \mathscr{C}(\sigma)/\sim_s} |S_{pq} \cap S_{rs}| \cdot |S_{pq} \triangle S_{rs}|,$$

where the $pqrs$ term in the summation represents the strength of the bond between cells S_{pq} and S_{rs}, this strength being the product of the social participation by the differentiation.

Disjoint cells do not contribute to cohesion while overlapping cells do, provided they differ somewhat. If the nonempty overlaps all have unit measure, i.e. $|S_{pq} \cap S_{rs}| = 1$ for all pairs of cells except those along the same column of the social structure matrix, then cohesion reduces to differentiation (Definition 5.26). If on the other hand all the local differences are the same, and in particular $|S_{pq} \triangle S_{rs}| = 1$, then cohesion coincides with participation (Definition 5.27).

Minimal differentiation accompanies minimal cohesion. In fact, if $|S_{ik}| = N$ and $n_k = |\mathscr{C}(\sigma) / \sim_k| = 1$ for every k, then $S_{pq} = \mathscr{C}(\sigma)$ regardless of q, whence $S_{pq} \triangle S_{rs} = \varnothing$ in all cases, whence $\kappa_s(\sigma, \sim) = 0$. Likewise maximal differentiation ensues in a total lack of cohesion. Indeed, if $|S_{ik}| = 1$ and $n_k = |\mathscr{C}(\sigma) / \sim_k| = N$ for every k, the overlaps are nonvoid just in case the cells coincide, which implies that the symmetric difference is nil. In sum, in this case too $\kappa_s(\sigma, \sim) = 0$. Finally consider the intermediate case of homogeneity: $|S_{ik}| = m$ and $n_k = |\mathscr{C}(\sigma) / \sim_k| = N/m$ for every k, and $|S_{pq} \cap S_{rs}| = t$ unless the cells are in the same column. It turns out that in this case

$$\kappa_s(\sigma, \sim) = Nt(m-t)/(N-1)m^2 \cong t(m-t)/m^2.$$

The maximal value of spontaneous cohesion for this homogeneous case occurs for $t = m/2$ and is

$$\kappa_s(\sigma, \sim) = \tfrac{1}{4}N/(N-1) \cong \tfrac{1}{4}.$$

In short, the spontaneous cohesiveness of a society consists in a moderate participation of its members in the various social groups of the community. If the latter is hardly differentiated, there is not much to participate in, and competition overpowers cooperation, whence cohesiveness is low. And if the differentiation is rigid, as is the case with hierarchical societies (e.g. caste systems), there is not much participation either: in this case stability does not result from spontaneous cohesiveness but is imposed by force of arms or of ideology.

In contrast to spontaneous cohesion, forced cohesion is related to cultural or political repression, i.e. to the action of organized ideologies (e.g. churches and parties) or to police forces and the like. Though forced cohesiveness is an output variable, it can be roughly indicated by the corresponding input, namely the strength of the repressive forces relative to the total population. Calling $\kappa_f(\sigma)$ whatever indicator of the degree of forced cohesion we may choose, we can define the *total cohesiveness* of σ as $\kappa(\sigma) = \kappa_s(\sigma) + \kappa_f(\sigma)$. (In disregarding the set $\sim$ of social equivalence relations used to compute $\kappa_s(\sigma)$ we are tacitly assuming that all "significant" relations of that kind are being taken care of.) In principle, $\kappa(\sigma)$ is comprised between 0 and 2.

There are a number of definitions of the concept of a free society. Since most of them are ideological rather than scientific, there is little hope of achieving consensus. Thus while X will rightly count the freedom of the press among the necessary conditions for a free society, Y will counter, equally correctly, that such freedom is illusory as long as it can be exercised only by those with the economic freedom to run the press. We can avoid this debate by proposing an objective and even quantitative criterion of *political freedom*, namely this: A society is *politically free* iff its spontaneous cohesiveness is much greater than its forced cohesiveness.

If a society is not politically free, i.e. if its social structure is kept or modified by force of arms or ideology, then this force overpowers, balances, or eventually elicits a rebellion on the part of some social groups. However, the strength of the disruptive forces in a society is not just proportional to its numbers. Indeed, a small band of enthusiastic and well organized revolutionaries espousing a popular cause, or else acting amidst popular indifference, may defeat a large army of demoralized supporters of an unpopular dictatorship, just as a large mass of disorganized rebels will be no match for a small but highly trained repressive force. Consequently the politologist, unlike the sociologist, must study not only the indicators of objective cohesiveness but also those of "perceived" cohesiveness, or adherence (Sec. 3.3).

4.4. *Social Relations and Structures: Real*?

While holists acknowledge the reality of social relations, they reduce them to two types: the whole–whole (or system to system) relations – exemplified by international relations – and the whole– part causal relation, allegedly exemplified by group pressure. They are interested more in mythical wholes hovering above individuals – such as the Nation, the State, and the Spirit of the Times – than in interpersonal relations. The latter are on the other hand stressed by individualists: "Repeated interactions between particular persons are the very guts and marrow of social life" (Homans, 1974, p. 57). The systemist gives full assent to this view but he also admits intersystem bonds (e.g., Postulate 5.8(ii)). And he insists that social bonds, whether interpersonal or intersystemic, are real as long as they accompany changes in the relata (individuals or sociosystems). Consequently social structures are perfectly real, and moreover the main concern of the sociologist.

Social scientists rarely espouse the extremes of holism and individualism: most of them study social systems and their subsystems. And, whether they sympathize with holism or with individualism, they rarely deny the reality of social relations and, as a consequence, the reality of social structures. (Structuralist anthropology is an exception: it claims that social structures are in the head of the anthropologist. See Lévi-Strauss, 1963. For a devastating criticism see Harris, 1968.)

On the other hand the philosophers of social science who favor ontological and methodological individualism tend to deny the reality of social bonds, hence of social structures, perhaps because of their eagerness to deny the reality of the wholes extolled by the holists. Thus Winch (1958, pp. 131 ff.) has claimed that *all* "social relations between men exist only in and through their ideas", and "fall into the same logical category as do relations between ideas", so that social science is a chapter of epistemology. And Popper (1974, p. 14) has stated that "social relations belong, in many ways, to what I have more recently called 'the third world', or 'world 3', the world of theories, of books, of ideas, of problems". We saw in Ch. 4, Sec. 4.5, that there is no such thing, because a genuine world – i.e. a system – cannot be composed of both concrete things, such as books, and fictions, such as ideas in themselves (i.e. apart from ideating brains). Here we shall note only how convenient it would be for all oppressors to be able to persuade their subjects that their bondage belongs in the world of fairy tales and mathematical formulas, so they could not possibly suffer from it, let alone rebel against it.

Sociologists, who study social relations and social structures, and politicians, who attempt to control them, cannot take seriously the claim that social bonds and the social organizations they constitute are fictions. They regard them as just as real as physical or chemical bonds. Hence, when intent on changing them they resort to social action rather than incantation.

5. Social change

5.1. *The Social Structure Approach*

A social change is a change in some of the social properties of a community, such as a change in cultural level, or a redistribution of economic or political power. In principle it is possible to account for social change with the help of the concept of state function: the instantaneous state of a society would be the value of its state function at the given time, and the history of the society over a certain period would be the list of all the instantaneous values of its state function. This approach to social change should work for small sociosystems. But even a rough determination of the overall economic, cultural and political state of a modern society would require a staggering amount of information. And an account of its evolution would require an equally staggering number of equations of evolution (which in the simplest case would be rate equations for the state functions). We shall therefore adopt a more feasible method.

We shall assume that every social change consists in the redistribution of the composition of a society among the various cells of the social matrix $\|S_{ik}\|$. (For example, the emergence or the disappearance of a given kind of work, or of a given power relation, will ensue in the formation or extinction of the corresponding social cells.) Hence tracing the evolution of a society boils down to taking a periodic census – which is of course what the census bureaus do. This procedure will give us a precise, albeit coarse, description, though not an explanation, of social change. More precisely, we can make

DEFINITION 5.29 Let σ be a society with social structure matrix $\|S_{ik}\|$ and corresponding population density matrix

$$D(\sigma, t) = (1/N(\sigma, t)) \cdot \|N_{ik}(\sigma, t)\|.$$

The *net structural change* in σ over the time interval $[t_1, t_2]$ is

$$\Delta(\sigma; t_1, t_2) = D(\sigma, t_2) - D(\sigma, t_1).$$

Example At t_1, σ is divided into two social classes: masters (M) and slaves (S). At time $t_2 > t_1$, σ is leveled to a society of free men (F) with conservation of the total population. (I.e. either the revolution has been bloodless or the losses in lives have been compensated for by births or immigration.) The net class change (or change in class structure) is then

$$(1/F) \cdot \begin{Vmatrix} 0 \\ 0 \\ F \end{Vmatrix} - (1/F) \cdot \begin{Vmatrix} M \\ S \\ O \end{Vmatrix} = (1/F) \cdot \begin{Vmatrix} -M \\ -S \\ F \end{Vmatrix}, \quad F = M + S.$$

A negative (positive) sign indicates population loss (gain) in a cell. It goes without saying that, by virtue of the interrelations among the various social properties, a full account of this (imaginary) social revolution would require an investigation of all the major social cells.

Whatever causes, natural or social, may influence the distribution of the total population of a society among its cells, may be lumped into a social evolution operator. We assume that this operator satisfies

POSTULATE 5.10 Let $D(\sigma, t_1)$ and $D(\sigma, t_2)$ represent the social structure of a society σ at times t_1 and t_2 respectively. Then there exists an operator $E(\sigma; t_1, t_2)$ such that

$$D(\sigma, t_2) = E(\sigma; t_1, t_2) D(\sigma, t_1)$$

representing the mode of change of σ over the period $[t_1, t_2]$ and called the *social evolution operator* for that period.

This is a substantive assumption that may be found empirically wanting, particularly over a long period: the actual social evolution may well be nonlinear for all we know. However, we may assume that the postulate holds at least for short periods.

From the above postulate and Definition 5.29 it follows that the net structural change is

$$\Delta(\sigma; t_2, t_1) = (E(\sigma; t_1, t_2) - I)\, D(\sigma, t_1),$$

where I is a suitable unit matrix. Given the initial state $D(\sigma, t_1)$ and the mode of evolution $E(\sigma; t_1, t_2)$, the final structure, i.e. $D(\sigma, t_2)$, remains uniquely determined. This is the direct or forecast problem. The inverse or explanation problem is this: Given the values of the density matrix at two different times, e.g. now and one decade ago, find the mode of evolution, i.e. compute the evolution operator. Being an inverse problem, in general it lacks a unique solution and is much harder than the forecast problem. In fact, consider again the example studied a moment ago. The following evolution operators, among others, account for the same structural change:

$$E = \begin{Vmatrix} -1 & 0 & 0 \\ 0 & -1 & 0 \\ 1 & 1 & 0 \end{Vmatrix} \quad \text{and} \quad E = \begin{Vmatrix} -1 & 0 & 1 \\ 0 & -1 & 1 \\ 1 & 1 & 1 \end{Vmatrix}.$$

It can be shown in general that the following holds (Bunge, 1974d):

THEOREM 5.1 Any given or prescribed social change in a society can be brought about in alternative ways, each represented by a different social evolution operator.

Further, since the populations along any given column of the social structure matrix must add up to the total population, we also have

THEOREM 5.2 The relative growth of any cell in a column of the social structure matrix occurs at the expense of the relative decay of some other cell(s) in the same column.

For example, if a column consists of just two cells and one of them grows exponentially, then the other must decay exponentially. Thus all growth has to be paid for by some decline.

Finally, note that social change is the likelier the more heterogeneous or varied a society is, and this if only because the more social cells there are, the more numerous the possible transitions among them. For the same reason, the more homogeneous a society the less changeable it is.

5.2. *History*

The main types of structural change of a society are listed in

DEFINITION 5.30 Let σ be a society with social matrix $\|S_{ik}\|$ and population matrix $\|N_{ik}\|$. Then

(i) σ remains *stationary* (or in a state of *static equilibrium*) in the kth respect (where $1 \leq k \leq n$) over the period $[0, \tau]$ iff $N_{ik}(t)$ = constant for all $1 \leq i \leq n$ and all $t \in [0, \tau]$;

(ii) σ is stable (or in a *state* of *dynamical equilibrium*) in the kth respect $(1 \leq k \leq n)$ over the period $[0, \tau]$ iff all the cells in the kth column of $\|S_{ik}\|$ expand or shrink at the same rate:

$$N_{ik}(t) = at \quad \text{with} \quad a \in \mathbb{R}^+ \quad \text{and} \quad a \neq 1 \quad \text{for all } 1 \leq i \leq m;$$

(iii) σ undergoes *cyclic changes* iff there is an evolution operator E and a natural number p such that $p = \tau$ and $E^p = I$;

(iv) σ undergoes an *irreversible* change iff σ changes in a noncyclic way;

(v) σ undergoes a *partial revolution* iff some cells S_{ik} change appreciably (in particular if they are either filled up or emptied);

(vi) σ undergoes a *total revolution* iff all the cells S_{ik} change appreciably.

Example A society consists initially of four social classes, 1, 2, 3, and 4, with populations N_1, N_2, N_3, and N_4 respectively. A revolution wipes out all but a fraction f_i of each class, and merges 1 with 2, and 3 with 4. A possible representation of the net change is this:

$$\begin{Vmatrix} f_1 & f_2 & 0 & 0 \\ 0 & 0 & 0 & 0 \\ 0 & 0 & f_3 & f_4 \\ 0 & 0 & 0 & 0 \end{Vmatrix} \begin{Vmatrix} N_1 \\ N_2 \\ N_3 \\ N_4 \end{Vmatrix} = \begin{Vmatrix} f_1N_1 + f_2N_2 \\ 0 \\ f_3N_3 + f_4N_4 \\ 0 \end{Vmatrix}$$

$$V = 1 - (1/N(0)) \textstyle\sum_i f_i N_i(0).$$

This index equals 0 for peaceful revolutions, $\frac{1}{10}$ for those that decimate each social stratum, and $\frac{1}{2}$ for those that take the lives of half the population in each group.

So far we have drawn no distinction between spontaneous and planned social change – hence between descriptive models and prescriptive ones. But all contemporary societies, even those with economic free enterprise, are subjected to planning to some degree or other, and it may be argued that they need more (and better) of the same if they are to survive. Take the rather obvious case of population growth. In the absence of any social or cultural constraints, and pretending that there are no environmental strictures either – in particular no interpopulational competition – the expected population size satisfies the rate law

$$\dot{N} = aN, \quad \text{with} \quad a = \text{birth rate} - \text{death rate}.$$

Here $a \in \mathbb{R}$ is a parameter summarizing biological, economic, cultural, and political characteristics, such as fertility, health level, food supply, and feeling of security. If $a > 0$, the population grows exponentially, if $a = 0$ it has zero growth, and if $a < 0$ it declines exponentially. Add now family planning on an individual basis, such as practised in all industrialized nations. Anything can happen depending on economic, cultural and political changes: the expected population size may increase, decrease, oscillate, remain stagnant, or neither.

Consider next family planning associated with the adoption of an explicit desideratum concerning the desirable family size. This norm will result in a total population goal G. A new law of population growth emerges, one that is controlled by the families themselves jointly with state organizations. In the simplest case this law takes the form

$$\dot{N} = b(G - N)\begin{cases} > 0 & \text{iff } N < G \text{ (underpopulation)} \\ = 0 & \text{iff } N = G \text{ (attained ideal)} \\ < 0 & \text{iff } N > G \text{ (overpopulation).} \end{cases}$$

Now population growth is under control. The total population grows exponentially (barring war, pestilence or famine) just in case the actual population is inferior to the goal value. It remains constant if the ideal has been attained. And it declines as soon as the goal is surpassed.

The above pattern of planned social change can obviously be generalized to an arbitrary social cell, such as a given occupational or educational group. As a matter of fact all modern nations do plan and control a number of activities, particularly agriculture and education: just think of the subsidies to the former and of the restrictions on entrance to medical schools.

Note the following points. First, mathematics can handle goals or norms; moreover, only with the help of mathematical models can social change be steered successfully. Second, knowledge and valuation can change the laws of social change. Such modifications consist, essentially, in the deliberate introduction of feedback terms of the form $G - X$, where X is the current value and G the corresponding goal value of a state variable. Third, controls are not restricted to temporal patterns but can be generalized to changes of all kinds. For example, if the spatial distribution of a social cell (e.g. farmers, craftsmen, or medical doctors) is to be controlled, the social planner will regard the density of each cell as a function of both spatial and temporal coordinates, and will set up diffusion equations containing feedback or control terms.

Whether spontaneous or planned, the social changes occurring in each social cell of a society form a time series or temporal sequence. And the whole bundle of such partial social histories, one per cell, constitutes the global social history of the given society. More precisely, we make

DEFINITION 5.31 Let σ be a society with instantaneous social structure (or density matrix) $D(\sigma, t)$. Then

(i) the *history* of the S_{ik} cell of σ throughout the period $[0, \tau]$ is the sequence

$$H_{ik}(\sigma; 0, \tau) = \langle D_{ik}(\sigma, t) \mid t \in [0, \tau] \rangle;$$

(ii) the *social* (or *anonymous*) *history* of σ over the period $[0, \tau]$ is the matrix $H = \|H_{ik}\|$ composed of all the social cell histories.

In other words, by collecting the values of the density of a cell over a period of time we obtain the history of that cell during that period. And doing the same for every cell in the social structure we come up with the social history of the society. (This may be dubbed the 'demographic view' of social history.) Obviously it will include the net effects of the contributions of the individual components of the society, though not their biographies. I.e. social history does not display the accomplishments of great men or great crooks: it is anonymous, therefore harder to reconstruct than biography. It attempts to describe, and eventually to explain, the evolution of entire cells, societies, and supersystems of such. It attacks big problems, in particular the coming into being and evolution of entire social cells, organizations, and institutions. For example, social history is concerned with the origin of agriculture and of capitalism, the origin of the state and of organized religion, the expansion of the world market and advanced technology.

A systemic approach to social history encourages studying the evolution of systems proper rather than artificial units such as many a kingdom resulting from war or political compromise. Such social systems may be circumscribed such as the Florentine banking system during the Renaissance, or vast like the Mediterranean basin during the same period (cf. Braudel, 1949). And, of course, a systemic approach will emphasize rather than ignore the interdependence among the various subsystems. For example it will study the evolution of the military–industrial complex, i.e. the symbiotic system composed of certain industries and the military. Or it will investigate the various facets of the Green Revolution, which started in the laboratory, changed the agrotechnics of India and elsewhere, and unexpectedly ended by worsening the lot of the poor peasants. So much for piecemeal social engineering, or the reform policy that ignores the interrelationships among the economic, the cultural, and the political subsystems of any society.

6. Concluding remarks

We have sketched a systemic view of society, which may be regarded as a synthesis of two traditional conceptions: individualism (or atomism), and collectivism (or holism). In fact we regard society or any of its subsystems neither as an aggregate of individuals nor as a mystic totality hovering above its humble membership, but rather as a system composed of persons bonded by social relations. These ties generate social equivalence relations, such as those of belonging to the same economic, cultural, or political group. And these relations in turn partition the composition of a society, at a given time, into social cells or equivalence classes – such as the group of destitutes, or of illiterates, or of people deprived of political rights. Every such partition, or set of social cells, constitutes a social structure, or rather a primary social structure. A secondary social structure rides on the former, as when the set of families is classed into equivalent income groups. The totality of social structures of a society, both primary and secondary, constitutes its social fabric.

The systemic view of society should not be mistaken for the *organismic* view, a variant of holism according to which sociosystems are organisms. Surely the components of sociosystems are alive; besides, sociosystems are born, develop, and become extinct. However, they are not alive: if they were then all of the peculiar laws of the living, and in the first place those of genetics, would hold for them. But these do not, hence sociosystems are not alive. Nor are they dead: the categories of life and death do not apply to sociosystems. This is not to say that sociosystems are suprabiological, in particular spiritual entities, or that they are infrabiological, in particular chemical, systems. Sociosystems are concrete systems with peculiar properties and laws. They are concrete because they are composed of concrete things interacting in concrete ways and because they transform energy and effect changes in their environment. But they are as little physical or chemical systems as they are biosystems. This view of sociosystems may be christened *systemic materialism*, provided 'systemic' is not mistaken for 'holist' and 'materialism' is not construed in the narrow physicalist way.

We have introduced a method for both describing the structure of a society and keeping track of its multifarious changes, namely the social structure approach. The individualist is bound to dislike it because it achieves what he holds to be impossible, namely describing the state, and the changes of state, of a society as a whole. But he cannot in good faith dismiss it as being "merely a holistic confusion", for it is based on a conception of society which, being systemic, is neither holist nor individualist. Besides, the method is mathematically transparent, so it cannot be branded as confused – in particular because it does not involve the individualistic confusion of a sociosystem with its composition.

Our structural approach to social change ensues in a view of social (or anonymous) history which is as precise as it is shallow: it only recounts how the social cells are populated or depleted. But at least this view of social history is superior to both the great hero myth and the great whole myth. It is moreover consistent with contemporary social history and social demography. And far from ignoring the individual it encompasses every individual contribution with a social impact – i.e. that influences the social behavior of other individuals. Finally, nothing prevents the social scientist from supplementing our descriptive account of social structure and social change with mechanismic models capable of explaining the current structure and its development.

Besides a general framework for the study of social systems, we have introduced definite measures of certain emergent social properties, such as social differentiation, participation, and cohesion. And we have submitted that every society, however primitive, is made up of four main subsystems: the kinship, economic, cultural, and political ones. These are not independent spheres but interdependent subsystems. Therefore the neglect or ill health of any of them will sooner or later lead to a deterioration of the others. And, because of the feedback links among the various subsystems, favoring either at the expense of the others will eventually result in a decay of the very favored subsystem.

Thus *economism* ensues in the wilting of culture, which may include a technology indispensable to keeping the economy up to date; and it will tend to leave politics in the hands of "those who know best", thus fostering a concentration of power that will eventually stifle both the economy and the culture. Likewise *politicism*, or the exaggeration of the role of politics, may lead to a disruption of the very economy and culture that politics is supposed to serve, to the point that not much worth being managed may remain. Analogously *culturalism* would end up by draining the economy and drying up the political life, which would backfire on culture itself by depriving it of sustenance and political influence. All this, which is obvious from our systemic point of view, entails a moral that has yet to be learned by most political, economic, and cultural leaders, to wit: In a healthy society there is a balance among its main subsystems, even during periods of vigorous growth. In particular, there is no unilateral (e.g. exclusively economic) development – nor is development the same as growth.

Although we have proposed a handful of substantive hypotheses on the nature of sociosystems, ours is mainly a framework for the study of things of that kind, in particular for the building of theories in social science. (See Bunge (1974d) for the difference between framework and theory.) And surely no framework, however suggestive, is a substitute for a full fledged testable theory. Yet while some frameworks block, others facilitate the conception of theories. Ours discharges its heuristic function in the following ways:

(i) It invites studying social systems rather than either outstanding individuals or amorphous sets of persons or totalities way above individuals. In this way it stimulates the search for societal properties and laws.

(ii) It directs one to identifying the coordinates of every social system – its composition, environment (natural and artificial), and structure. In this way neither the people nor the environment nor the social reslationships can be lost sight of.

(iii) It stimulates the analysis of extremely large and complex systems into more manageable subsystems with their own inner dynamics as well as their mutual actions. Without such analyses hardly any scientific study of society is possible.

(iv) It suggests spotting and characterizing the three chief artificial subsystems of every society – the economy, the culture, and the polity. Thus it helps avoiding the pitfalls of economism, culturalism, and politicism.

(v) It guides the search for relationships among the components of each subsystem as well as among the subsystems of every system. Hence it avoids the fallacies of the lone wolf and the isolated organization.

(vi) It clarifies and simplifies the study of the so-called cross relations, or relations among relations, by constructing them as links among system components or subsystems – with considerable gain in clarity and economy.

(vii) It makes room for cultural, in particular ideological, influences in every walk of social life. But it avoids the reification of cultural activities and is itself ideology-free.

(viii) Although it is ideology-free, our framework does suggest certain social policies, such as that of combining equity with participation in the interests of social cohesion and therefore stability.

This concludes our study of sociosystems and at the same time that of all other genera of concrete system. We return now to the general concept of a concrete system, with the advantage of our realization of the diversity of kinds of system, hence of change.

CHAPTER 6

A SYSTEMIC WORLD VIEW

In this, the last chapter, we shall generalize some of the results of the previous chapters. We shall thus come up with the core of a systemic world view. This *Weltanschauung* is a continuation of certain philosophical traditions and is also in tune with contemporary science. But it should not be mistaken for the popular "systems philosophy," a new version of holism according to which every thing is a system (false) and the patterns of being and becoming are basically the same at all levels (false). Our systemist philosophy is neither holistic nor atomistic: it acknowledges the variety of properties, kinds and patterns found in the world and, by using certain elementary formal tools, it avoids the obscurities of traditional philosophy. Ours is, in short, a kind of scientific ontology. (Cf. Vol. 3, Introduction.) Let us review some of its assumptions.

6.1. *A World of Systems*

An assumption of this book and its companion, *The Furniture of the World* (Vol. 3), is that there are no stray things: that every thing interacts with other things, so that all things cohere forming systems. That is, we propose

POSTULATE 6.1 Every concrete thing is either a system or a component of one.

Another assumption is that systems come in Chinese boxes or nested systems. Hence, given any system but the world, one may expect to find that it is part of some supersystem. (The converse is false: not every system component is itself a system. There seem to be basic or simple things, such as electrons and photons.) Thus we lay down

POSTULATE 6.2 Every system except the universe is a subsystem of some other system.

Finally, the universe itself is neither a mere thing nor a loose aggregate of things but a system of systems:

POSTULATE 6.3 The universe is a system – namely the system such that every other thing is a component of it.

From the previous assumptions, together with some of our axioms on things and change (Vol. 3), a number of further sweeping generalizations follow. One such theorem is that every system is engaged in some process or other. Second: Every change in any system in lawful. Third: Because every subsystem acts upon or is acted on by other subsystems, it exists somehow *per aliud* (by another) rather than autonomously. Fourth: The universe as a whole exists *per se* (by itself) – and moreover it is the only absolute (autonomous) existent. Fifth: The universe has neither beginning nor end in time: it endures eternally although no part of its does.

6.2. *System Genera*

Philosophers are not equipped to ascertain what kinds of system there are or could be in the world: they can only take note of the variety of systems that science discovers, and help categorize them. Current science, pure and applied, seems to accept the existence of five system genera. (Recall that a kind, such as a species or a genus, is said to exist just in case it is not empty. Strictly speaking only the members of a kind can be said to exist really. Cf. Vol. 3, Ch. 3, Sec. 4.3.) The system genera characterized in the previous chapters are:

S_1 = *Physical*
Microphysical, e.g. atoms and small scale fields
Mesophysical, e.g. bodies and large scale fields
Megaphysical, e.g. galaxies
S_2 = *Chemical*
Microchemical e.g. chemical reactors involving only monomers
Mesochemical, e.g. chemical reactors involving polymers
Megachemical, e.g. compost piles
S_3 = *Biological*
Microbiological, e.g. single cells
Mesobiological, e.g. individual multicellular organisms
Megabiological, e.g. ecosystems
S_4 = *Social*
Microsocial, e.g. families
Mesosocial, e.g. villages and concerns
Megasocial, e.g. large cities and nations
S_5 = *Technical*
Microtechnical, e.g. this book

Mesotechnical, e.g. farms and industrial plants
Megatechnical, e.g. a lumber-paper manufacturing-and-commercialization system.

Remark 1 We have not lumped physical and chemical systems into a single category for the following reasons. Firstly, not all physical entities are systems. Secondly, chemical systems are never at rest: if all chemical reactions in a system come to a halt, the system becomes a physical one. Thirdly, chemosystems have hardly any inertia, as shown by their quickness to respond to external inputs and to the cessation of the latter. (Mathematically: the equations of chemical kinetics are of the first order in the rates, whereas most physical equations of evolution are of the second order or higher.) Fourthly, controllers, such as catalyzers, play an important role in chemosystems, for even the H^+ and OH^- ions are catalyzers. On the other hand the only physical systems endowed with control mechanisms seem to be artificial. *Remark 2* We might have distinguished a system genus between biosystems and sociosystems, namely psychosystems. We have refrained from doing so from fear of encouraging the myth of disembodied minds. In our view psychosystems belong in the biosystem genus because they are animals endowed with a highly evolved nervous system. *Remark 3* We have not grouped technical systems, i.e. artifacts, together with any others, because they have ontological traits that set them apart: they are a product of human work and, as such, they bear the stamp of human intelligence and purposiveness as well as of social organization.

We compress the above remarks into

POSTULATE 6.4 At the present stage of the evolution of the universe there are five system genera: physical, chemical, biological, social, and technical. I.e. the family of system genera is $\mathcal{S} = \{S_1, S_2, S_3, S_4, S_5\}$.

The cautious clause about the present is a reminder that the higher kinds have not always existed (so far as we know), and may either become extinct or give rise to further system genera. Thus the exploration of outer space might reveal the existence of further system genera, and a nuclear holocaust would wipe out S_3, S_4 and S_5 on our planet.

6.3. *Novelty Sources*

Even assuming that the universe has always contained things of certain kinds, such as protons, there is little doubt that things of new kinds come into being every so often. I.e. at least part of the marvelous variety of the world is a product of change. Now, there are two main kinds of change: quantitative (e.g. motion) and qualitative (e.g. chemical combination). In the former case all of the axes of the state space of the thing are used; in the latter some new axes pop up and others drop out as change proceeds. However, even quantitative change may end up in qualitative newness, as in the case of high speed collisions of atomic nuclei, the reshuffling of the components of a system, and growth by accretion or multiplication. Let us take a look at reshuffling and accretion, which would seem to be unlikely emergence mechanisms.

A *reorganization* or *restructuring* is of course a change in the structure of a system, i.e. a change modifying some of the links among the system components even though it may not alter appreciably the intrinsic properties of the said components. Two simple examples are isomeric transitions (in an atomic nucleus or in a molecule) and institutional reorganizations without changes in personnel. In a closed system, repeated shuffling ends up in a state of equilibrium and disorder where no novelty can occur. On the other hand in an open system certain reorganizations accompanied by qualitative novelty and increase of order can occur. *Examples* The cooling of a liquid until it solidifies forming a crystal; the cooling of a metallic ring until it becomes superconducting; the magnetization of an iron bar; and the formation of patterns in fluids and in chemical systems. (Cf. Turing, 1952; Glansdorff and Prigogine, 1971; Haken, 1975.)

Another mechanism that may ensue in qualitative novelty is accretion or assembly. Thus stars and planets are believed to have formed by gravitational accretion of cosmic gas and dust. However, novelty is more likely to emerge along processes where heterogeneous things come together to form systems. This may occur even if "mistakes" – i.e. departures from the common pattern – are possible, that jeopardize the completion of the assembly process. Now, in principle assembly can proceed either in one stage – whether instantly or gradually – or in several. For example, a one stage assembly process may consist in the addition of one component per unit time; and a multiple stage assembly process may start with the formation of decamers out of monomers, proceeding with the formation of hectomers out of decamers, and so on. If there is a nonvanishing probability that each item (whether a monomer, a decamer, etc.) joins another in the "wrong" way – e.g. forming an unstable system – then the direct or one stage process is unlikely to yield systems of high complexity. As the number of intermediate stages in the synthesis of a complex system increases, the probability of departure from normality decreases and so the probability of success increases. (For the Crane-Rosen mathematical model of multiple stage assembly, see Rosen, 1970.) Since this result is independent of the nature of the units that get assembled, as well as of the bonds among them, we may assume that all complex systems, whether living or nonliving, have assembled by stages. This general principle of system architectonics deserves being incorporated into our ontology:

POSTULATE 6.5 The more complex a system, the more numerous the stages in the process of its assembly.

The dual of this hypothesis concerns the breakdown or dismantling of systems: it states that complex systems can decay in different ways, each way corresponding to the weakening of one set of bonds among the components. (Even comparatively simple things such as mu mesons can have multiple disintegration schemata.) In short we lay down

POSTULATE 6.6 The more complex a system, the more numerous its possible breakdown modes.

6.4. *Emergence*

Our next assumption is that every concrete system has assembled from, or with the help of, things in the same or lower order genera. In order to formulate this assumption with some precision we need

DEFINITION 6.1 Let $S = \{S_1, S_2, S_3, S_4, S_5\}$ be the family of system genera. Then S_i *precedes* S_j, or $S_i < S_j$ for short, iff things of genus S_i take part (either as components or as agents) in the assembly of every system in S_j.

Our hypothesis is

POSTULATE 6.7 For any $x \in S_j$ there is at least one $y \in S_i$, where $S_i < S_j$, such that y has taken part in the assembly of x.

Thus far we have only asserted that every system has been preceded by pre-existing things – which is the reciprocal of the holistic thesis that the whole precedes its parts. But we have not said exactly what precedes what. This we do now:

POSTULATE 6.8 The precedence relations among the system genera are

$$S_1 < S_2 < S_3 < S_4, S_5.$$

That is, the family of system genera is a tree rooted in S_1 and branching out at S_3 into S_4 and S_5: see Figure 6.1.

The precedence relation $<$ is a partial order relation. Hence the ordered $\mathcal{S} = \langle S, < \rangle$ is a partially ordered set. It may be called the *system structure of the world.* This is often mistaken for a hierarchy. Yet the difference between the two concepts is clear: whereas $\mathcal{S} = \langle S, < \rangle$ is the set of system genera ordered by the precedence (or emergence) relation, a hierarchy is a set of things (or of sets of things) ordered by a dominance relation. Besides, the thesis that the world has a hierarchical structure is usually a form of supernaturalism, and is not countenanced by science. On the other hand the thesis that the concrete systems in the world are distributed among S, and that higher genus systems have emerged from lower genera ones, is a component of a naturalist, systemist, pluralist, and dynamicist ontology.

6.5. *Systemism Supersedes Atomism and Holism*

The eight postulates proposed in the preceding subsections match with the assumptions made in the preceding chapters and crown our version of systemism. This ontology must not be mistaken for holism, or the romantic extolling of wholeness and emergence criticized in Ch. 1, Sec. 4.2. For, while holism claims that wholes and emergents must be accepted with "reverence" (Goethe) or "natural piety" (Alexander), systemism encourages attempts to analyze systems into their composition, environment, and structure, as well as to disclose the mechanisms of their formation and breakdown.

The demand for explanations of the way of systems should not be mis-

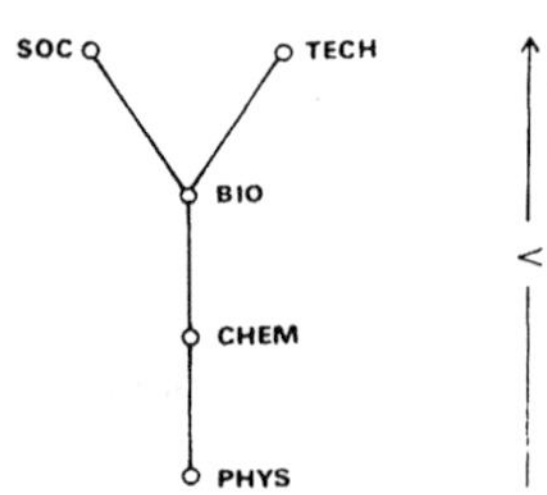

Fig. 6.1. The precedence relations among system genera.

taken for naive reductionism, i.e. the belief that all it takes to account for a whole is a knowledge of its parts. This belief is mistaken because systems possess properties absent from their components, hence not to be discovered by studying the latter in isolation from one another. Analysis is necessary but insufficient to account for the formation and persistence of systems: we must also study systems on their own level and interaction with things of different genera if we wish to understand them. This is what the theories of systems of various kinds – physical, chemical, biological, social, and technical – are all about.

The philosophy that combines an acknowledgment of emergence with the thesis that emergence is explainable and predictable within bounds may be called *rational emergentism*. This philosophy, foreshadowed by Sellars (1922), supersedes both atomism (though not its allegiance to science) and holism (though not its insistence on emergence), and it incorporates a critical realist theory of knowledge – of which more in the next two volumes of this *Treatise*.

6.6. *Synopsis*

It is closing time. We close by characterizing the ontology presented in this volume and its companion, *The Furniture of the World*, in terms of a few traditional *isms*. (A single *ism* is bad because it is one-sides and rigid. Many-sidedness and flexibility can be achieved only by a system of matching *isms*.)

Our ontology endorses

(i) *naturalism* or *materialism*, for it countenances only material existents and discounts autonomous ideas, ghosts, and the like; but *not physicalism* (or *mechanism*), as this denies that all things are physical entities;

(ii) *systemism*, for it holds that every thing is either a system or a component of one; but *not holism*, as it rejects the myths that the whole is incomprehensible, prior to its components, and superior to them;

(iii) *pluralism* as regards the variety of things and processes, hence the plurality of kinds of thing and laws; and also *monism* as regards the substance that possesses properties and undergoes change (namely matter) as well as the number of worlds (just one);

(iv) *emergentism* with regard to novelty, for it holds that, while some bulk properties of systems are resultant, others are emergent; but *not irrationalism* with regard to the possibility of explaining and predicting emergence;

(v) *dynamicism*, for it assumes that every thing is in flux in some respect or other; but *not dialectics*, for it rejects the tenets that every thing is a unity of opposites, and that every change consists in or is caused by some strife or ontic contradiction;

(vi) *evolutionism* with regard to the formation of systems of new kinds, for it maintains that new systems pop up all the time and are selected by their environment; but *neither gradualism nor saltationism*, for it recognizes both smooth changes and leaps;

(vii) *determinism* with regard to events and processes, by holding all of them to be lawful and none of them to come out of the blue or disappear without leaving traces; but *not causalism*, for it recognizes randomness and goal striving as types of process alongside causal ones;

(viii) *biosystemism* with regard to life, for it regards organisms as material systems that, though composed of chemosystems, have properties not found on other levels; but *neither vitalism nor machinism nor mechanism*;

(ix) *psychosystemism* with regard to mind, for it holds that mental functions are emergent activities (processes) of complex neural systems; but *neither eliminative nor reductive materialism*, for it affirms that the mental, though explainable with the help of physical, chemical, biological, and social premises, is emergent;

(x) *sociosystemism* with regard to society, for it claims that society is a system composed of subsystems (economy, culture, polity, etc.), and possessing properties (such as stratification and political stability) that no individual has; hence *neither individualism nor collectivism, neither idealism nor vulgar materialism.*

The reader accustomed to dwell in a single *ism* or in none is likely to throw up his hands in despair at the multiplicity of *isms* embraced by our ontology. Let this be said in defense of such multiplicity. First, it is possible to synthesize a variety of philosophical *isms* provided they are not mutually inconsistent – i.e. provided the result is a coherent conceptual system rather than an eclectic bag. (We have tried to secure consistency by adopting the axiomatic format.) Second, it is necessary to adopt (and elaborate) a number of philosophical *isms* to account for the variety and mutability of reality – provided the various theses harmonize with science. Third, tradition can be avoided only at the risk of unfairness and ignorance: rather than dismiss our philosophical legacy altogether, we should try and enrich it.

APPENDIX A

SYSTEM MODELS

This Appendix is a review and philosophical analysis of some simple yet serviceable models of concrete systems regardless of the nature of their components, hence of the kind of bonds among them. The aims of this presentation are (*a*) to clarify some of the key concepts employed in the text, (*b*) to supply a stock of examples, and (*c*) to discuss some problems in the as yet immature philosophy of systems science. For details see Ashby (1956), Athans and Falb (1966), Dorf (1974), Harrison (1965), Klir (1969), Klir and Valach (1967), Lange (1965), Padulo and Arbib (1974), and Zadeh and Desoer (1963).

1. INPUT-OUTPUT MODELS

1.1. *The Black Box*

The simplest (i.e. poorest) model of a thing one can think of is a black box that does not interact with its environment. ('Black' because it has no structure or we do not care to disclose it.) This, the bare black box model, ignores both the internal structure of the thing modeled and its actions upon other things as well as the actions its environment exerts on it. Despite its poverty, such a model can be endowed with a number of attributes assumed to represent properties of the real thing. The classical particle in free motion and the totally isolated thermostatic system are cases in point.

Next in order of complexity comes the black box proper, fitted with terminals connecting it to its environment – which is in turn treated globally or as composed of other black boxes. The simplest black box has just one input terminal, representing either the action (in a given respect) of the environment on the thing, or the thing's reaction (in the same or a different respect) upon the environment. The former model represents a passive thing, the latter an active one, and both are called *monopoles*. Since in reality there are neither totally passive nor fully active systems, the monopole model is adequate only in extreme cases and when a single property is considered.

A more realistic model of a thing is of course the black box fitted with both input and output terminals, i.e. sensitive to (a part of) its environment and in turn reacting on (some features of) the latter. We call this the *complete black box*, or *black box* for short. The simplest complete black box is sensitive to stimuli of just one kind and accepts just one stimulus at a time – i.e. has a single input terminal; and it also has a single output terminal, i.e. acts just in one fashion on its environment and performs one action at a time. This, the simplest black box, is called a *dipole*. Any more complex black box is called a *multipole*: it is equipped with a certain number of input terminals of the same or different kinds. See Figure A1. We shall concentrate our attention on the dipole, which is the building block of every other input-output model of a system.

From a strictly formal point of view a dipole is just a double entry table of items (e.g. numbers), or set of ordered pairs $\langle u, v \rangle$, where $u \in U$ is an input value and $v \in V$ the corresponding output value. In other words, a black box of this kind boils down to a binary relation f (perhaps a function) with graph or extension $\mathscr{E}(f) \subset U \times V$, where U is the collection of possible inputs and V that of possible outputs. (This is what Mesarović (1968) calls a *general system*. But it is a model of a system and moreover not completely general, for it neglects the internal structure of the system.)

The simplest of all such model systems, i.e. the dipole, has inputs and outputs of the yes or no type: i.e. the terminals are either on (value 1) or off (value 0). That is, $U = V = \mathbf{2} = \{0, 1\}$, and $f\colon \mathbf{2} \to \mathbf{2}$ is the *transfer function*. *Example 1* The conformist box: $f(0) = 0$, $f(1) = 1$. *Example 2* The nonconformist box: $f(0) = 1$, $f(1) = 0$. *Example 3* The capricious (yet not chaotic) or stochastic box: $Pr(0 \mid 0) = p > 0$, $Pr(1 \mid 0) = 1 - p$, $Pr(0 \mid 1) = q > 0$, $Pr(1 \mid 1) = 1 - q$, where p is the probability of response 0 to stimulus 0, and similarly for q. (Note the difference between the stochastic box, characterized by constant values of p and q, hence statistically reliable, and a totally chaotic box, i.e. one with no transfer function.) All three boxes are coarse models of a specialist; the third models the creative type.

Other black boxes accept continuous streams of stimuli and deliver

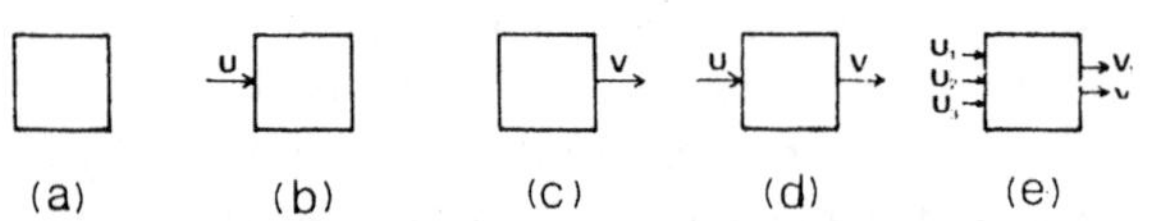

Fig. A1. (*a*) The bare black box. (*b*) The passive monopole, or system without outputs. (*c*) The spontaneous monopole, or system without inputs. (*d*) The dipole, or simplest complete black box. (*e*) The reactive multipole.

continuous streams of outputs. That is, their transfer functions are smooth functions of time. *Example 1* Differentiator. The state of the system at time $t \in T$ is the ordered pair $\langle u(t), \dot{u}(t)\rangle$, where $u : T \to \mathbb{R}$ and $\dot{u}$ is the rate of change of u. The state space of the system is $S = U \times V$, where U is the image of u and V that of $v = \dot{u}$. *Example 2* Delay line. Here $v(t) = u(t - \delta)$ with $\delta > 0$. I.e. the current output equals the input at the earlier time $t - \delta$. The state space is $S = U \times \{u(t - \delta) \mid t \in T\}$.

In the last two examples the input and output values were parametrized by time – i.e. U and V were construed as the images of time-dependent functions. In general the output value $v(t)$ at time t will depend not only on the current value of the input (as in Example 1) but also on some or even all of the inputs accepted earlier by the system (as in Example 2). I.e. in general the system will have *memory* (or be *hereditary*). A fairly general input–output relation covering all such cases for the dipole, and easily generalizable to the multipole, is

$$v(t) = \int_{-\infty}^{t} d\tau\, M(t, \tau)F(u(\tau)),$$

where M summarizes the structure of the box (Bunge, 1963a). The domain of integration, which stops at t, has been chosen to conform with the antecedence principle (Postulate 6.12 in Vol. 3, Ch. 6, Sec. 5.3). If M is a delta function, then the box is memoriless. If F is a linear function then the above represents a linear dipole. In a great many cases M does not depend on t but only on the past (if at all), and $F(u(\tau)) = u(t - \delta)$, where δ is a time lag. In such cases the integral is the convolution of the two functions, and the mathematics is greatly simplified.

1.2. *Connecting Black Boxes*

Strictly speaking a black box is a suitable model not of a system but rather of a system component. In many cases a system can be adequately modeled as an assembly of black boxes connected in a suitable way. The basic modes of coupling are as follows.

(i) *Series coupling*: see Figure A2a. If the two component boxes are characterized by the transfer functions $f : U \to V$ and $g : V \to W$, then the resulting box is characterized by the composition $h = f \circ g$ with values $w = g(f(u))$. Note that, in general, function composition is not commutative; thus g might filter out all stimuli except those in a certain set, while

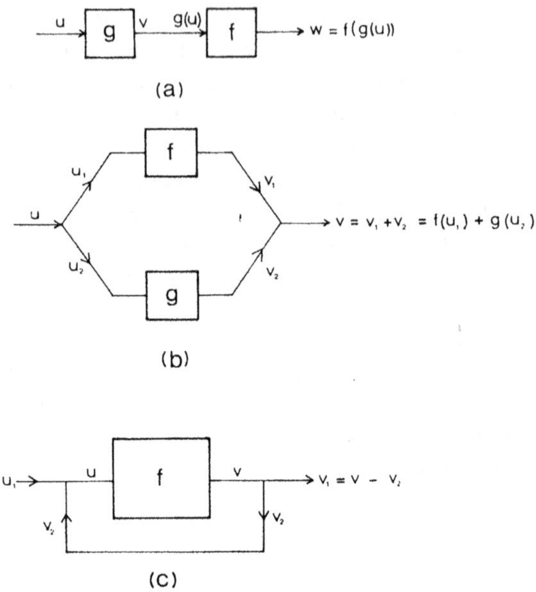

Fig. A2. (*a*) Series combination of two dipoles. Example: cloth manufacturer *g* supplies clothing manufacturer *f*. (*b*) Parallel connection of two dipoles. Example: two branches of a cloth manufacturing concern. (*c*) Feedback coupling (or self interconnection). Example: blood pressure control system.

f could just delay the responses. Example: a (living or industrial) plant *f* that manufactures materials produced by another plant *g*.

(ii) *Parallel coupling*: see Figure A2b. The input is shared by the two components and the output is the sum of the partial outputs. I.e. $u = u_1 + u_2$ and $v = f(u_1) + g(u_2)$. Example: two industrial plants that consume and produce similar goods at the same time.

(iii) *Feedback coupling*: see Figure A2c. The system is coupled to itself in such a way that part of its output is fed back as an input. That is, $v = f(u) = v_1 + v_2$ and $u = u_1 + v_2$, where now u_1 is the external input (stimulus) and v_2 the net output (response) whereas v_2 is the correcting feedback.

By combining tandem, parallel and feedback couplings one can synthesize any complex system of the input–output type. An assemblage of interconnected black boxes is not a black box itself but has emergent properties. One way of representing the resulting system is by means of the list of all the transfer functions characterizing its components when inter-connected, i.e.

$$\mathbb{F} = \langle f_1, f_2, \ldots, f_n \rangle, \quad \text{where} \quad f_i\colon U_i \longrightarrow V_i.$$

While some of these functions represent external terminals (or *exogenous* variables), others represent internal terminals (or *endogenous* variables).

Input–output models are used in a number of fields, from thermodynamics and electrical engineering (the two historical sources) to biology, psychology, and social science. (Most of what goes under the name of *systems analysis* consists in building such models.) Their success has fostered the beliefs that (*a*) every system can be fully modeled by an input–output model, and (*b*) the world as a whole is a supersystem composed of multipoles. But these are illusions. (The technologist's sweet dream is the metaphysician's nightmare.) Firstly, there are plenty of systems without any recognizable inputs and outputs: just think of a radio wave. Secondly, in many cases we need to know the spatial locations of the parts of a system: time is not enough. (Think again of a radio wave, or of a brain, or of a city.) An input-output analysis, though practicable and useful in many areas – particularly in technology – is much too poor and is bound to be eventually supplemented or even superseded by a deeper and more detailed model. Thus thermodynamics is supplemented by statistical mechanics, electric network theory by electrodynamics, stimulus-response psychology by psychobiology, and so on. (For an evaluation of black boxes and a criticism of blackboxist philosophy see Bunge (1964).)

1.3. *Control System*

A control system is composed of two subsystems: an ordinary or controlled system and a controlling system, i.e. one that controls or regulates the output of the former. The controlling system acts by absorbing part of the total output and feeding it back into the system in such a way that the effective input itself is modified, and the net output differs from that of the uncontrolled system. The controlling system has two main components: the *imbalance* (or *error*) *detector* and the *response mechanism*. The latter is driven by the former in proportion to the deviation of the current state of the controlled system from the set point or ideal state or "goal". In other words, the detector is activated (has an output) just in case the controlled system is in a state other than the set point. See Figure A3.

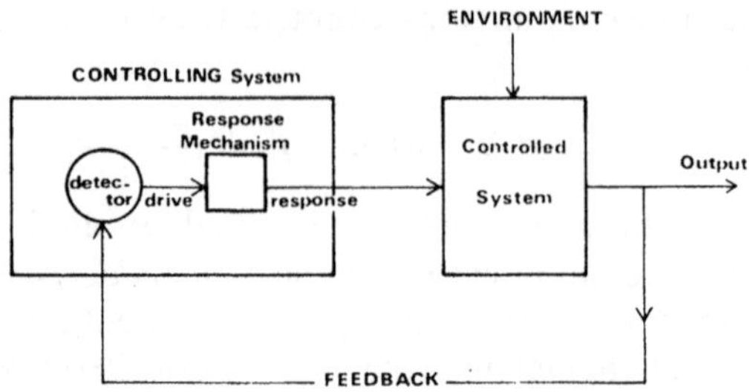

Fig. A3. Feedback control system. The detector compares the output (actual state) of the controlled system and, if it differs from the set point, drives the response or corrective device, which acts on the controlled system to correct the imbalance.

The study of specific control systems belongs to the special sciences, such as mechanical and electrical engineering, biochemistry, physiology, economics, and politology. And the scientific study of the features common to all control systems, regardless of their composition, is of course cybernetics (Wiener, 1948). Cybernetics was born in advanced engineering but, since it is stuff-free (i.e. independent of the kind of material), it assumes no special physical, chemical, biological, or social laws, so it covers all levels. It is therefore a chapter of the general theory of systems as well as of scientific ontology.

The supersystem composed by the controlled and the controlling systems can be described as follows. The total state function of the supersystem is

$$\mathbb{F} = \langle F_1, F_2, \ldots, F_p, F_{p+1}, F_{p+2}, \ldots, F_n \rangle : T \longrightarrow \mathbb{R}^n,$$

where the first p coordinates are the so-called *state variables* and the remaining $n - p$ components the *control variables*. These components are subjected to three sets of conditions:

(i) *constraints* on the state variables (e.g. their values must lie within a certain region of the state space);

(ii) *control* or *steering conditions* specifying the manner in which the control variables affect the state variables, e.g.

$$\dot{F}_i = g_i(\mathbb{F}, t) \quad \text{for } 1 \leq i \leq p;$$

(iii) *optimization conditions* on certain state variables (e.g. minimization of energy consumption, or maximization of the rate of production of some substance).

A good way to get a feel for cybernetics is to study a simple control system such as an integrator (e.g. area calculator) with feedback control. In the absence of a controller and if the system is initially inert, the input-output equation is

$$\dot{v} = u, \quad \text{with} \quad v(t) = 0 \quad \text{for } t \leq 0.$$

This entails

$$v(t) = \int_0^t d\tau \, u(\tau),$$

so that the system integrates the input. In particular, a constant input

$$u(t) = H(t) = \begin{cases} a \in \mathbb{R}^+ & \text{for } t \geq 0 \\ 0 & \text{for } t < 0 \end{cases}$$

is converted into $v(t) = at$. Obviously this output, unless checked, gets eventually out of hand. This can be prevented by coupling the integrator to a controller that diverts part of the output and sends it back to the input terminal, so that a closed loop results: see Figure A4.

In the simplest case the output fraction $kv(t)$, where $|k| < 1$, is fed back diminished but without distortion, so that the input–output equation reads now

$$v = u + kv, \quad \text{with} \quad -1 < k < 1.$$

If $k > 0$, the feedback is *positive*, i.e. it enhances the effect of the input, so that the system may become uncontrollable. The feedback is *negative*, and results in a stable system, if it checks the input, i.e. $k < 0$. The solutions for the case of the constant input $u = H(t)$ applied for $t > 0$ are

(i) for $k > 0$ (*positive feedback*)

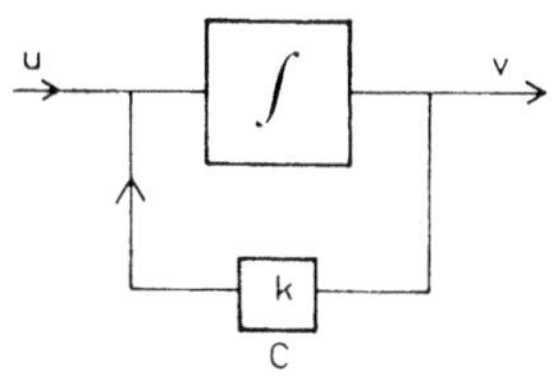

Fig. A4. Integrator with feedback control. The effect of the control subsystem C is to convert a constant input into either an exponentially growing response (for $k > 0$) or an exponentially decaying one (for $k < 0$).

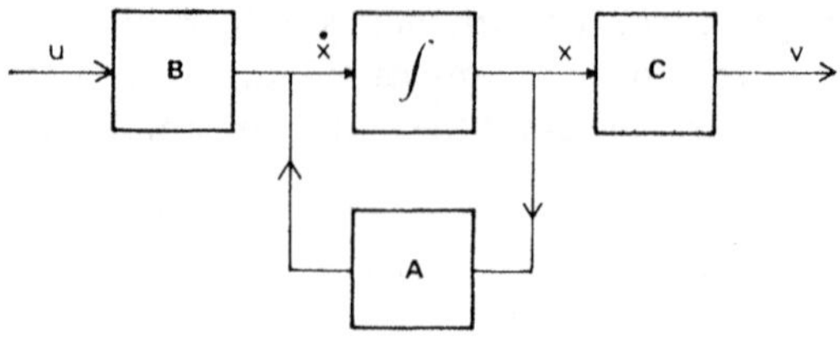

Fig. A5. Linear control system.

$$v(t) = \frac{a}{k}(\exp(kt) - 1) \quad \text{exponential growth}$$

(ii) for $k < 0$ (*negative feedback*)

$$v(t) = \frac{a}{|k|}(1 - \exp(-|k|t)) \quad \text{decelerated growth approaching the target value } a/|k|$$

Note that both the controlled and the controlling subsystems have been treated as black boxes: their mechanisms have not been specified. Nor could they without roping in definite law statements (physical, biological, or what have you).

The above example illustrates the black box model of a control system. In most cases such a model is insufficient and a grey box model is required, i.e. one involving not only an input vector u and an output vector v but also an (internal) state vector x. The simplest example of such a model of a control system is, of course, the linear system

$$\dot{x} = Ax + Bu, \qquad v = Cx,$$

where the state matrix A, the input matrix B, and the output matrix C are constant. The corresponding block diagram is shown in Figure A5.

1.4. *Stability and Breakdown*

As we saw in Sec. 1.3, feedback can be positive (enhancing) or negative (checking). If negative, feedback may correct misalignments. If it does, the corresponding system is said to be *self-stabilizing*, for in such case all the components of its state vector remain bounded all the time. See Figure A6.

Equilibrium can of course be static or dynamic. In the former case nothing changes, so that the representative point in the state space remains

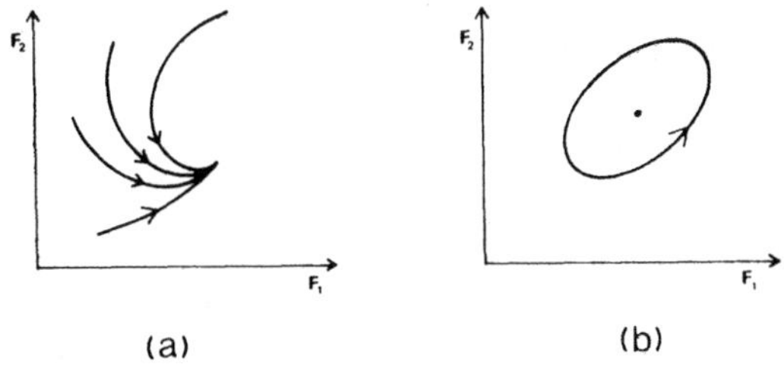

Fig. A6. Two kinds of stable orbits in state space. (a) Spiral approach to static equilibrium. (b) Stationary orbit.

fixed. (But of course no system remains forever in a state of static equilibrium for, even if it is internally stable, every system is subjected to disruptive external influences.) In the case of dynamic equilibrium the trajectories remain within a certain limited region of the state space. The departures from equilibrium may be regular (e.g. sinusoidal oscillations about a point) or irregular – in which case they are called *fluctuations* or *stochastic oscillations*. A system may oscillate around an equilibrium position, whether regularly or irregularly, either spontaneously (self-excited oscillations) or under the action of its environment (forced oscillations). If the oscillation amplitude increases in the course of time instead of remaining bounded or fading out (damping), the system is unstable and eventually breaks down.

If the ongoing existence of a system depends critically on its keeping its properties within certain bounds, then it will have more chances of "survival" if it is self-stabilizing than in any other case. But of course even a system of this kind may be subjected to such strong external disturbances that they drive it permanently far from equilibrium and eventually into breakdown. If the parameters of the controlling components (e.g. thresholds, delays, stiffness) are rigid (constant), then the system may not be able to withstand any large inputs – i.e. may not adapt to the new circumstances. The only chance a system has to withstand strong environmental impacts, within bounds, is by varying some of its parameters as required by the occasion – e.g. to raise some of its thresholds, to lengthen some of its delays, or to decrease its stiffness, or to vary whatever other parameters in such a manner that its wholeness or integrity be preserved. A "plastic" system of this kind is capable of evaluating its own performance and adjusting some of its properties in such a way that its outputs be kept within the stability limits. Far from being a passive toy of its environment, such a system adapts itself actively to it and thus manages to endure. Because such systems are of paramount importance in the life sciences and in the social sciences, not to mention technology, they deserve a name of their own:

DEFINITION A1 A control system of the negative feedback type that changes some of its parameters in such a way that it remains stable in the presence of a changing environment is called a *self-governed, self-controlled, autonomic, adaptive*, or *plastic* system.

What happens if a system becomes unstable? There are two possibilities: if rigid it suffers *structural breakdown*, i.e. destruction. Nuclear disintegration and fission, the blowing up of a section of a network of any kind, the death of an organism, and the extinction of a biopopulation or an ecosystem, illustrate structural breakdown. On the other hand if the system is plastic it is said to suffer *functional breakdown*, i.e. it becomes a stable system of another kind. For example, an insulator subjected to a voltage above a certain point becomes a conductor; a man in a state of shock may lose some haunting memories; a society may become viable by modernizing its technology or getting rid of its parasites. In any case instability, usually dreaded by the engineer, is of interest to the scientist and the philosopher because it is often at the root of radical novelty. In other words, functional breakdown may result in structural buildup or morphogenesis.

2. Grey box models

2.1. *Generalities*

Take a black box and observe or conjecture its internal states without however disclosing the mechanism that makes it tick: you have a grey box. Psychologists say that an *intervening variable* has been added to the stimulus-response pairs, and that the new variable is not a "hypothetical construct" but a mere formal link between inputs and outputs. But if the intervening variable is interpreted as representing the mechanism that transforms inputs into outputs, then instead of a grey box we have a dynamical box (Appendix B, Sec. 1). A grey box is then midway in depth between a black box and a translucid box or mechanismic model. We may picture it as a box equipped with a ladder every rung of which represents a state of the system: see Figure A7.

Of all the grey box theories the simplest is perhaps automata theory (Arbib, 1969; Ginsburg, 1962; Guinzburg, 1968; Harrison, 1965; McNaughton, 1961; Rabin and Scott, 1959; Rabin, 1963). As idealized by this theory, an automaton is a one component system admitting inputs of some kind (e.g. punched cards) and producing outputs of the same or of a different kind, such as printed symbols. The automaton jumps from one state to another in response to, or at least under the influence of, the inputs it accepts. All three – inputs, internal states, and outputs – are assumed to be denumerable. In short, an automaton is a discrete and sequential system. If lacking in spontaneity, i.e. acting only under external compulsion, the automaton is called *deterministic.* If inputs and internal states determine only output probabilities, the automaton is called *probabilistic.* Deterministic automata are of interest mainly in technology, where highly reliable systems are desirable. Probabilistic automata are of greater scientific and philosophical interest, for natural systems, such as brains and communities, are endowed with some spontaneity and seem to have strong stochastic components.

Automata are of greater interest than black box models because, unlike the latter, they take into account not only the behavior but also the internal states of the system. Consequently an automaton model can succeed (i.e. be true) where a black box model must fail (i.e. be false), namely (*a*) when a stimulus elicits different responses depending on the state of the system, or (*b*) when the system has spontaneous (uncaused) outputs. But of course automata models share with black box models the lack of concern with the specific nature of the thing modeled and even the spatial configuration of its components. In other words automata theory is a global (but not holistic), stuff-free (but not ghostly), atopic (nonspatial) and achronic (nontemporal) theory. Being so extremely poor, it is applicable to all

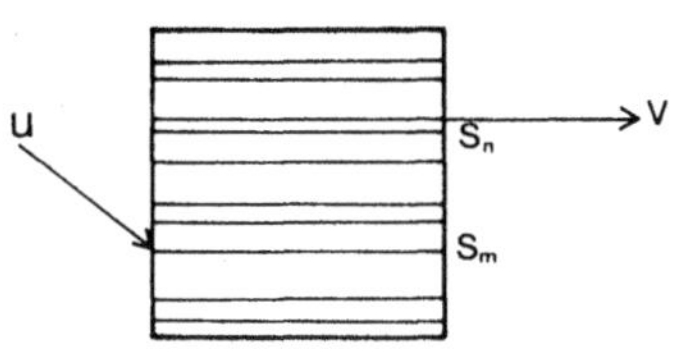

Fig. A7. Transition of system from state s_m to state s_n under the influence of input u.

kinds of system. (Perhaps for this reason von Neumann, 1951, mistook it for a chapter of formal logic.)

All information processing systems may be regarded, at least to a first approximation, as automata: nervous systems, servomechanisms, computers, TV networks, and so on. Therefore the referent of automata theory is not a narrow species of system but the entire genus of information processing systems, whether physical, chemical, living or social. And because the theory is concerned with certain traits of concrete systems, it is not a formal theory but a factual one. Indeed it provides an exact and simple (hence also superficial) model of a system interacting with its environment regardless of any specific features of interest to the special sciences – such as the kinds of material it is built from and the way it is energized. For this reason automata theory belongs not only in advanced technology but also in ontology. Let us therefore take a quick look at it.

2.2. *Deterministic Automata*

Automata theory presupposes only ordinary logic, the concepts of set and function, and some abstract algebra: it does not rest on any body of factual knowledge; in particular it contains no law statement borrowed from other theories. The specific defining (primitive) concepts of the theory of deterministic automata are collected in Table A1.

The above concepts are interrelated and made precise by the axiomatic

TABLE A1
Basic concepts of the theory of deterministic automata

Symbol	Mathematical nature	Factual interpretation
Σ	Set	Set of distinct unit inputs
Ω	Set	Set of distinct unit outputs
σ_0	Individual	Null input or pause
$\circ$	Binary operation	Concatenation
S	Set	State space
M	Function	Transition (next state) function
N	Function	Output function
s_0	Individual	Initial state
F	Set	Set of final states

DEFINITION A2 The structure $\mathscr{A} = \langle \Sigma, \Omega, \sigma_0, \circ, S, M, N, s_0, F \rangle$ represents a *finite deterministic automaton* (or *sequential machine with output*) A iff

(i) $\Sigma = \{\sigma_0, \sigma_1, \ldots, \sigma_k\}$, called the *input alphabet*, is a nonempty set with $k + 1$ elements called *letters*, every one of which represents a unit input on A from its environment;

(ii) Ω, called the *output alphabet*, is a set with two elements: 0 and 1, where 0 represents the absence of an output and 1 a unitary output;

(iii) let Σ^* be the set of finite concatenations, called *tapes*, of elements of Σ (i.e. letters), and let $\sigma_0 \in \Sigma$ be such that, for every tape $x \in \Sigma^*$, $\sigma_0 x = x\sigma_0 = x$. Likewise let Ω^* be the set of finite concatenations of elements of Ω, with $0 \in \Omega$ such that, for every $y \in \Omega^*$, $0y = y0 = y$. The structures $\langle \Sigma^*, \circ, \sigma_0 \rangle$ and $\langle \Omega^*, \circ, 0 \rangle$ are free monoids; $\circ$ represents the combination or concatenation of successive inputs to or outputs of A;

(iv) S, called the *state space* of A, is a finite nonempty set with n elements, where $n > 1$; every member of S represents an internal state of A;

(v) the function $M\colon S \times \Sigma^* \to S$ is called the *transition* (or *next state*) function; if $s \in S$ and $x \in \Sigma^*$, then $M(s, x) \in S$ represents the state A goes into when A, while in state s, admits input x;

(vi) the function $N\colon S \times \Sigma^* \to \Omega^*$ is called the *output function* and is such that $N(f, x) = 1 \in \Omega^*$ for every $f \in F \subset S$ and every $x \in \Sigma^*$, where $N(s, x)$ represents the output of A when acted on by the input $x \in \Sigma^*$ while in state s;

(vii) s_0 is a distinguished element of S representing the initial internal state of A;

(viii) F is a nonempty set included in the state space S, and every member f of F represents a final state of A;

(ix) A makes no spontaneous transitions; i.e. the null input has no effect: for every s in S, $M(s, \sigma_0) = s$;

(x) the internal states form sequences, i.e. the effect of a compound input $x \circ y$, where $x, y \in \Sigma^*$, equals the effect of the second input acting on the automaton in the state to which the first input carried it:

$$M(s, x \circ y) = M(M(s, x), y).$$

The last two assumptions are the essential components of Definition A2. They characterize an entire class of transition or next state functions, and state that a deterministic automaton operates only under external stimulation, sequentially, and in a determinate fashion. Nothing is said about the precise structure of the system, let alone its physics and chemistry: this is why automata theory has such a wide range of applications.

Examples The tables and state transition diagrams in Figure A 8 characterize two different automata with the same internal states reacting differently to the same unit inputs σ_1 and σ_2. They may thus represent some trait of the behavior of two different insects confronted with the same circumstances.

	s_0	s_1	s_2
$M_1(s, \sigma_1)$	s_1	s_2	s_1
$M_1(s, \sigma_2)$	s_2	s_0	s_2
$M_2(s, \sigma_1)$	s_0	s_2	s_2
$M_2(s, \sigma_2)$	s_1	s_0	s_0

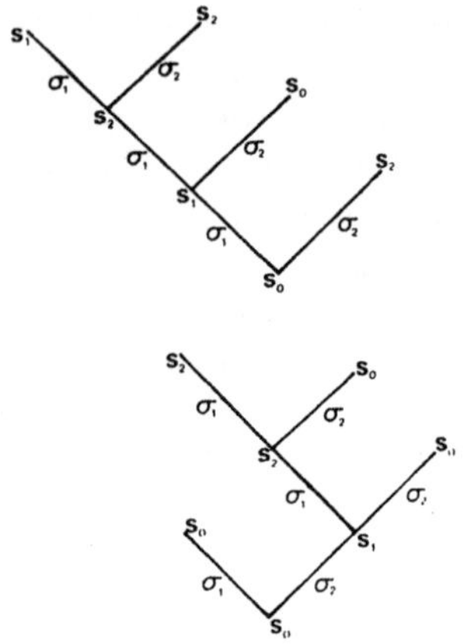

Fig. A8. Two different three state automata, each characterized by a different transition function (M_1 and M_2 respectively), subjected to the same stimuli.

In order to exhibit some consequences of the axiomatic definition of a deterministic automaton we need a couple of conventions. First,

DEFINITION A3 Let $\mathscr{A} = \langle \Sigma, \Omega, \sigma_0, \circ, S, M, N, s_0, F \rangle$ represent an automaton A. Then

(i) the restriction of the transition function M to s_0 is called the *response function*: $rp(x) = M(s_0, x)$, for $x \in \Sigma^*$;

(ii) a state $s \in S$ of A is *accessible* iff there is some input $x \in \Sigma^*$ such that $s = rp(x)$ – otherwise s is *inaccessible*;

(iii) a tape $x \in \Sigma^*$ is *accepted* (or *recognized*) by A iff $rp(x) \in F$ (i.e. if x has an overt effect when acting on A in s_0);

(iv) the *behavior* of A is the set of all the tapes accepted (or recognized) by A: $\beta(A) =_{df} \{x \in \Sigma^* \mid rp(x) \in F\}$;

(v) two automata A and A' are *behaviorally identical* iff $\beta(A) = \beta(A')$.

Note the difference between the above notion of behavior and that used in the so-called behavioral sciences.

And now a few typical consequences. Firstly

THEOREM A1 If two inputs $x, y \in \Sigma^*$ have the same effect, then the effects of the compound inputs xz and yz are the same for an arbitrary tape $z \in \Sigma^*$. I.e. for all states $s \in S$ and all inputs $x, y, z \in \Sigma^*$,

$$M(s, x) = M(s, y) \Rightarrow M(s, x \circ z) = M(s, y \circ z).$$

COROLLARY A1 For any tapes $x, y, z \in \Sigma^*$,

$$rp(x) = rp(y) \Rightarrow rp(xz) = rp(yz).$$

THEOREM A2 Every accessible state of an automaton which is the response to an input whose length is at least equal to the number of states of the system, can also be produced by a shorter input. In other words: If A has n states and if $s = rp(x)$ is an accessible state of A for some tape x composed of $m \geq n$ unit letters, then there exists another tape y such that $rp(y) = s$ and whose length is smaller than m.

THEOREM A3 Every accessible state of an automaton with n states is the response to some input of length smaller than n.

Let us now define the concept of automata analogy:

DEFINITION A4 Let $\mathscr{A} = \langle \Sigma, \Omega, \sigma_0, \circ, S, M, N, s_0, F \rangle$ and $A' = \langle \Sigma', \Omega', \sigma_0', \circ, S', M', N', s_0', F' \rangle$ represent finite automata A and A' respectively. Then

(i) A is *homomorphic* to A' iff there exists a mapping φ from $\mathscr{A}$ to $\mathscr{A}'$ such that

(*a*) $\varphi: \Sigma^* \cup S \to \Sigma'^* \cup S'$;

(b) for every u in $\Sigma^* \cup S$: if $u \in S$ then $\varphi(u) \in S'$;

(c) $\varphi(s_0) = s_0'$;

(d) for every t in $\Sigma^* \cup S$: if $t \in F$ then $\varphi(t) \in F'$;

(e) for every s in S and every σ in Σ: $\varphi(M(s, \sigma)) = M'(\varphi(s), \varphi(\sigma))$, $\varphi(N(s, \sigma)) = N'(\varphi(s), \varphi(\sigma))$;

(ii) A and A' are *isomorphic* iff φ is bijective;

(iii) A and A' are *structurally equivalent* iff they isomorphic;

(iv) the set of all automata isomorphic to a given automaton A is called the *structure type* of A.

It can be proved that, if there exists a homomorphism from one automaton to another, then it is unique. (The same holds, *a fortiori*, for isomorphisms.) The necessary and sufficient conditions for homomorphism and isomorphism are formulated in terms of

DEFINITION A5 An *equiresponse* relation $\sim_A$ on the set Σ^* of tapes is a binary relation on Σ^* such that, for any $x, y \in \Sigma^*$,

$$x \sim_A y \text{ iff } rp(x) = rp(y).$$

By using Definition A4 of homomorphism it is possible to prove

THEOREM A4 Let A and A' be automata. Then there is a homomorphism from A into A' iff the extension of $\sim_A$ is included in that of $\sim_{A'}$ and the behavior of A is included in that of A'.

COROLLARY A2 Two automata are isomorphic iff the extensions of their equiresponse relations coincide and their behaviors are the same.

2.3. *Probabilistic Automata*

The theory of probabilistic automata has the same formal background as that of deterministic automata plus elementary (discrete) probability theory. Its foundations are compressed into

DEFINTION A6 The structure $\mathscr{P} = \langle \Sigma, \Omega, \sigma_0, \circ, S, M, N, s_0, F \rangle$ represents a *finite probabilistic automaton P* iff all the coordinates of $\mathscr{P}$ but M satisfy the clauses of Definition A2, and M is now a function from $S \times \Sigma$ into the real interval $[0, 1]^{n+1}$, called the *table of transition probabilities*, such that, for every state $s \in S$ and letter $\sigma \in \Sigma$,

$$M(s, \sigma) = \langle p_0(s, \sigma), p_1(s, \sigma), \ldots, p_n(s, \sigma)\rangle$$

where

(i) the $p_i(s, \sigma)$, for $i = 0, 1, \ldots, n$, are probabilities, subject to the condition

$$\Sigma_i\, p_i(s, \sigma) = 1 \quad \text{for all } \sigma \in \Sigma;$$

(ii) $p_i(s, \sigma)$, for $i = 0, 1, \ldots, n$, measures the strength of the disposition or propensity for the automaton P, when in state s, acted on by a unit stimulus σ, to jump into the state $s_i \in S$.

Since the transition function is now a probability, we have relinquished not only the determinate behavior of the system but also the serial or sequential character of its operations. The theory describes then the possible rather than the actual behavior of the system. How this is done can be gathered from what follows.

So far we have dealt only with unit inputs or letters. In order to deal with tapes or words we need one more item of notation:

DEFINITION A7 The *stochastic matrix* associated with a unit input $\sigma \in \Sigma$ is

$$P(\sigma) = \|p_j(s_i, \sigma)\|, \quad \text{with } i, j = 0, 1, \ldots, n.$$

The main result concerning the probability associated with a tape or word expresses the statistical independence of the effects of its components:

THEOREM A5 The stochastic matrix of an input $x = \sigma_p \sigma_q \cdots \sigma_r$ equals the product of the stochastic matrices of the unit components of the tape:

$$P(\sigma_p \sigma_q \cdots \sigma_r) \equiv \|p_j(s_i, \sigma_p \sigma_q \ldots \sigma_r)\| = P(\sigma_p) \cdot P(\sigma_q) \cdots P(\sigma_r).$$

DEFINITION A8 Let $\mathscr{P} = \langle \Sigma, \Omega, \sigma_0, \circ, S, M, N, s_0, F \rangle$ represent an automaton P and let $I = \{i_0, i_1, \ldots, i_r\} \subset \mathbb{N}$ be an index set used to label the final states, i.e. set $F = \{s_{i_0}, s_{i_0}, \ldots s_{i_r}\}$. The *probability of tape* $x \in \Sigma^*$ *being accepted* (recognized) by P when in s_0 is

$$p(x) = \sum_{i \in I} p_i(s_0, x).$$

Example Let $\Sigma = \{\sigma_0, \sigma_1, \sigma_2\}$, $S = \{s_0, s_1\}$, $F = \{s_1\}$, and assume the stochastic matrices

$$P(\sigma_1) = \begin{Vmatrix} 1 & 0 \\ \frac{1}{2} & \frac{1}{2} \end{Vmatrix} \qquad P(\sigma_2) = \begin{Vmatrix} \frac{1}{2} & \frac{1}{2} \\ 0 & 1 \end{Vmatrix}.$$

Then the table of transition probabilities for the tape $x = \sigma_1 \sigma_2 \sigma_2$ is

$$P(\sigma_1 \sigma_2 \sigma_2) = P(\sigma_1) \cdot P(\sigma_2) \cdot P(\sigma_2) = \begin{Vmatrix} \frac{1}{4} & \frac{3}{4} \\ \frac{1}{8} & \frac{7}{8} \end{Vmatrix}$$

$$= \begin{Vmatrix} p_0(s_0, x) & p_1(s_0, x) \\ p_0(s_1, x) & p_1(s_1, x) \end{Vmatrix}.$$

The corresponding diagram in the state space is

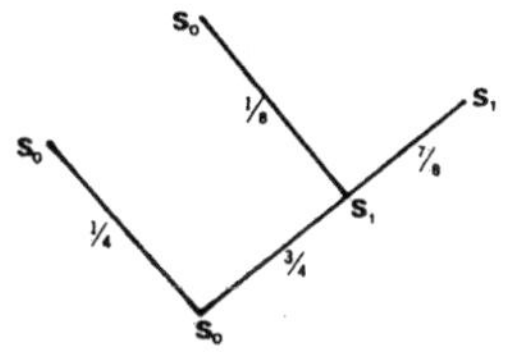

Only the probable set of behaviors of a probabilistic automaton can be computed: its actual behavior can only be described after the event. However, if the various transition probabilities are markedly different from one another, it is possible to anticipate the most probable behavior of an automaton by taking into account only those inputs that have a fair chance of being accepted, and ignoring all others. More precisely, one can choose a real number comprised between 0 and 1, called the *cutpoint*, below which all probabilities can be neglected – at a risk of course. In other words, one can use

DEFINITION A9 Let P be a probabilistic automaton and let r be a real number such that $0 \leq r \leq 1$. Then *the most probable behavior* above (the cutpoint) r is

$$\beta_r(P) = \{x \in \Sigma^* | p(x) > r\}.$$

The relation between probabilistic and deterministic automata, as regards their behavior, is given by

THEOREM A6 P is a deterministic automaton iff $\beta_r(p) = \beta(P)$ for any r such that $0 \leq r < 1$.

This is but an instance of an important metatheorem that has no known proof, namely: *Every stochastic theory includes a deterministic subtheory.*

2.4. *Information Systems*

Our last example of grey box models will be the information system. The signals impinging upon an automaton, whether deterministic or probabilistic, drive it from one state to the next, either invariably or with a certain probability, without altering the set of its possible final states. On the other hand if certain inputs can alter – e.g. expand or shrink – the subset of final states in the state space of a system, then the latter does not qualify as an automaton but does qualify as an *information system*. That is, an information system is one that can gain or lose final states in response to certain external stimuli. If the latter have such an effect, they are called *signals* or *messages*, and each such message can be assigned a definite amount of information measuring the effect of the message on the system.

Needless to say, every message is carried by some information carrier or channel, and every information gain or loss involves some energy. However, the general theory dealing with information systems is not interested in the physical characteristics of either system or message, not even in the energetic aspect of the transaction: it focuses instead on the net change in the number of final states (or outcomes) of a system affected by a message, whatever its nature. So much so that it regards source, channel and receiver as atomic parts of the system and does not include any physical variables. See Figure A9. (In the popular literature on information it is often stated that an input to a system is called a *signal* iff it elicits an output whose energy is much greater than the input energy. However, information theory does not contain the concept of energy, so the popular characterization is inadequate.)

Consider then a system with n possible final states and a message arriving at the system, the effect of which is to reduce that number of alternatives from n to some number $m \leq n$. The *amount of information* carried by the message is taken to equal $\log_2(n/m)$ bits. (The *bit* is of course the unit of information, i.e. the quantity of information gained by the system when the number of alternatives is halved. In fact, in this case $\log_2(n/(n/2)) = \log_2 2 = 1$.) In short, if a message shrinks the number of final states of a system from n to $m < n$, it is said to carry the quantity of information $I = \log_2 (n/m)$. There need be nothing cognitive, let alone subjective, about this concept of information.

If the signals are random, either because they are generated at random or, more commonly, because the information channel is subject to random disturbances, then the statistical theory of information (Shannon and Weaver, 1949) applies. The basic notion of this theory is that of a random binary physical signal appearing at the receiving end of a communication system. The central postulate of the theory is that the quantity of information of such a signal is $I = \log_2(1/p)$, where p is the probability of the signal's reaching the receiver. If the latter happens to be an intelligent being then I is interpreted as the degree of unexpectedness of the message. But

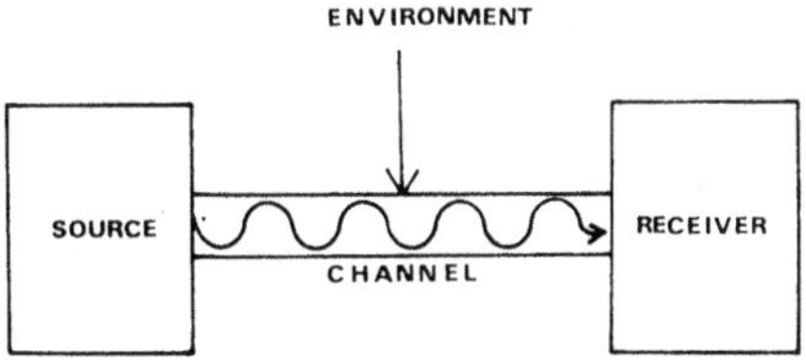

Fig. A9. The components of an information system. The effects of the environment on the channel, when random, are lumped under the category of noise.

this interpretation, legitimate as it is in some applications of statistical information theory to psychology, is irrelevant to communications engineering, molecular biology, and semantics.

One final note of caution. Some claim that information measures organization. The argument runs like this: a deterministic system is predictable, so watching its behavior provides no information; on the other hand a probabilistic system is unpredictable, so whatever it does is surprising or informative. Hence the more disorganized a system the more informative the observation of it. This argument contains several important mistakes. Firstly, the information function is not defined on a set of systems (or of states of a system) but on system-message pairs. Consequently a quantity of information, far from being an intrinsic property of a system (e.g. a transfer RNA molecule), is a mutual property of system and message. Hence it cannot measure the degree of organization of a system nor any other intrinsic property of it. Secondly, systems are neither predictable nor unpredictable: whatever can be predicted can be so with the help of theories and data. The predicate "predictable" is not defined on the set of systems but on the set of pairs $\langle$body of knowledge, system$\rangle$. Thirdly, prediction must not be equated with deterministic prediction. If a system is stochastic, as is the case with an atom, a learning organism, or a social group, then its behavior may be predicted statistically – provided we have succeeded in building a true probabilistic theory about it. In this case we may be able to predict probabilities (of e.g. transitions) and a number of properties derived from them, such as averages and standard deviations. Fourthly, a stochastic system – such as a probabilistic automaton – is not chaotic: being probabilistic it is "governed" by probabilistic laws. Thus, though stochastic, an atom is not chaotic like the top of my desk, over which no probability distribution can be defined. (See Vol. 3, Ch. 4, Sec.6.4 for the difference between randomness and chaos.)

In conclusion, information does not measure organization. And there seems to be no better measure, perhaps because organization is an extremely complex property that defies lumping into a single number. What we do have are measures of efficiency, such as thermodynamic efficiency, output/input ratios, or signal/noise ratios. Knowing that a system is highly (or fairly or poorly) efficient we can infer that it is well (or fairly or poorly) organized to attain a certain goal or to help us perform a certain task. But efficiency is an indicator of organization not a measure of it.

APPENDIX B

CHANGE MODELS

In this Appendix the accent will be on change. We shall exhibit both kinematical and dynamical models, first of quantitative, then of qualitative change. And we shall study only extremely general change models, i.e. such that are portable from one field of inquiry to another. As in the case of Appendix A, most of the material to be presented is standard or has at least appeared before. But the interpretation and the philosophical discussion is often novel.

1. KINEMATICAL MODELS

1.1. *Global kinematics*

A kinematical model of a system is a theory representing the change, step by step, in some of the properties of the system. The simplest kinematical model assumes that the properties of the system are representable by smooth functions, and that the collection of all of these, i.e. the state function, is a time dependent function with values in $\mathbb{R}^n$ obeying an equation of evolution of the form

$$\dot{\mathbb{F}} = \mathbb{G}(\mathbb{F}, t).$$

Example The rate of change of the state function is a linear combination of its own components, i.e. $\mathbb{F} = G \cdot \mathbb{F}$, where G is a square matrix with time independent coefficients. The solution is $\mathbb{F}(t) = e^{Gt} \cdot \mathbb{F}(0)$, where $\mathbb{F}(0)$ is the value of the state function at $t = 0$.

Some of the components of $\mathbb{G}$ may remain constant throughout the evolution of the system. In this case they are called the (time) *invariants*, or *constants of the motion*, of the system. (The system as a whole need not move through space.) If all the n components of $\mathbb{G}$ vanish, the system has n constants of motion and is always in a steady state even though its components may not be in such a state.

Another systemic or global property that can be studied with the help of a kinematical model is stability. A system represented by a state equation of the form $\dot{\mathbb{F}} = \mathbb{G}(\mathbb{F}, t)$ is (*a*) in a *stationary state* (or in *equilibrium*) during a certain time interval iff $\dot{\mathbb{F}} = 0$ during that interval (so that $\mathbb{F}$

equals a constant n-tuple); (b) *asymptotically stable* iff $\mathbb{F}$ approaches an equilibrium value for increasing values of t.

Example Suppose a system is characterized by just two properties, represented by F_1 and F_2, so that its equations of evolution are

$$\dot{F}_1 = G_1(F_1, F_2, t), \qquad \dot{F}_2 = G_2(F_1, F_2, t).$$

The equilibrium property can be found by going over to the state space, namely by dividing one equation by the other to obtain

$$\frac{dF_1}{dF_2} = \frac{G_1}{G_2},$$

which represents the possible processes in the state space. The system is stable iff dF_2/dF_1 approaches a definite number as F_1 increases. This number is the slope of a straight line in the state space, namely the equilibrium trajectory. If on the other hand the system diverges from the (imaginary) equilibrium line, it is unstable. See Figure Bl. (The equilibrium trajectory may be a finite segment or just a point.)

1.2. *Analytical Kinematics*

In the preceding subsection we dealt with the global kinematics of an arbitrary system. In contrast, analytical kinematics pays attention to the components of a system and the links among them. Assume, for the sake of simplicity, that a system has just two components characterized by one property each. Call F_1 and F_2 the functions representing the peculiar property of component 1 and component 2 respectively. Suppose that these functions vary smoothly in the course of time and that the law of evolution is

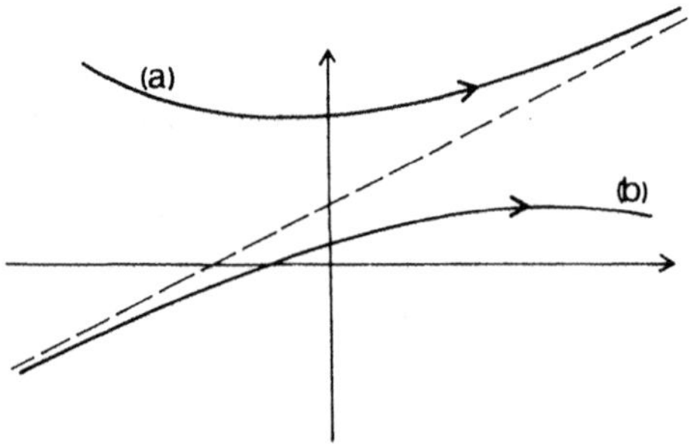

Fig. B1. (a) Stable system. (b) Unstable system.

$$\dot{\mathbb{F}} = \mathbb{G}(\mathbb{F}) \quad \text{with} \quad \mathbb{F} = \langle F_1, F_2 \rangle, \quad \mathbb{G} = \langle G_1, G_2 \rangle.$$

Assume further that the system is linear, i.e. that

$$\dot{\mathbb{F}} = A \cdot \mathbb{F} \quad \text{with} \quad A = \begin{Vmatrix} a_{11} & a_{12} \\ a_{21} & a_{22} \end{Vmatrix}, \quad \text{where } a_{ij} \in \mathbb{R}.$$

If the off-diagonal coefficients vanish or are very small compared with the diagonal elements, then each component evolves separately or nearly so and, instead of a system proper, we have just an aggregate. Otherwise the evolutions of the system components are coupled.

Example Let $F_i(t)$ represent the amount of stuff of some kind in the ith component of the system at time t. (Such stuff can be matter, information, goods, capital, or what have you.) The system will be in a stable state, and remain in it, provided each component gives back as much stuff as it receives, i.e. iff $a_{12} = a_{21}$. Any departure from this equilibrium state will ensue in an uneven distribution of the stuff measured by F_i – i.e. there will be wind, a leak, an imbalance of payment, a migratory flow, or some other diffusion or transport process. If this trend continues unchecked – i.e. if the system lacks control mechanisms – the entire stuff may end up by concentrating in one of the components. At this point of maximal heterogeneity the exchange may cease and thus the system itself will come to an end *qua* system: each component starts evolving independently from the other. In short, depending on the control mechanisms, whatever disparity there was will either level out (as is the case with heat transfer) or be greatly enhanced (as is the case with unchecked capital transfer). In short, if balance is required then checks are called for.

1.3. *Balance Equations*

So far in this Appendix we have not singled out any state variable representing the transactions of the system with its environment. We shall now remedy this shortcoming. To this end we shall have to introduce the concept of ordinary (physical) space.

Example In $\dot{\mathbb{F}} = \mathbb{G}(x, t)$ set $\mathbb{G} = -\nabla \cdot j$, where j is a vector valued function of position and time representing the flux of something. This law schema occurs so frequently under different guises that it is called the *general balance* (or *conservation* or *continuity*) equation, and is sometimes regarded as a purely mathematical statement without any factual content. Let us examine this matter.

Let σ be a system with volume V and surface S in Euclidean three-space. (We are assuming that σ is not a megasystem, so it does not require Riemannian geometry.) Call ρ an intensive (nonextensive) property of ε, and assume that σ interacts with its environment. By definition, the extensive property ε associated with ρ is

$$\varepsilon(\rho, t) = \int_V dV \rho(\sigma, x, t).$$

For the sake of definiteness think of ρ as the mass density and of ε as the total mass of the system.

We now assume the following factual hypothesis:

POSTULATE B1 For every extensive property ε of a system interacting with its environment, ε may change in the course of time in two (mutually compatible) ways: spontaneously (internally) or as a consequence of system-environment interactions. More precisely, the rate of change of ε is composed of two parts: a source s and a flux j, or

$$\dot{\varepsilon} = \int_V \mathrm{d}V s + \int_S \mathrm{d}S\, j_n,$$

where j_n is the projection of j onto the interior normal to S. See Figure B2.

Using now Green's theorem, a purely mathematical result, we derive

THEOREM B1 If ρ is the intensive property associated with the extensive property ε occurring in Postulate B1, then

$$\dot{\rho} + \nabla \cdot j = s.$$

This is the local (rather than integral or global) balance equation. It is not a special law statement but a *law schema*: it holds for every intensive

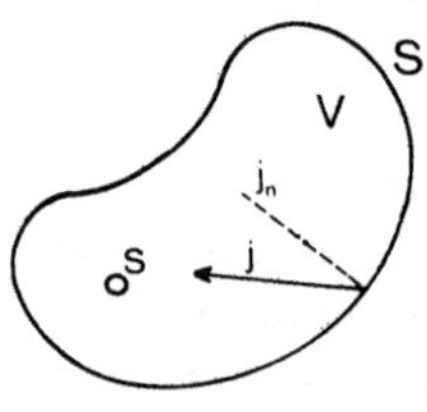

Fig. B2. Transactions of a system with its environment as reflected in the rate of change of the former's extensive property ε. This rate consists of the (positive or negative) contribution of the source (or sink s) and the (positive or negative) contribution of the flow j.

property of an arbitrary system, ergo it will occur as often as intensive properties are treated, whether in quantum mechanics or in continuous models of traffic flow. (The only limitation on it is that it is local, i.e. it holds only in the neighborhood of a reference frame.)

Example Let $j = -a(\partial\rho/\partial x) - b\rho$, with a and b real numbers, and $s = 0$. Then

$$\dot{\rho} = a\,\partial^2\rho/\partial x^2 + b\,\partial\rho/\partial x$$

represents the superposition of a diffusion process, that may be random, and a drift caused by the force $b\,\partial\rho/\partial x$. But even this is just a law schema unless ρ is interpreted in factual terms.

Theorem B1 allows us to give a quantitative definition of the notion of system closure:

DEFINITION B1 Let a system σ interacting with its environment possess n intensive properties ρ_i with the corresponding fluxes j_i, where $1 \leq i \leq n$. Then σ is *isolated* from (*closed* to) its environment during a given time interval

(i) *in the ith respect* iff the flux j_i vanishes on the system boundary for all time in the given interval;

(ii) *in all respects* iff all the fluxes j_i vanish on the system boundary for all time in the given interval.

COROLLARY B1 For any system isolated (closed) in the ith respect, $\dot{\rho}_i = s$.

Moral: Before blaming any changes in the system on its environment, search for the system's internal sources of change. This holds not only in the natural sciences but also in the social ones.

1.4. *Lagrangian Framework*

Consider a system of any kind with n components each of which is characterized by just three basic state functions (called *generalized coordinates*), the values of which depend only on a phase variable interpretable as the time coordinate of a certain reference frame. A Lagrangian model of such a system boils down to a certain function L of the $3n$ generalized coordinates, their rates, and time, that satisfies a certain extremality condition. Because the nature of the components is left unspecified, the model fits any system with a finite number of distinct parts, be it physical, biological, or social. This referential generality explains the wide use of lagrangian kinematics in a variety of scientific and technological fields.

A possible formalization of the notion of a lagrangian model is given by

DEFINITION B2 Let σ be a system composed of n parts and moreover one with a constant composition, i.e. such that $\mathscr{C}_t(\sigma) = \mathscr{C}_{t'}(\sigma)$ for any t, $t' \in T$. Then the structure $\mathscr{L} = \langle T, Q, L \rangle$ is a *lagrangian model* of σ iff

(i) T is a subset of the real line ordered by the relation $\leq$ of temporal priority or simultaneity, and every $t \in T$ represents an instant of time;

(ii) $Q = \{q_i \mid q_i: T \to \mathbb{R} \;\&\; 1 \leq i \leq 3n\}$ is a set of time dependent, real valued and once differentiable functions [the basic state functions of the system];

(iii) the $6n$-tuple $\langle q_1(\sigma, t), \ldots, q_{3n}(\sigma, t), \dot{q}_1(\sigma, t), \ldots, \dot{q}_{3n}(\sigma, t)\rangle$ represents the state of σ at time t;

(iv) L is a real valued function of the state variables, their rates, and t – i.e. $\langle q_i(t), \dot{q}_i(t), t\rangle \mapsto L(q_i, \dot{q}_i, t)$ – such that, for any $t \in [t_1, t_2] \subset T$ and fixed end points $q_i(t_1)$ and $q\,(t_2)$, the integral

$$I =_{df} \int_{t_1}^{t_2} \mathrm{d}t\, L\,(\sigma, q_i, \dot{q}_i, t)$$

is either minimal or maximal.

By performing the indicated first order variations (i.e. by computing δI and setting it equal to 0) it is easily seen that the above axiomatic definition entails the following system of $3n$ equations

$$\frac{\mathrm{d}}{\mathrm{d}t}\frac{\partial L}{\partial \dot{q}_i} - \frac{\partial L}{\partial q_i} = 0, \quad 1 \leq i \leq 3n.$$

If there are constraints, i.e. if the generalized coordinates q are not mutually independent, the equations of "motion" are more general: the right hand side equals a function Q_i representing the net effect of the constraints.

Sometimes these kinematical equations can be assigned a dynamical interpretation, namely as representing a balance of forces. In particular $\partial L/\partial q_i$ represents the intensity of the combined action of the environment and the rest of the system upon the component characterized by three values of the index i, and Q_i the forces amounting to the constraints. The latter are particularly important in the case of social systems, where no variable can be said to be independent of the others.

Note that so far we have not specified the Lagrangian function L. A suitable choice of L and of the semantic assumptions concerning the generalized coordinates q_i will yield the laws of particle mechanics. An alternative interpretation yields the theory of electrical networks. Still another interpretation ensues in a certain theory of population dynamics – and so on. In sum, the lagrangian formalism is subject-matter independent. For this reason it belongs not only to science and technology but also to ontology (Bunge, 1974c).

In many fields of inquiry the equations of evolution are of the form

$$\dot{\mathbb{F}} = \mathbb{G}(\mathbb{F}, t)$$

(Sec. 1.1). These equations are just a particular case of the lagrangian formalism with constraints. In fact regarding the components F_i of the state function $\mathbb{F}$ as the generalized coordinates, and introducing the definitions

$$L =_{df} \sum_{k=1}^{n} F_k \dot{F}^k - \sum_{k=1}^{n} G_k F^k$$

$$Q_i =_{df} \mathbb{F} \cdot \frac{\partial \mathbb{G}}{\partial F_i},$$

the two sets of equations coincide. (And, because $\partial L/\partial \dot{F}_i = F_i$, the generalized momenta $p_i = \partial L/\partial \dot{F}_i$ are seen to coincide with the corresponding generalized coordinates. This does no harm since, in general, the generalized coordinates are not interpretable as position coordinates. Thus in the case of a society F_1 may represent the population, F_2 the energy consumption per capita, F_3 the GNP per capita, and so on.) So much for the Lagrangian models of discrete systems.

In the case of a continuous system, such as an elastic body, an electric field, or a sufficiently dense population of organisms, one starts from a lagrangian density and obtains L as the volume integral of that density. As with the discrete case, a few rather mild conditions on the Lagrangian suffice to yield a theory schema with a wide coverage – so wide indeed that, when enriched with suitable subsidiary conditions or hypotheses, it yields every one of the classical field theories (Edelen, 1962). Such conditions specify then a specific Lagrangian model.

What is usually called *axiomatic field theory* intends to discharge a similar function with respect to the quantum field theories, in particular quantum electrodynamics. However, unlike the general classical field theory mentioned a moment ago, its quantum counterpart does not yield quantum electrodynamics as a special case. "It can at most give a framework for a concrete theory of existing particles and their corresponding fields" (Jost, 1965, p. xi). In addition, one hopes that this approach may avoid the inconsistencies that mar the quantum field theories. Its role in metaphysics is yet to be investigated.

We close this subsection with an historical remark. The Lagrangian framework was foreshadowed by Maupertuis as a synthesis of dynamics and metaphysics. It was brought to fruition by Lagrange, who freed it from metaphysics, and it was generalized beyond dynamics by Maxwell. The axiom contained in Definition B2 (iv) is one of the most secure and yet misinterpreted principles of science. Maupertuis took the (action) integral I to be always a minimum and interpreted this as a sure sign of divine wisdom: Nature always chose the paths that minimized I. It was later discovered that in some cases (notably in optics) I is not a minimum but a maximum, so the principle was rechristened from that of minimal to that of extremal action. Nevertheless a number of physicists, among them Planck and Margenau, continued to regard Maupertuis' principle as an expression of purpose and design. This is mistaken if only because the principle, together with some lemmas borrowed from the variational calculus, entails the equations of motion – in which no traces of purpose or design have ever been detected. It is not that the system behaves in order to minimize or maximize its action: it is just a law, and an extremely pervasive one, that all systems behave that way. Because of this generality the Lagrangian framework should be recognized as belonging to the intersection of science and metaphysics.

1.5. *Kinematical Analogy*

We close this section with some remarks on the notion of kinematical analogy. Any systems which, regardless of their detailed composition and structure, have the same kinematics – hence are described adequately by the same equations of evolution – can be said to be *kinematically analogous* (or *equivalent*). For example some periodic oscillations, whether mechanical, electrical, chemical, biological, or economic, happen to satisfy the same differential equations, hence they are kinematically analogous. Thus the relaxation oscillations in a vacuum tube amplifier are described by the equation $\ddot{x} - a(1 - x^2)\dot{x} + bx = 0$, which has been said to describe also the noise made by scratching a piece of chalk on a blackboard, the waving of a flag in the wind, shivering from cold, and the periodic recurrence of epidemics.

There can be no doubt that the finding of kinematical analogies can play a heuristic role and save time. But they have hardly any ontological import except insofar as they show that different mechanisms may result in similar overall behavior patterns, whence the study of behavior does not exhaust that of reality. For example brownian particles, domestic flies, and drunkards move randomly, but the underlying mechanisms are very different, so that the observation that all three are kinematically analogous has no metaphysical significance. What does have ontological import is the finding of *dynamical* analogies among systems of different kinds, i.e. the discovery that certain *mechanisms* – such as transport, diffusion, merging, cooperation, competition, random shuffling, and casual mutation – are conspicuous on all levels. So, let us turn our attention to mechanisms.

2. Dynamical Models

2.1. *Generalities*

A dynamical model of a system is a kinematical model (Sec. 1) some variables of which are interpretable as "driving" others. In other words, a dynamical model involves a change *mechanism*, i.e. either a subsytem or a process whereby the changes in some properties of the system effect changes in some other properties of the same system. Morever all such changes are spatiotemporal or reducible to changes in time (such as rates) or in space (such as gradients).

Example 1 The geographic distribution of a biospecies can be explained as the outcome of the action of three separate mechanisms: random diffusion, drift, and environmental resistance (or selection pressure). A dynamical model of the spatial pattern of a species (or rather a family of populations of the given species) is this. The rate of change of the population density ρ of the given collection of populations at some point x on the surface of the planet is

$$\frac{\partial \rho}{\partial t} = \underbrace{\alpha\, \partial^2\rho/\partial x^2}_{\text{Diffusion}} - \underbrace{2\beta\, \partial\rho/\partial x}_{\text{Drift}} + \underbrace{\gamma\rho(\rho - 1)}_{\text{Environmental pressure}}.$$

Example 2 Much the same holds if ρ is interpreted, not as the actual population density, but as the probability density, so that $\rho\, \mathrm{d}S$ represents the probability that $\rho\, \mathrm{d}S$ individuals live in the area $\mathrm{d}S$. In this case all three forces are stochastic, like the pushes exerted by molecules on a brownian particle. *Example 3* Field models are a kind of dynamical model. A field is a thing spread in a continuous manner over an expanse of space. (Think of the gravitational or of the magnetic field of our planet.) In principle a field does not qualify as a system any more than a particle does, for neither has atomic parts. However, a collection of fields superposed in a given region of space does qualify as a system. An electromagnetic wave is an example of a pure field system. Also, a field may play a part in a system not made up exclusively of fields: its role is that of glue that keeps the nonfield components together. We call a *field theoretic model* of a system any model in which field concepts occur essentially, as is the case with the (theoretical) model of a molecule. The "fields" invented by some biologists, psychologists, and social scientists, are phoney, for they are not substantial continua and are not described by field equations: they are just metaphors.

A field theoretic model is mechanismic, not phenomenological, but its mechanism is nonmechanical. (Mechanical mechanisms are a proper subset of mechanisms.)

2.2. *Formalities*

We repeat in a slightly more exact way the definition of a mechanism given at the beginning of the last subsection:

DEFINITION B3 Let σ be a system. Then a *mechanism* of σ is either
(i) a component m of σ such that m acts in σ, or
(ii) a process m in σ such that other processes in σ depend upon m.

And now we assume

POSTULATE B2 Every system has some mechanism.

Clearly this postulate (*a*) overreaches experience, (*b*) goes against the grain of phenomenalism and descriptivism (e.g. behaviorism), and (*c*) encourages theorizing. The kind of theorizing our postulate stimulates is, of course, the building of dynamical models. A *dynamical model* of a system may be defined as a model representing one or more mechanisms of the system.

A standard formulation of a dynamical model is within the general framework of hamiltonian dynamics. Just as every kinematical model can be formulated in Lagrangian terms (Sec. 1.4), so nearly every dynamical model can be formulated in a Hamiltonian fashion. It will be recalled that the former has just one set of $3n$ basic state functions, namely the generalized coordinates q_i. A Hamiltonian model, on the other hand, rests on two sets of basic state functions: the very same generalized coordinates q_i and the corresponding generalized momenta p_i, which are logically independent of the former. Just as the solutions q_i of a Lagrangian problem may represent the spatial configuration of the system, so the corresponding p_i represent the dynamics of the system. In particular, the rates $\dot{p}_i$ may represent the forces acting on the various parts of the system. If the Lagrangian L of the system is known, then the p_i are given by $p_i = \partial L/\partial \dot{q}_i$, and the hamiltonian of the system by $H = \Sigma p_i \dot{q}_i - L$. But if L is unknown then the problem must be formulated afresh by conveniently choosing the basic variables and guessing the adequate Hamiltonian. The basic equations of motion are

$$\dot{p}_i = -\partial H/\partial q_i, \qquad \dot{q}_i = \partial H/\partial p_i.$$

So, the solution of a hamiltonian problem is compressed into the state function $\mathbb{F} = \langle q_1, q_2, \ldots, q_{3n}, p_1, p_2, \ldots, p_{3n}\rangle$. Every value of this function represents a possible state of the system. And as time goes by the tip of $\mathbb{F}$ traces a trajectory in the state space of the system, which in this case is called its *phase space*.

Although the general hamiltonian theory – namely that in which H is left unspecified and the basic state variables uninterpreted – belongs to scientific ontology, we shall not go into any details of it. (For this and a few other general dynamical theories, see Bunge, 1967b, Ch. 2.) Our purposes in mentioning the theory here are (*a*) to emphasize the richness of the ontology that originates in science, and (*b*) to elucidate the concept of dynamical analogy. The latter may be defined as follows. Let σ and σ' be two systems modeled by the Hamiltonians $H_\sigma(q, p, t)$ and $H_{\sigma'}(q', p', t')$ respectively. Then σ and σ' are said to be *dynamically* analogous iff H_σ and $H_{\sigma'}$ are formally the same up to a similarity transformation of its dependent variables – i.e. if $q' = aq$, $p' = bp$, and $t' = ct$, where a, b and c are complex numbers. Nothing is required of the interpretation of these variables.

Let us next discuss the dynamics of processes of two special yet rather widespread kinds, namely competition and cooperation, which will be treated jointly in a single dynamical model.

2.3. *The Pervasiveness of Cooperation and Competition*

When two or more things get together and form a system, they may be said to cooperate with one another even if they do not do so deliberately.

For example, each of the syntheses (physical, chemical, biological, or social) $A + B \rightharpoonup C$ and $A + D \rightharpoonup E$ may be regarded as a cooperative process: the A's cooperate with the B's to form C's, and likewise for the second reaction. However, if the two processes occur simultaneously at the same place, then the two cooperative processes compete with each other: indeed, in this case the B's and the D's compete for the A's. So, in this case there is both cooperation (in some respects) and competition (in others). Likewise two families (or other sociosystems) may compete with one another while cooperating internally.

A possible definition of the notions of interest is this:

DEFINITION B4 Let x, y and z be things. Then

(i) x *cooperates* with y iff there is a system z such that x and y are in the composition of z;

(ii) x *competes* with y for z iff the formation of a system composed of x and z interferes with that of a system composed of y and z.

Note that, according to this definition, competition is not the opposite of cooperation. Moreover two things may cooperate in certain respects and compete in others.

Cooperation has received far less attention than competition. In particular, most philosophers since Heraclitus have stressed war at the expense of cooperation, ignoring that intersystemic competition calls for intrasystemic cooperation. On the other hand biologists, pharmacologists and physicians are fond of emphasizing that the components of the normal cell act in cooperation, that every healthy organism is synergic, and that only sickness and senility cut the cooperative ties.

Of late physicists have joined life scientists in studying a number of cooperative processes. One striking discovery along this line marked the birth of a whole new branch of science, namely plasma physics. Collective or molar modes of oscillation and propagation were discovered in such lowly systems as bodies constituted by very hot and highly ionized gases – which, according to classical ideas, should be the paradigm of chaos and homogeneity. Plasma waves, both stationary and traveling, as well as the emergence of complex shapes of plasma bodies (plasmons), gave the lie to the belief that in every gas all a molecule can do is to move haphazardly and collide with other molecules in a random fashion without ever giving rise to a new system characterized by novel overall modes of behavior. At about the same time chemists discovered the hydrogen bond, which can give rise to marvels of cooperation such as the biomolecules. In short, cooperation is nowadays just as "in" as competition was one century ago. Which is just as well, for unless cooperation is acknowledged alongside competition, the very existence of systems cannot be accounted for. For one thing the pervasiveness of assembly processes would not have been recognized if scientists had adhered to the Heraclitean ontology of strife.

We shall stick our neck out and assume

POSTULATE B3 Cooperation is about as frequent as competition.

We shall presently atone for the vagueness of this formulation by reviewing a mathematical model of processes wherein cooperation intertwines with competition (Bunge, 1976).

2.4. *The Dynamics of Competitive–Cooperative Processes*

Suppose that two things or systems, a and b, consume or produce a third thing c. We make no assumption concerning the nature of these things, let alone the mechanism of consumption or production. But we do assume something about the modes of cooperation and competition between a and b with regard to c. In fact we assume that each component, when on its own, consumes or produces c at a constant rate (possibly nil), and that each may or may not succeed in making ends meet. We also assume that the components are sure to attain the optimum level (of consumption or of production) if they cooperate with one another. Finally, we suppose that this cooperation becomes intenser the more it is needed, that it ceases when no longer needed, and that from then onwards it turns into hindrance, so that growth remains under control. Such a process can be formalized as follows.

Let $A(t)$ and $B(t)$ be the instantaneous amounts of thing c consumed or produced by the components a and b respectively, and call α and β the optimum or satiety levels. Further, let

$$A(0) = c_1 < \alpha, \qquad B(0) = c_2 < \beta$$

be the initial values. In the absence of interaction the rates are assumed to be constant (possibly zero):

$$\dot{A} = a_{11}, \qquad \dot{B} = a_{22} \quad \text{with} \quad a_{11}, a_{22} \in \mathbb{R}^+.$$

If these rates are nil, no processes occur. If they are positive, each individual eventually attains and surpasses its optimum level. Furthermore if cooperation sets in from the start, a will help b in proportion to the latter's needs, i.e. in such a way that the imbalance $\beta - B$ will be offset; and b in turn contributes in proportion to the imbalance $\alpha - A$, though not necessarily with the same intensity. This mutual help, we have assumed, is positive as long as there is a deficit (negative balance); it becomes nil on attaining satiety, and negative as soon as there is superavit (positive balance). The simplest formulas representing such a cooperative process are the nonlinear equations

$$\dot{A} = a_{11} + a_{12}(\alpha - A)B, \qquad \dot{B} = a_{22} + a_{21}(\beta - B)A$$

with $a_{ij} \geq 0$ for $i, j = 1, 2$.

The coefficient a_{12} measures the strength of the help a gets from b, while a_{21} measures the intensity of the help a lends b. At the optimum point $\langle\alpha, \beta\rangle$ there is neither help nor hindrance: each component fends for itself, so the system breaks down momentarily.

We spare the reader the treatment of the general case in the vicinity of the optimum point, showing that, in effect, no sooner does a attain the α level than its partner b turns against it forcing it to slow down its rate of consumption or of production. (Cf. Bunge, 1976.) Intead, we shall study the particular case when the two partners behave in exactly the same way. In this case, i.e. for $c_1 = c_2 = c$ (no initial advantage), $a_{11} = a_{22}$ (same individual rates), $a_{12} = a_{21}$ (symmetric help) and $\alpha = \beta$ (same requirements), we are left with a single equation of evolution, namely

$$\dot{X} = a_{11} + a_{12}(\alpha - X)X, \qquad X = A, B.$$

The general solution to this equation is found to be

$$X(t) = a_{11}t + \frac{\alpha c}{c + (\alpha - c)\exp(-a_{12}\,\alpha t)}.$$

If each partner is self-sufficient ($a_{11} = a_{22} > 0$), cooperation only speeds up the process up to the optimum point, and curbs it after that point. Otherwise (i.e. if $a_{11} = a_{22} = 0$), mutual help makes the process possible. Indeed, the second term of the last equation represents the growth of A (which is equal to B) from the initial value c towards the asymptote $A(\infty) = \alpha > c$. Rivalry from this point on prevents unlimited growth in either consumption or production. See Figure B3. This limitation takes place whether or not there is an external limitation such as scarcity of resources: the control is strictly internal. Not that there are no external limitations; only, our model does not take them into account.

General dynamical models of this kind have at least three possible uses:

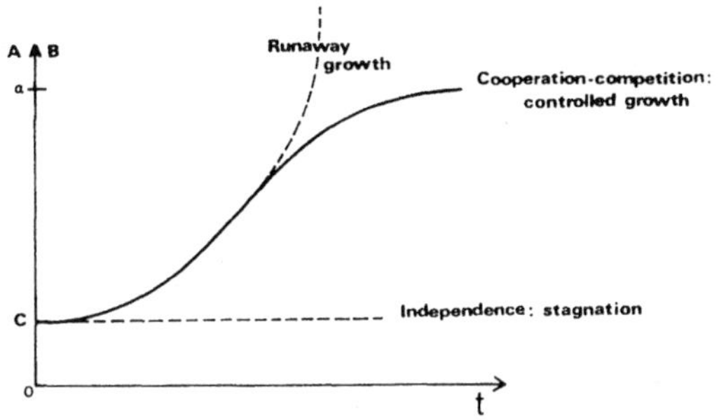

Fig. B3. Full line of a system composed of partners that cooperate and compete ($a_{12} \neq 0$) but are powerless in isolation (i.e. when $a_{11} = a_{22} = 0$). Dotted line: stagnation in the absence of interaction (i.e. when $a_{12} = 0$).

(*a*) they may suggest the search for real systems (e.g. in morphogenesis, neurology, ecology, or sociology) that approach a zero growth state in some variable as a result of their cooperative-competitive structure; (*b*) they may stimulate the building of specific theories covering real systems behaving in the way outlined by the model; (*c*) they may help in planning social controls. In particular, our model suggests redesigning society in such a way that each member helps his/her neighbor in proportion to his/her own ability and the other party's needs, and also controls his/her partner's growth and is in turn controlled by it, thus preventing scarcity as well as waste, and thus combining the good features of competition with those of cooperation.

3. Qualitative Change Models

3.1. *Kinematical: Birth and Death Operators*

All sciences deal with qualitative changes and even deep ones such as changes in kind, or transmutations of things of one kind into things of another. Such changes in kind can be conceptualized in several ways. The simplest is by treating each kind as a graph node, and the transformation as a graph edge, of a directed graph. The latter is then a directed and connected graph – e.g. one representing a lineage such as a disintegration schema or a phylogenetic tree. If a quantitative, yet still purely phenomenological, representation of such qualitative changes is wanted, then one has to devise a way of saying how many things of each kind emerge or disappear in a given process, without suggesting any transmutation mechanisms. We shall exhibit one such model, inspired in quantum field theory.

Consider a system composed of things of different species (natural kinds) K_m, where m is a natural number greater than 2. Call n_m the instantaneous population of kind K_m in the system. The state of the system such that there are n_1 things of kind K_1, n_2 of kind K_2, and in general n_m of kind K_m, can be represented by a state function $u(n_1 n_2 \ldots n_i \ldots)$. Thus

$u(0000\ldots)$ represents the no-system state,

$u(1000\ldots)$ represents the state such that the system has a single component of kind K_1 and none of the others;

$u(0n0\ldots)$ represents the state of the system when it is composed of n units of kind K_2.

We assume that the totality of such state function values spans the state space S of the system. Moreover we postulate that S is a vector space, so that every linear combination of u's represents a possible state of the system. The unit vectors $\varepsilon_1 = (100\ldots)$, $\varepsilon_2 = (0100\ldots)$, $\varepsilon_3 = (00100\ldots)$ constitute a basis of the vector space.

We now introduce the *birth operators* $B_i: S \to S$ defined by

$$B_i u(n_1 \ldots n_i \ldots) = (n_i + 1)^{1/2} u(n_1 \ldots n_i + 1 \ldots).$$

B_i represents the change of state of the system brought about by the increase by one thing of kind K_i in the composition of the system.

Similarly we define the disappearance of one unit of kind K_i from the composition of the system:

$$D_i u(n_1 \ldots n_i \ldots) = (n_i)^{1/2} u(n_1 \ldots n_i - 1 \ldots),$$

and call $D_i: S \to S$ the *destruction operator* of kind K_i. It can be shown (e.g. Mandl, 1959) that the two operators are mutually adjoint, i.e. $D_i = B_i^{\dagger}$, where the dagger designates transposition and conjugation of the entries in the matrix representations of the operators. It follows that $B_i D_i$ leaves things in the state they were to begin with, i.e.

$$B_i D_i u(n_1 \ldots n_i \ldots) = n_i u(n_1 \ldots n_i \ldots).$$

The eigenvalues of $B_i D_i = B_i B_i^{\dagger}$ are the possible populations of kind K_i in the system. Therefore $B_i D_i = N_i$ is an operator that can be interpreted as the population (or occupation number) operator for the kind K_i.

The basic birth operators can be combined to yield operators representing arbitrary changes in the composition of the system. Thus $(B_i^{\dagger})^p B_j{}^q$

represents the destruction of p individuals of kind K_i and the birth of q individuals of kind K_j. In other words,

$$(B_i^\dagger)^p B_j^q \text{ represents the reaction } pK_i \longrightarrow qK_j,$$

whereas $(B_j^\dagger)^p B_i^q$ represents the converse reaction. Mathematically speaking, $(B_i^\dagger)^p B_j^q$ transforms the vector $u(n_1 \ldots n_i \ldots n_j \ldots)$ into $(n_i!(n_i + q)!/p!n_j!)^{1/2}\ u(n_1 \ldots n_i - p \ldots n_j + q \ldots)$. Finally, the combinations of birth operators generate the entire state space out of the no-system (or zero) state $u(000\ldots)$ Thus $B_i{}^p B_j{}^q u(000\ldots)$ is proportional to the state vector representing the situation where there are p components of kind K_i and q of kind K_j in the system, and nothing else.

Because we have made no assumption other than that the changes under consideration are discrete changes in kind, the above is a framework for a kinematical (phenomenological) theory rather than a theory proper. But this is what we want in ontology: a more specific formalism would belong in science.

3.2. *Dynamical: Random Hits*

We shall presently sketch an extremely simple theory, both dynamical and stochastic, suggestive of a very general ontological hypothesis. The theory deals with the mutagenic effect of external agents impinging randomly upon a system. (See Maynard Smith (1968)). Suppose a system contains m components (*targets*) sensitive to certain discrete transmuting agents (*bullets*). The sensitive components must be a critical part of the whole thing even if they are only a small part: they may be atomic or cellular nuclei, organelles, key sociosystems, or what not. The disrupting agents can be ionizing particles, radiations, terrorists, bureaucrats, or what have you. Every hit or collision of a bullet with a target is assumed to either mutate or destroy the latter and, in any case, to effect a change in kind. The problem is to compute the probability that n random bullets will produce a change in kind (or transmutation) in a system with m sensitive components.

Assume that each bullet can act on a single target rather than being able to kill several birds at one stroke: i.e. let us restrict our attention to single hits. Call p the probability that an arbitrary bullet will mutate a target, and suppose that a total of n bullets fire at the system in an aimless and unconcerted way. Suppose also that the effect each successful bullet produces has no consequences on the other targets of the system – i.e. that the effects are mutually independent, so that only their cumulative effect is of interest. Since the probability that a bullet fails to hit a target is $q = 1 - p$, the probability that the target will fail to mutate under the action of n random bullets is

$$q^n = (1 - p)^n \cong e^{-pn} \quad \text{for large } n.$$

And the probability that all m sensitive components will remain unaltered after a round of n firings will be the m-fold product of the above, i.e.

$$Q \cong (e^{-pn})^m.$$

Consequently the probability that all m sensitive components will be hit and mutate is

$$P = 1 - Q \cong 1 - e^{-pmn},$$

which approaches 1 as the number of bullets increases. Therefore the average number of mutations induced by the n bullets is Pm, which approaches m with increasing n. In words: the probability that the bullets be ineffectual decreases exponentially as the number of bullets increases, while the probability that the system undergoes a radical transmutation increases monotonically with the number of bullets: see Figure B4.

Now, every system is subjected to some mutagenic agent or other – fast charged particles, slow neutrons, hard radiation, predators, critics, saboteurs, or what have you. Moreover, whether the missiles are aimed and concerted, or random, they are likely to have some effects – not necessarily destructive but in any case transmuting. Therefore we can make the sweeping generalization

POSTULATE B4 Every system except the world is subject to random actions that eventually elicit its breakdown or its mutation (change in kind).

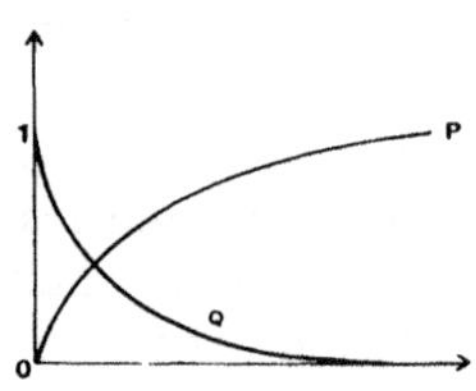

Fig. B4. Probable effects of random hits.

Note that we are not assuming that all mutagenic agents are external to the system concerned: it suffices that they be external to its sensitive components. For example, chromosomes are subjected not only to external radiation but also to thermal and chemical actions of the surrounding protoplasm. And every cultural system is the beneficiary or victim not only of economic and political actions but also of its own creative and destructive members.

BIBLIOGRAPHY

Alston, William P. (1974). Conceptual prolegomena to a psychological theory of intentional action. In Brown, Ed., 1974, pp. 71–101.

Anderson, James A. (1972). A simple neural network generating an interactive memory. *Math. Biosci.* **14**: 197–220.

Anderson, James A. (1973). A theory for the recognition of items from short memorized lists. *Psychol. Rev.* **80**: 417–438.

Arbib, Michael A. (1969). Automata theory. In Kalman *et al.*, 1969.

Apter, M. J., and Lewis Wolpert (1965). Cybernetics and development. *J. Theoret. Biol.* **8**: 244–257.

Aris, Rutherford (1965). Prolegomena to the rational analysis of systems of chemical equations. *Arch. Rational Mech. Anal.* **19**: 81–99.

Aristotle, *The Basic Works of A.* Richard McKeon, Ed. New York: Random House, 1941.

Armstrong, David M. (1968) *A Materialist Theory of the Mind.* London: Routledge & Kegan Paul.

Ashby, W. Ross (1956). *An Introduction to Cybernetics.* New York: Wiley.

Athans, M. and P. L. Falb (1966). *Optimal Control.* New York: McGraw-Hill.

Ayala, Francisco J. (1974). The concept of biological progress. In Ayala and Dobzhansky, Eds., 1974, pp. 339–355.

Ayala, Francisco J., and Theodosius Dobzhansky, Eds. (1974). *Studies in the Philosophy of Biology.* Los Angeles & Berkeley: University of California Press.

Bergson, Henri (1907). *L'évolution créatrice*, 77th ed. Paris: Presses Universitaries de France.

Berlinski, David (1976). *On Systems Analysis.* Cambridge, Mass.: MIT Press.

Bertalanffy, Ludwig von (1950). An outline of general system theory. *Brit. Jo. Philos. Sci.* **1**: 139–164.

Bertalanffy, Ludwig von (1968). *General System Theory.* New York: Braziller.

Bindra, Dalbir (1976). *A Theory of Intelligent Behavior.* New York: Wiley.

Birkhoff, Garrett, and Thomas C. Bartee (1970). *Modern Applied Algebra.* New York: McGraw-Hill Book Co.

Bogen, Joseph (1969). The other side of the brain: an appositional mind. *Bull. Los Angeles Neurol. Soc.* **34**, 3: 135–162.

Bonner, John Tyler (1974). *On Development. The Biology of Form.* Cambridge, Mass.: Harvard University Press.

Borst, C. V., Ed. (1970). *The Mind-Brain Identity Theory.* New York: St. Martin's Press.

Boulding, Kenneth (1956). General systems theory – the skeleton of science. *Management Sci.* **2**: 197–208.

Bowler, Peter J. (1976). *Fossils and Progress.* New York: Science History Publications.

Brandt, Richard, and Jaegwon Kim (1967). The logic of the identity theory. *J. Philos.* **64**: 515–537.

Braudel, Fernand (1949). *La Méditerranée et le monde méditerranéen à l'époque de Philippe II.* Paris: Armand Colin. Engl. transl.: *The Mediterranean.* New York: Harper and Row, 1973.
Braudel, Fernand (1969). *Écrits sur l'histoire.* Paris: Flammarion.
Brown, S. C., Ed. (1974). *Philosophy and Psychology.* London: Macmillan.
Bunge, Mario (1959a). *Causality: The Place of the Causal Principle in Modern Science.* Cambridge, Mass.: Harvard University Press, Rev. ed.: New York: Dover Publ., 1979.
Bunge, Mario (1959b). *Metascientific Queries.* Springfield, Ill.: Charles C. Thomas.
Bunge, Mario (1960). Levels: a semantical preliminary. *Rev. Metaphys.* **13**: 396–406. Repr. with slight changes in Bunge, 1963b.
Bunge, Mario (1963a). A general black box theory. *Philos. Sci.* **30**: 346–358.
Bunge, Mario (1963b). *The Myth of Simplicity.* Englewood Cliffs, N.J.: Prentice-Hall.
Bunge, Mario (1964). Phenomenological Theories. In M. Bunge, Ed., *The Critical Approach: Essays in Honor of Karl Popper.* Glencoe: The Free Press.
Bunge, Mario (1967a). *Scientific Research,* 2 vols. New York: Springer-Verlag.
Bunge, Mario (1967b). *Foundations of Physics.* New York: Springer-Verlag.
Bunge, Mario (1971). Is scientific metaphysics possible? *J. Philos.* 68: 507–520.
Bunge, Mario (1973a). *Method, Model, and Matter.* Dordrecht: D. Reidel Publ. Co.
Bunge, Mario (1973b). *Philosophy of Physics.* Dordrecht: D. Reidel Publ. Co.
Bunge, Mario (1974a). *Sense and Reference.* Dordrecht & Boston: D. Reidel Publ. Co.
Bunge, Mario (1974b). *Interpretation and Truth.* Dordrecht & Boston: D. Reidel Publ. Co.
Bunge, Mario (1974c). Les présupposés et les produits métaphysiques de la science et de la technique contemporaines. *Dialogue* **13**: 443–453.
Bunge, Mario (1974d). The concept of social structure. In W. Leinfellner and E. Köhler, Eds., *Developments in the Methodology of Social Science,* pp. 175–215. Dordrecht: D. Reidel Pub. Co.
Bunge, Mario (1976). A model for processes combining competition with cooperation. *Appl. Math. Model.* **1**: 21–23.
Bunge, Mario (1977a). *The Furniture of the World.* Dordrecht & Boston: D. Reidel Publ. Co.
Bunge, Mario (1977b). States and events. In William E. Hartnett, Ed., *Systems: Approaches, Theories, Applications.* Dordrecht & Boston: D. Reidel Publ. Co.
Bunge, Mario (1977c). The GST challenge to the classical philosophies of science. *Intern. J. General Systems* **4**: 29–37.
Bunge, Mario (1977d). General systems and holism. *General Systems* **22**: 87–90.
Bunge, Mario (1977e). Emergence and the mind. *Neurosci.* **2**: 501–509
Bunge, Mario (1977f). Levels and reduction. *Am. J. Physiol.* **R2**: 75–82.
Bunge, Mario (1977g). The philosophical richness of technology. In F. Suppe and P. D. Asquith, Eds., *PSA* **2**: 153–172.
Bunge, Mario (1977h) A systems concept of the international system. In *Bunge et al.,* Eds., pp. 291–305.
Bunge, Mario (1978a). A model of evolution. *Appl. Math. Model.* **2**: 201–204.
Bunge, Mario (1978b) A systems concept of society: beyond individualism and holism. Theory and Decision **10**: 13–30.

Bunge, Mario, and Máximo García Sucre (1976). Differentiation, participation and cohesion. *Quality and Quantity* **10**: 171–178.

Bunge, Mario, Johann Galtung, and Mircea Malitza, Eds. (1977) *Mathematical Approaches to International Relations.* Bucharest: Romanian Academy of Social and Political Sciences.

Bunge, Mario, and Rodolfo Llinás (1978). The mind-body problem in the light of neuroscience. *Proc. 16th World Congress of Philosophy.*

Calvin, Melvin (1969). *Chemical Evolution.* New York & Oxford: Oxford University Press.

Chiaraviglio, Lucio (1965). Sequential machines and the coding of polypeptides. *J. Theoret. Biol.* **8**: 130–140.

Chomsky, Noam (1968). *Language and Mind.* New York: Harcourt, Brace & World.

Cowan, Jack (1976). Are there modifiable synapses in the visual cortex? In Rosenzweig and Bennett, 1976, pp. 133–143.

Craik, Kenneth J. W. (1966), *The Nature of Psychology.* Stephen L. Sherwood, Ed. Cambridge: University Press.

Crick, Francis (1970), Central dogma of molecular biology. *Nature* **227**: 561–563.

Delgado, José M. R. (1969), *Physical Control of the Mind.* New York: Harper & Row.

Deutsch, Karl W. (1968), *The Analysis of International Relations.* Englewoods Cliffs, N. J.: Prentice-Hall.

Dobzhansky, Theososius (1970), *Genetics of the Evolutionary Process.* New York & London: Columbia University Press.

Dorf, Richard C. (1974), *Modern Control Systems,* 2nd ed. Reading, Mass.: Addison-Wesley Publ. Co.

Du Bois-Reymond, Emil (1872), Ueber die Grenzen des Naturerkennens. In *Reden,* 2nd ed. Leipzig: Veit und Co.

Dunbar, Max (1960). The evolution of stability in marine ecosystems: natural selection at the level of the ecosystem. *Am. Natur.* **94**: 129–136.

Edelen, Dominic G. B. (1962). *The Structure of Field Space.* Berkeley: University of California.

Eigen, Manfred (1971). Selforganization of matter and the evolution of biological macromolecules. *Naturwissenschaften* **58**: 465–523.

Feigl, Herbert (1958), The 'mental' and the 'physical'. In H. Feigl, M. Scriven, and G. Maxwell, Eds., *Minnesota Studies in the Philosophy of Science* II, pp. 370–497. Minneapolis: University of Minnesota Press.

Feinberg, Martin (1972). On chemical kinetics of a certain class. *Arch. Rat. Mech. Anal.* **46**: 1–41.

Florkin, M. , Ed. (1960). *Aspects of the Origin of Life.* Oxford: Pergamon Press.

Fox, Sidney W., and K. Dose (1972). *Molecular Evolution and the Origin of Life.* San Francisco: Freeman & Co.

Freeman, Walter J. (1973). A model of the olfactory system. In M. A. M. Brazier, D. O. Walter, and D. Schneider, Eds., *Neural Modeling,* pp. 41–62. Los Angeles: UCLA Brain Research Institute.

Fritsch, Bruno (1974), *Wachstum Begrenzung als Machtinstrument.* Stuttgart: Deutsche Verlags-Anstalt.

Frondizi, Risieri (1953). *The Nature of the Self. A Functional Interpretation.* New Haven, Con.: Yale University Press. Repr.: Carbondale, Ill.: Southern Illinois Press, 1971.

Gall, John (1977) *Systemantics: How Systems Work and Especially How They Fail.* New York: Quadrangle.

Gardner, R. A., and B. T. Gardner (1969). Teaching sign language to a chimpanzee. *Science* **165**: 664–672.

Gazzaniga, Michael S. (1967). The split brain in man. *Sci. Am.* **217**, 2: 24–29.

Gierer, H., and H. Meinhardt (1972). A theory of biological pattern formation. *Kybernetik* **12**: 30–39.

Ginsburg, S. (1962). *An Introduction to Mathematical Machine Theory.* Reading, Mass.: Addison-Wesley Publ. Co.

Glansdorff, P., and I. Prigogine (1971). *Thermodynamic Theory of Structure, Stability and Fluctuations.* London: Wiley-Interscience.

Glass, Leon (1973). Stochastic generation of regular distributions. *Science* **180**: 1061–1063.

Goldin-Meadow, Susan, and Heidi Feldman (1977). The development of language-like communication without a language model *Science* **197**: 401–403.

Gray, J. A. (1972a). The psychophysiological nature of introversion-extroversion: A modification of Eysenck's theory. In Nebylitsyn and Gray, 1972, pp. 182–205.

Gray, J. A. (1972b). Learning theory, the conceptual nervous system, and personality. In Nebylitsyn and Gray, 1972, pp. 372–399.

Gray Eaton, G. (1976). The social order of Japanese macaques. *Sci. Am.* **235**, 4: 96–106.

Griffin, Donald R. (1976). *The Question of Animal Awareness.* New York: Rockefeller University Press.

Gross, Charles G., and H. Philip Zeigler, Eds. (1969). *Readings in Physiological Psychology. Motivation.* New York: Harper & Row.

Guinzburg, A. (1968). Algebraic Theory of Automata. New York: Academic Press.

Haken, H. (1975). Cooperative phenomena in systems far from equilibrium and in non-physical systems. *Rev. Mod. Phys.* **47**: 67–121.

Hamilton, W. D. (1964). The genetical theory of social behavior. *J. Theoret. Biol.* **7**: 1–16, 17–32.

Hamilton, W. D. (1971), Geometry of the selfish herd. *J. Theoret. Biol.* **31**: 295–311.

Harris, Marvin (1968). *The Rise of Anthropological Theory.* New York: Thomas Y. Crowell Co.

Harrison, M. A. (1965). *Introduction to Switching and Automata Theory.* New York: McGraw-Hill.

Hebb, Donald O. (1949). *The Organization of Behavior.* New York: Wiley.

Hebb, Donald O. (1968). *A Textbood of Psychology.* Philadelphia: W. B. Saunders Co.

Hebb, Donald O. (1974).What psychology is about. *Am. Psychol.* **29**: 71–79.

Hegel, Georg Wilhelm Friedrich (1830). *Encyklopädie der philosophischen Wissenschaften im Grundrisse,* 3rd ed. Heidelberg: Verwaltung des Oswaldschen Verlags.

Henderson, Lawrence J. (1913). *The Fitness of the Environment.* Boston: Beacon Press, 1958.

Hinde, Robert A. (1974). *Biological Bases of Human Social Behaviour.* New York: McGraw-Hill.

Holland, John H. (1975). *Adaptation in Natural and Artificial Systems.* Ann Arbor: University of Michigan Press.

Homans, George C. (1974). *Social Behavior: Its Elementary Forms.* New York: Harcourt, Brace, Jovanovich.

Hubel, D. H., and T. N. Wiesel (1963), Receptive fields and functional architecture of monkey striate cortex. *J. Physiol.* **165**: 559–568.

Hull, David (1974), *Philosophy of Biological Science*. Englewood Cliffs, N.J.: Prentice-Hall.

Hume, David (1739). *A Treatise of Human Nature*. L. A. Selby-Bigge, Ed. Oxford: Clarendon Press, 1888.

Jacob, François (1970). *La logique du vivant*. Paris, Gallimard.

Jacob, François (1977). Evolution and tinkering. *Science* **196**: 1161–1166.

Jost, R. (1965). *The General Theory of Quantized Fields*. Providence: American Mathematical Society.

Kalman, R. E., P. L. Falb, and M. A. Arbib (1969). *Topics in Mathematical System Theory*. New York: McGraw-Hill.

Kawai, Masao (1965). Newly acquired pre-cultural behavior of the natural troop of Japanese monkeys on Koshima Islet. *Primates* 6: 1–30.

Kemeny, John, and Paul Oppenheim (1956). On reduction. *Philos. Stud.* **7**: 6–19.

Kendrew, John C. (1966). *The Thread of Life*. London: G. Bell and Sons Ltd.

Keosian, John (1968). *The Origin of Life*, 2nd ed. New York: Van Nostrand Reinhold Co.

King, M. -C., and A. C. Wilson (1975). Evolution at two levels in humans and chimpanzees. *Science* **188**: 107–116.

Klir, George (1969). *An Approach to General System Theory*. New York: Van Nostrand Reinhold Co.

Klir, Jiri, and Miroslav Valach (1967). *Cybernetic Modelling*. London: Iliffe Books Ltd.; Prague: SNTL.

Klir, George, and Gary Rogers (1977). *Basic and Applied General Systems Research: A Bibliography*. Binghampton, N.Y.: SUNY-Binghampton.

Kripke, Saul (1971). Identity and necessity. In Milton K. Munitz, Ed., *Identity and Individuation*, pp 135–164 New York: New York University Press.

Krohn, Kenneth, Rudolph Langer and John Rhodes (1967). Algebraic principles for the analysis of a biochemical system. *J. Comp. Syst. Sci.* **1**: 119–136.

Küppers, Bernd (1975). The general principles of selection and evolution at the molecular level. *Prog. Biophys. Molec. Biol.* **30**: 1–22.

Lange, Oscar (1965). *Wholes and Parts: A General Theory of Systems Behavior*. Oxford: Pergamon Press; Warsaw: PWN.

Lehninger, Albert L. (1975). *Biochemistry*, 2nd ed. New York: Worth Publishers Inc.

Lepper, M. R., D. Greene, and R. E. Nisbett (1973). Undermining children's intrinsic interest with extrinsic rewards. *J. Pers. Soc. Psychol.* **28**: 129–137.

Levins, Richard (1968). *Evolution in Changing Environments*. Princeton: Princeton University Press.

Lévi-Strauss, Claude (1963). *Structural Anthropology*. New York: Basic Books.

Lewontin, Richard C. (1974). *The Genetic Basis of Evolutionary Change*. New York & London: Columbia University Press.

Llinas, Rodolfo, and Mario Bunge (1978). Restricted applicability of the concept of command in neuroscience. *Behav. Brain Sci.* **1**: 30–31.

Lloyd Morgan, C. (1933). *The Emergence of Novelty*. London: Williams and Norgate.

Luria, A. R. (1966) *Human Brain and Psychological Processes*. New York: Harper & Row.

Luria, A. R. (1969) Speech development and the formation of mental processes. In Michael Cole and Irving Maltzman, Eds., *A Handbook of Contemporary Soviet Psychology*, pp. 121–162. New York: Basic Books.
Lwoff, André (1962). *Biological Order*. Cambridge, Mass.: MIT Press.
MacGregor, Ronald J., and Edwin R. Lewis (1977). *Neural Modelling*. New York & London: Plenum Press.
Malcolm, Norman (1973). Thoughtless brutes. *Proc. Add. Am. Philos. Assoc.* **46**: 5–20.
Malcolm, Norman (1964). Scientific materialism and the identity theory. *Dialogue* **3**: 115–125.
Mandl, F. (1959). *Introduction to Quantum Field Theory*. New York: Interscience.
Malsburg, C. von der (1973). Self-organization of orientation sensitive cells in the striate cortex. *Kybernetik* **14**: 85–100.
Margolis, Joseph (1978) *Persons and Minds: The Prospects of Nonreductive Materialism*. Dordrecht & Boston: D. Reidel Publ. Co.
Marr, David (1970). A theory for cerebral cortex. *Proc. Roy. Soc.* (London.) **B176**: 161–234.
Maynard Smith, J. (1968). *Mathematical Ideas in Biology*. Cambridge: University Press.
McNaughton, Robert (1961). The theory of automata: a survey. *Adv. Comp.* **2**: 379–421.
Menzel, E.W., Jr., R. K. Davenport, and C. M. Rogers (1972). Protocultural aspects of chimpanzee's responsiveness to novel objects. *Folia primatol.* **17**: 161–170.
Mesarović, M. D., Ed. (1964). *Views on General Systems Theory*. New York: Wiley.
Miller, Stanley, and Leslie Orgel (1974). *The Origins of Life on the Earth*. Englewood Cliffs, N.J.: Prentice-Hall.
Miller, Neal E. (1969). Learning of visceral and glandular responses. *Science* **163**: 434–445.
Milsum, John H., Ed. (1968). *Positive Feedback*. Oxford: Pergamon Press.
Monod, Jacques (1970). *Le hasard et la nécessité*. Paris: Ed. du seuil. Engl. transl.: *Chance and Necessity*. New York: Knopf, 1971.
Moore, Robert Y. (1976). Synaptogenesis and the morphology of learning and memory. In Rosenzweig and Bennet, Eds., pp. 340–347.
Morris, Charles W. (1938) Foundations of the theory of signs. *International Encyclopaedia of Unified Science*, vol. I., No. 2. Chicago: University of Chicago Press.
Mostow, G. D., Ed. (1975). *Mathematical Models for Cell Rearrangement*. New Haven, Conn.: Yale University Press.
Mountcastle, Vernon (1957). Modality and topographic properties of single neurons of cat's somatic sensory cortex. *J. Neurophys.* **20**: 408–434.
Nass, Menasche M., and Leon N. Cooper (1975). A theory for the development of feature detecting cells in visual cortex. *Biolog. Cybernet.* **19**: 1–18.
Nebylitsyn, V. D., and A. J. Gray, Eds. (1972). *Biological Bases of Individual Behaviour*. New York & London: Academic Press.
Neumann, John von (1951). The general and logical theory of automata. In L. A. Jeffress, Ed., *Cerebral Mechanisms in Behavior*. New York: Wiley.
Neumann, John von (1966). *Theory of Self-Reproducing Automata*. Ed. and completed by A. W. Burks. Urbana: University of Illinois Press.

Nicolis, G., and I. Prigogine (1977). *Self-Organization in Nonequilibrium Systems.* New York: Wiley-Interscience.

O'Connor, John, Ed. (1969). *Modern Materialism: Readings on Mind-Body Identity.* New York: Harcourt, Brace & World.

Olby, Robert (1974). *The Path to the Double Helix.* Seattle: University of Washington Press.

Omenn, Gilbert S., and Arno Motulsky (1972). Biochemical genetics and the evolution of human behavior. In L. Ehrman, G. S. Omenn, and E. Caspari, Eds., *Genetics, Environment and Behavior*, pp. 129–172. New York: Academic Press.

Oparin, A. I. (1968). *Genesis and Evolutionary Development of Life.* New York: Academic Press.

Oppenheim, Paul, and Hilary Putnam (1958). Unity of science as a working hypothesis. In H. Feigl, G. Maxwell, and M. Scriven, Eds., *Minnesota Studies in the Philosophy of Science*, Vol. II, pp. 3–36. Minneapolis: University of Minnesota Press.

Oró, J., S. L. Miller, C. Ponnamperuma, and R. S. Young, Eds. (1974). *Cosmochemical Evolution and the Origins of Life*, I. Dordrecht & Boston: D. Reidel Publ. Co.

Padulo, Louis, and Michael A. Arbib (1974). *System Theory.* Philadelphia: W. B. Saunders Co.

Paivio, A. (1971). *Imagery and Verbal Processes.* New York: Holt, Rinehart and Winston.

Penfield, Wilder (1975). *The Mystery of the Mind: A Critical Study of Consciousness and the Human Brain.* Princeton: University Press.

Penfield, Wilder, and T. Rasmussen (1950). *The Cerebral Cortex of Man.* New York: Macmillan.

Pirie, N. W. (1960). Chemical diversity and the origins of life. In Florkin, Ed., 1960, pp. 55–62.

Polanyi, Michael (1968). Life's irreducible structure. *Science* **160**: 1308–1312.

Popper, Karl R. (1968). Epistemology without a knowing subject. In B. van Rootselaar and J. F. Staal, Eds., *Logic, Methodology and Philosophy of Science* III, pp. 333–373. Amsterdam: North-Holland.

Popper, Karl R. (1972). *Objective Knowledge. An Evolutionary Approach.* Oxford: Clarendon Press.

Popper, Karl R. (1974). Autobiography. In P.A. Schilpp, Ed., *The Philosophy of Karl R. Popper*, I, pp. 1–181. La Salle, Ill.: Open Court.

Popper, Karl R., and John C. Eccles (1977). *The Self and its Brain.* Berlin, Heidelberg, London, New York: Springer International.

Powell, T.P.S., and Vernon Mountcastle (1959). Some aspects of the functional organization of the cortex of the postcentral gyrus of the monkey: a correlation of findings obtained in a single unit analysis with cytoarchitecture. *Bull. Johns Hopkins Hosp.* **105**: 133.

Premack, David (1971). Language in chimpanzee? *Science* **172**: 808–822.

Puccetti, Roland (1977). Sperry on consciousness: A critical appreciation. *J. Med. Philos.* **2**: 127–144.

Putnam, Hilary (1969). On properties. In Nicholas Rescher, Ed., *Essays in Honor of Carl G. Hempel*, pp. 235–254. Dordrecht: D. Reidel Publ. Co.

Rabin, Michael O. (1963). Probabilistic automata. *Information and Control* **6**: 230–245.

Rabin, Michael O., and Dana Scott (1959). Finite automata and their decision problems. *IBM J. Res. & Devel.* **3**: 114–125.

Rashevsky, Nicholas (1972). Some remarks on the central nervous system. *Bull. Math. Biophys.* **34**: 231–242.

Ratcliffe-Brown, A. R. (1935). On the concept of function in social science. *Am. Anthr.* **37**: 397–402.

Reynolds, Vernon (1976). *The Biology of Human Action*. Reading & San Francisco: W. H. Freeman and Co. Ltd.

Rosen, Robert (1959). The DNA-protein coding problem. *Bull. Math. Biophys.* **21**: 71–95.

Rosen, Robert (1970). Subunits and subassembly processes. *J. Theoret. Biol.* **28**: 415–422.

Rosenblueth, Arturo, Norbert Wiener, and Julian Bigelow (1943). Behavior, purpose and teleology. *Philos. Sci.* **10**: 18–24.

Rosenblueth, Arturo, and Norbert Wiener (1950). Purposeful and non-purposeful behavior. *Philos. Sci.* **17**: 318–326.

Rosenzweig, Mark R., and Edward L. Bennett, Eds. (1976). *Neural Mechanisms of Learning and Memory*. Cambridge, Mass.: MIT Press.

Rumbaugh, Duane M., Timothy V. Gill, and Ernst von Glasersfeld (1973). Reading and sentence completion by a chimpanzee (*Pan*). *Science* **182**: 731–733.

Ruse, Michael (1979). *Sociobiology: Sense or Nonsense*? Dordrecht & Boston: D. Reidel Publ. Co.

Rutledge, L. T. (1976). Synaptogenesis: effects of synaptic use. In Rosenzweig and Bennett, Eds., 1976, pp. 329–339.

Schlick, Moritz (1925). *Allgemeine Erkenntnislehre*, rev. ed. Transl. A. E. Blumberg, *General Theory of Knowledge*. Wien-New York: Springer-Verlag, 1974.

Sellars, Roy Wood (1922). *Evolutionary Naturalism*. Chicago: Open Court.

Sellers, Peter H. (1967). Algebraic complexes which characterize chemical networks. SIAM *J. Appl. Math.* **15**: 13–68.

Sellers, Peter H. (1970). Combinatorial analysis of a chemical network. *J. Franklin Inst.* **290**: 113–130.

Sellers, Peter H. (1971). An introduction to a mathematical theory of chemical reaction networks I. *Arch. Rational Mech. Anal.* **44**: 22–40.

Shaffer, Jerome A. (1977). Personal identity: the implications of brain bisection and brain transplant. *J. Medicine and Philosophy* **2**: 147–161.

Smart, J. J. C. (1959). Sensations and brain processes. *Philos. Rev.* **68**: 141–156.

Smith, Emil (1970). Evolution of enzymes. In Paul D. Boyer, Ed., *The Enzymes*, Vol. I, pp. 267–339. New York: Academic Press.

Smith, W. John (1965). Message, meaning, and context in ethology. *Am. Natur.* **99**: 405–409.

Smythies, J. R., Ed. (1965). *Brain and Mind*. London: Routledge & Kegan Paul.

Sommerhoff, Gerd (1974). *Logic of the Living Brain*. New York: Wiley.

Spencer, Herbert (1862). *First Principles*. Final ed.: London: Watts & Co., 1937.

Sourkes, Theodore (1962). *Biochemistry of Mental Disease*. New York: Harper & Row.

Sperry, Roger W. (1964). Neurology and the mind-body problem. In Robert Isaacson,

Ed., *Basic Readings in Neuropsychology*, pp. 403–429. New York: Harper & Row.

Sperry, Roger W. (1966). Brain bisection and mechanisms of consciousness. In John C. Eccles, Ed., *Brain and Conscious Experience*. Berlin, Heidelberg & New York: Springer-Verlag, 1966.

Sperry, Roger W. (1977). Forebrain commissurotomy and conscious awareness. *J. Med. Philos.* **2**: 101–126.

Spinoza, Benedict (1677). *Ethics*. Oxford: University Press, 1927.

Steinberg, Malcolm S. (1963). Reconstruction of tissues by dissociated cells. *Science* **141**: 401–408.

Thom, René (1972). *Stabilité structurelle et morphogenèse*. Reading, Mass.: W. A. Benjamin Inc.

Toulmin, Stephen (1972). The mentality of man's brain. In A.G. Karczmar and John C. Eccles, Eds., *Brain and Human Behavior*, pp. 409–422. New York, Heidelberg & Berlin: Springer-Verlag.

Turing, Alan (1950). Computing machinery and intelligence. *Mind* NS**59**: 433–460.

Turing, Alan (1952). The chemical basis of morphogenesis. *Philos. Trans. Roy. Soc. (Lond.)* **B237**: 7–94.

Tyson, John J. (1976). *The Belousov–Zhabotinskii Reaction*. Berlin, Heidelberg, New York: Springer-Verlag.

Vaucouleurs, Gérard de (1970). The case for a hierarchical cosmology. *Science* **167**: 1203–1213.

Vesey, G. N. A., Ed. (1964. *Body and Mind*. London: George Allen & Unwin.

Vol'kenshtein, Mikhail V. (1970). *Molecules and Life*. New York: Plenum Press.

Vygotskii, L. S. (1962). *Thought and Language*. Transl. E. Hanfman and G. Vakar. Cambridge, Mass.: MIT Press.

Waddington, Conrad Hall (1959). Evolutionary systems – animal and human. *Nature* **183**: 1654–1658.

Wallerstein, Immanuel (1974). *The Modern World-System: Capitalist Agriculture and the Origins of the European World Economy in the Sixteenth Century*. New York: Academic Press.

Watson, James (1976). *Molecular Biology of the Gene*, 3rd ed. Menlo Park, Cal.: W. A. Benjamin, Inc.

Wiener, Norbert (1948). *Cybernetics*. New York: John Wiley & Sons.

Williams, George C. (1966). *Adaptation and Natural Selection*. Princeton: Princeton University Press.

Wilson, Edward O. (1975). *Sociobiology: The New Synthesis*. Cambridge, Mass.: Harvard University Press.

Wilson, H. R., and J. D. Cowan (1973). A mathematical theory of the functional dynamics of cortical and thalamic nervous tissue. *Kybernetic* **13**: 55–80.

Winch, Peter (1958). *The Idea of a Social Science*. London: Routledge & Kegan Paul.

Worden, Frederic G., Judith P. Swazey, and George Adelman, Eds. (1975). *The Neurosciences: Paths of Discovery*. Cambridge, Mass.: MIT Press.

Zadeh, Lofti A., and Charles A. Desoer (1963). *Linear System Theory: The State Space Approach*. New York: McGraw-Hill.

INDEX OF NAMES

INDEX OF SUBJECTS

CONTENTS OF *ONTOLOGY I*

GPSR Compliance

The European Union's (EU) General Product Safety Regulation (GPSR) is a set of rules that requires consumer products to be safe and our obligations to ensure this.

If you have any concerns about our products, you can contact us on ProductSafety@springernature.com

In case Publisher is established outside the EU, the EU authorized representative is:

Springer Nature Customer Service Center GmbH
Europaplatz 3
69115 Heidelberg, Germany

Batch number: 10584501

Printed by Printforce, the Netherlands